1.700

D1338518

BIRD VOCALIZATIONS

BIRD VOCALIZATIONS

THEIR RELATIONS TO CURRENT PROBLEMS IN BIOLOGY AND PSYCHOLOGY

Essays presented to W. H. Thorpe

EDITED BY R. A. HINDE

CAMBRIDGE
AT THE UNIVERSITY PRESS 1969

Published by the Syndics of the Cambridge University Press
Bentley House, 200 Euston Road, London N.W.1.
American Branch: 32 East 57th Street, New York, N.Y. 10022
© Cambridge University Press 1969
Library of Congress Catalogue Card Number: 69-19376
Standard Book Number: 521 07409 6

Printed in Great Britain at the Aberdeen University Press

CONTENTS

EDITOR'S PREFACE

The sounds of birds have long been a source of pleasure to mankind, but only recently have scientists recognized that they provide a wealth of material pertinent to many problems of biology and psychology. For those interested in the ontogeny of behaviour, the subtle interplay of genetic and experiential factors in the development of bird song provides exceptionally favourable material for study, and the principles which emerge are applicable in many other contexts. The physiologist studying the causal bases of behaviour is often hampered by difficulties of measurement: vocal patterns are susceptible to quantitative treatment and provide useful end-points for causal analyses. In addition, vocalizations are themselves causes, influencing the behaviour of, and physiological changes in, other individuals: the function of sounds in social communication is proving to be an exceptionally fertile field since, with the use of a tape recorder, their effects on other individuals can be assessed under controlled conditions more readily than those of visual signals. As social signals, the vocalizations of each species form part of an adaptive complex of characters of structure, physiology and behaviour—a complex whose intricacies comparative studies help to reveal. Furthermore, since song plays an important role in promoting reproductive isolation, studies of individual responsiveness to song, and of inter- and intra-species differences in song patterns, can throw light on the mechanisms which act to reduce hybridization in incipient species. In addition, song can be used as a taxonomic character. Finally, the special part which bird song plays in man's appreciation of nature demands both historical perspective and analysis, promising rich fruits for those interested in the nature of aesthetics.

Providing material for such diverse disciplines, the study of bird song could easily become fragmented: the specialists might steal away with their recordings each to his own laboratory, blinded by their analyses to the interrelations between their problems. That this has not so far happened is, one may hope, not merely a consequence of the youth of the subject, for at least two other factors are operating.

First, many of the biologists with a professional interest in bird song remain amateur ornithologists, retaining an interest in the whole animal. Second, in the study of bird song it is especially clear that questions of form and function, development and causation, evolution and appreciation, are all inextricably interwoven. Topics such as this, which can run as rivets through the centre of biology, have a special role to play at a time when interdepartmental barriers are becoming a brake on progress.

The recent growth of interest in bird song has been due in large part to the development of refined methods of recording and sound analysis. Such methods would, however, have been valueless without the men to exploit them, and these essays are presented to a pioneer in this field—W. H. Thorpe. Realizing that birds, with their highly developed stereotyped patterns of behaviour and marked learning ability, provide exceptional material for the study of behaviour, he founded the Ornithological Field Station (subsequently the Sub-department of Animal Behaviour) at Cambridge in 1950. It was natural that the combination of his interest in ornithology and his love of music should lead him to the study of bird song, and he was quick to exploit the new techniques which became available in the early fifties. His own work on the chaffinch has been a model for later studies, and his book on *Bird Song* (1961) an important review of the field.

This collection of essays is intended to serve two purposes. First, the essays themselves review some of the more important areas of advance since Thorpe's book. They thus help to bring the subject up to date. The essays have been grouped according to the biological and psychological problems to which they are pertinent, but it will be apparent that the divisions are little more than an editorial device, for each essay contains material pertinent to many sections. Second, the volume is intended as a tribute to W. H. Thorpe. Nine of the contributors have worked in Thorpe's department, another (E. A. Armstrong) is a friend of long-standing, and all the others would, I believe, acknowledge that he had had a strong influence on their work. Many other friends and colleagues would have wished to join in had the volume been planned on a basis sufficiently broad to represent Thorpe's diverse interests, but it seemed preferable to present him with a volume having a central theme of its own. It is suitable that those others should be represented here by Konrad Lorenz, a long-term friend whose work was partly responsible for stimulating Thorpe to found the Madingley department.

R. A. HINDE

University of Cambridge,
Sub-department of Animal Behaviour,
High Street,
Madingley,
Cambridge.

FOREWORD

by **KONRAD Z. LORENZ**

It is a great honour to be allowed to write a little foreword to the essays on bird vocalizations presented to William H. Thorpe: since my claim to that honour is so small and so largely personal, I feel I must justify it. That the growth of a scientist's mind can involve something resembling convergent evolution was never brought home to me more forcibly than when I read Thorpe's book 'Science, Man and Morals'. He started life studying insects, which I know less about than any other comparable group of animals, yet from the start we were interested in very much the same problems, centreing about the evolution and the ontogeny of behaviour. Independently of each other we hit on the phenomenon of imprinting: he in parasitic wasps, I in birds. To this day, the complicated and still enigmatical interaction between the genetical basis of species-specific behaviour and the adaptively modifying influence of individual experience is still in the very centre of both our interests. One might say that the main object of study for each of us is the phylogeny of the ontogeny of behaviour. It is the subject of Thorpe's book 'Learning and Instinct in Animals' as well as that of my volume on 'Evolution and Modification of Behaviour'.

If and when a biologist is successful in his research and produces valuable results, well-wishing contemporaries very often remark that he was 'lucky in his choice of object'. This may be an altogether unjust depreciation of his merits. Often the successful discoverer knew beforehand that the particular object offered a great chance to solve the problem he was attacking. It was in fact Thorpe's genius which led him to choose the sounds produced by song-birds as the object of a long-term research programme—a programme whose successful prosecution was largely responsible for the recent award by the British Ornithologists' Union of the Godwin-Salvin medal to him. The great advantages of bird song as an object of comparative, physiological, sociological and even aesthetic investigation, have already been mentioned in the Preface, and I need not repeat what has been said there. Indeed, all of these possibilities for fruitful research are exploited most successfully in the contributions to this volume. However, I want to emphasize the unique importance of the sound utterances of birds as an object of *ontogenetical* study. It is no exaggeration to say that there is no type of animal behaviour whose ontogeny is so accessible to experimental research, nor one that has been investigated with equal success.

To cite one example: M. Konishi made the surprising discovery that, in some species of American sparrows, the inexperienced young individual possesses a phylogenetically preformed auditory template, containing a rather detailed 'blueprint' of what its species-song ought to sound like, and that hearing its own experimental subsong, the bird learns by trial and error to perform motor coordinations producing a pattern of sounds matching this innate template. Deafened birds never achieve more than an amorphous chirping noise. At a certain time, the acquired motor pattern becomes irreversibly fixated and independent of further auditory feedback. In both respects, the process is closely akin to inductive determination in embryology, as well as to imprinting. In the same birds all sounds other than those of song—such as flight calls and warning calls—are determined as simple fixed motor patterns, as are all sound utterances in Anatidae, Gallinaceous birds and others.

I have mentioned these examples from the work of Konishi because they furnish good illustrations of the extremely important results which are achieved by the logical continuance of a line of research initiated by W. H. Thorpe many years ago. They demonstrate that the conceptualizations as well as the methods of experimental embryology are fully applicable to the study of the ontogeny of animal behaviour. Thus the importance of these results is of an extremely general nature, far exceeding the narrower field of knowledge in which they were originally gained. They open up tremendous possibilities for future research in the directions of the all-important problems of the relationship of the phylogenetically acquired, inherited 'blueprint' or 'programme', and the epigenetic modifications achieved during ontogeny; in other words of the relationship between instinct and learning, the subject of W. H. Thorpe's book.

What I have said here is just one example of the way in which the work of W. H. Thorpe has initiated lines of investigation and stimulated the research done by his pupils and his friends. There are few scientists who have an equal number of good pupils, and fewer still with an equal number of friends—all of whom join in the sincere wish that the seemingly much-too-early retirement of so youthful a man may provide him with more time to pursue further what he has so successfully begun.

EDITORIAL ACKNOWLEDGEMENTS

I would like to thank P. Bateson, Joan Stevenson, R. White, and other colleagues at the Sub-Department of Animal Behaviour, Madingley, for helpful discussion during the preparation of this volume. I also wish to acknowledge a very great debt to Rachel Pollard-Urquhart for her conscientious attention to detail at all stages from the initial planning to the final proofs.

R. A. H.

PART A
THE PHYSICAL ANALYSIS OF AVIAN VOCALIZATIONS

INTRODUCTION

Progress in the study of bird vocalizations was for long restricted by difficulties of description and measurement. Onomatopoeic renderings of the 'Little-bit-of-bread-and-no-cheese' variety gave at most a crude description, and conveyed little idea of either pitch changes or the quality of the notes. Some idea of quality can be conveyed by the use of terms such as 'click' or 'cheep' but, while these are capable of reasonably precise definition (Broughton in Busnel, 1963), many authors have used them loosely.

There are in fact two problems here—first that of recording, and then that of description. Field recordings of bird song have been possible for over 50 years and, by the use of modern directional microphones and light-weight tape recorders, results of high quality can now be obtained. But good recordings are not in themselves sufficient, for they can have but limited circulation, cannot be included in printed papers, and are not amenable to measurement. A means of description is thus essential.

It might be thought that musical notation could be used at least for the more 'musical' bird calls, but this is rarely the case. For one thing, musical notation is designed for series of sounds possessing both temporal (rhythmic) and frequency (scale) structures different from those of most animal sounds. For another, musical notation is not intended as a means of representing the output of an instrument: it is a shorthand method of indicating the way in which the instrument should be manipulated. To interpret the score, a knowledge of the sorts of sounds the instrument produces is essential.

Pictorial representation of bird vocalizations was first made possible by the use of the oscillograph: this provides a record of amplitude as a function of time, but frequency is poorly represented. More generally valuable has been the sound spectrograph, which provides a record of frequency against time and can also be used to obtain amplitude sections across frequencies over particular short intervals in time (Thorpe, 1954; Andrieu in Busnel, 1963): many examples are shown in later chapters of this book. The last two decades have seen advances in the study of animal sounds which would have been quite impossible without this instrument.

Spectrograms have an additional merit. A trained observer can sometimes infer from them perceptual characteristics of the original sound: this implies that they portray physical features which are 'perceptually important'. In describing a spectrogram it would be desirable to concentrate on these, but it is not always clear how this should be done. Not all features of every spectrogram are perceptually significant, and the physical precision of the spectrograph can be misleading unless this is recognized. Thus to consider the three physical attributes of a sound:

(a) *Temporal characteristics*. The extent to which notes are repeated or continuous, start and end gradually or suddenly, and so on, is in large part responsible for the peculiar properties of such sounds as trills, chirps, buzzes. The relative lengths of notes, and of the intervals between them, give the phrases their rhythmic qualities. With experience, such perceptual features can easily be read from a spectrogram.

(b) *Frequency*. This also is well represented on the spectrogram, and is the main parameter determining subjective pitch. The relationship, however, is not one to one. In the first place, pitch perception takes time, so that short notes sound like clicks whatever their frequency. Second, it depends to some extent on physical intensity: for a low tone, pitch decreases with intensity, but for a high tone, it increases. Third, with some complex sounds, pitch depends on differences between frequencies rather than on their absolute values. For example when a tone is made up of frequencies with a constant difference greater than 100 Hz, the reported pitch is equal to the constant difference. Thus since the harmonics of a note differ by an amount equal to the fundamental frequency, they alone can be sufficient to determine pitch, even if the fundamental is filtered out (in Thorpe and Lade, 1961). A further discussion of subjective correlates of such auditory stimuli is given by Licklider (1951).

(c) *Distribution of energy between frequencies.* This is the major parameter determining tonal quality. Pure-sounding notes, for instance, have a small frequency range, or a strong fundamental and a series of harmonics above it. With harsh sounds, by contrast, the energy is distributed over a wide frequency range. Tonal quality also depends on phase relationships between harmonics: these are not measured by the sonagraph, though for most purposes the loss of information is not serious.

Valuable as they are, the interpretation of spectrograms is thus far from straightforward. Indeed, the relations between the measured physical properties of bird sounds and their perceived qualities are still not completely understood. The following contribution by Marler constitutes an important advance in this field.

REFERENCES

BUSNEL, R. G. (1963) *Acoustic Behaviour of Animals.* Elsevier, Amsterdam.

LICKLIDER, J. C. R. (1951) Basic correlates of the auditory stimulus. In *Handbook of Experimental Psychology* (S. S. Stevens, ed.), pp. 985–1039. Wiley, New York.

THORPE, W. H. (1954) The process of song-learning in the Chaffinch as studied by means of the sound spectrograph. *Nature* **173**, 465.

THORPE, W. H. and LADE, B. (1961) The songs of some families of the Passeriformes. I. Introduction: the analysis of bird songs and their expression in graphic notation. *Ibis* **103a**, 231–45.

1. TONAL QUALITY OF BIRD SOUNDS

by **P. MARLER**

The Rockefeller University and The New York Zoological Society, New York

INTRODUCTION

Many animal sounds share characteristics with the music of man. They form patterns of sound in time reminiscent of our own melodies. The units from which the patterns are constructed are sometimes simple noise, with a minimum of internal structure. Often however, they are tones, and we are reminded of the varying quality of sounds produced by our own musical instruments. Nowhere are these parallels more prominent than among birds. What function does this musicality serve? Some investigators regard bird song as exercises in aesthetics, with perhaps a minimum of significance to other individuals. Analogies with human aesthetics may or may not be valid but some of the structure of bird songs seems to serve their functions as signals in a system of communication.

Many birds, particularly oscines, rely extensively on sounds as a means of social communication. The complexity of their social behaviour, of their life histories, and of the communities in which they live, conspire to create an unusual need for a variety of signals. The individual bird needs a repertoire of easily distinguishable calls to elicit different responses from companions. The existence of bonds between two or more individuals to form pairs, families, and still larger groups of individuals, creates a need for variation of signal pattern between individual members of the species. Sounds uttered for long periods of time by one individual to another may gain effectiveness by variety, preventing habituation of the recipient. Cohabitation of many species in the same area favours further evolution of signal variety, to satisfy the requirements that some signals be species-specific. The fragmentation of plant communities by song birds into spatially overlapping niches creates unusual risks of interspecific confusion.

As with music, the sounds of birds vary in structure in two basic ways. First, the pattern in time may be varied. Many birds' songs consist of trains of sounds uttered in a regular and distinctive pattern. Their temporal patterns have been extensively studied by ornithologists. Second, tonal quality may be varied. This has been rather less studied (Thorpe and Lade, 1961*a,b*). This paper discusses some bird songs whose tonal quality may be important to the species, and some problems in analysing the physical basis of tonal quality.

ANALYSIS OF TONAL QUALITY

In a classical treatise on the psychology of music, Seashore (1938) distinguishes between the timbre of a sound and what he calls sonance. He makes an analogy with moving pictures. The timbre of a tone corresponds to the impression made by the sequence of frames projected at normal speed. Timbre refers to the spectral structure of the sound at one instance. Sonance is manifest in the changes in intensity or spectral structure, not on the long-time scale over which a melody develops, but from instant to instant within a single note. Sonance is especially notable at the beginning, or 'attack', and ending, or 'release', of a note, as well as in internal variations or 'portamento'. Thus, 'in order to give a complete description of the quality of a tone, it is evident that we must know the spectrum of each wave at representative stages and the character of the changes in spectra which takes place for the duration of the 'tone' (Seashore, 1938, p. 110).

Recordings of several different musical instruments playing the same sustained note can readily be told apart. This is true even if we cut the tape and discard the attack and release portions so that our only clue is the timbre of the notes themselves. Our ability to distinguish between them is based on differences among the instruments in the overtones accompanying a note of the same pitch. The fundamental frequency, by which we identify the frequency of the note, is accompanied by varying numbers of harmonics, and the timbre depends on how many harmonics there are and their relative amplitudes (Figure 1).

Addition of a second harmonic to the fundamental is said to add clearness and brilliance to the tone. A third again adds brilliance, but also contributes a hollow, throaty, or nasal quality. A fourth adds more brilliance and even shrillness; a fifth, a rich somewhat horn-like quality; a sixth adds a delicate shrill or nasal quality (Wood, 1962).

We can find similar harmonic patterns in bird songs (Thorpe, 1961). For example, the sounds of many thrushes have a rich tonal quality to our ears as a result of the wide array of harmonics. Plate 1 shows songs of a thrush and two other North American species that are famous for the rich timbre of their notes.

There is evidence of something more subtle here than just emphasis of the first few harmonics of the fundamental. The distinctive tone

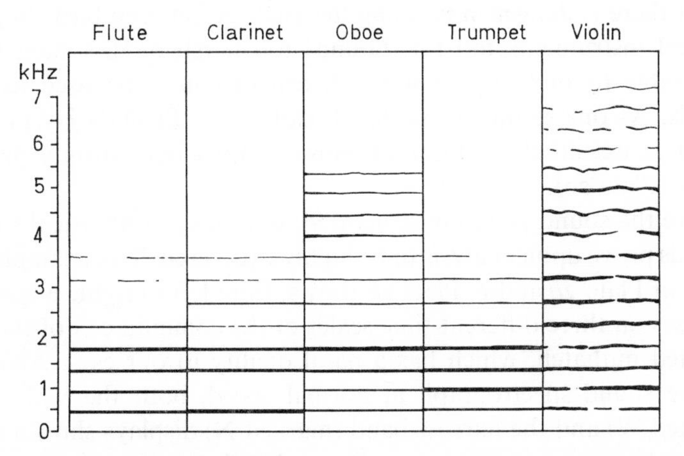

FIGURE 1. Sound spectrograms of a flute, clarinet, oboe, trumpet, and violin playing the same note, middle A, 440 Hz. These are narrow band analyses. (After Bergeijk, Pierce and David, 1960.)

of the clarinet results from emphasis on odd-numbered harmonics. Some notes of the hermit thrush have a similar mellow quality and analyses show that here, too, there is emphasis on the fundamental and third harmonic, with the second and fourth much weaker (Plate 1*c* left). The rich tone of many other bird sounds is undoubtedly attributable to a similar spectral organization.

Although as yet there is little proof, it is reasonable to suppose that such a distinctive timbre is relevant to the function of the sounds as signals (e.g. Dilger, 1956). However, the number of alternative types of tonal quality is not great. The size of the bird may also restrict variety. The smaller the vocal tract, the higher will be the natural frequency of its resonating parts. Harmonics are multiples of the fundamental frequency and, as the fundamental of a song is increased in frequency, the harmonics become more widely separated.

Eventually they begin to exceed the upper limit of the bird's hearing. High frequencies also attentuate very rapidly, and are inaudible at only a short distance from the singer. A song with a fundamental frequency of 4 or 5 kHz may have a second or even a third harmonic that is audible to other birds. It is unlikely that any above this will be heard, and even the second and third will be faint. Thus, one might suppose that small birds have more occasion to exploit sonance than timbre, as a means of achieving signal diversity.

But there is danger in pressing the analogy between bird song and musical instruments too far. Sounds too harsh or dissonant to be acceptable to our ears as musical, can of course be used as bird signals. As one example, the harsh call of the Steller's jay in California is essentially a blast of noise, emphasized around $3\frac{1}{2}$ and 7 kHz (Plate 2*a*).

Here the sound is barely structured, but many nonmusical sounds of birds prove upon analysis to be highly structured. Two examples are given in Plate 2*b* and *c*. Plate 2*b* shows, from left to right, a series of analyses on three different time scales of the common call of the red-breasted nuthatch, which has a nasal quality to our ears. Analysed on the sound spectrograph at normal speed, both the wide band (marked W) and the narrow band (marked N) displays show a series of simultaneous overtones across a wide frequency spectrum. We are reminded of the harmonically structured sounds of some of our own instruments. However, this is a small bird, and thus an apparent exception to the notion that only large birds should be able to develop an extensive array of harmonics. Further analysis reveals complications. When the recording is slowed to half- and then quarter-speed, these sounds are revealed in wide band analysis not as combinations of pure tones but as rapid trains of clicks or pulses.

The call of the blue-gray gnatcatcher (Plate 2*c*), which also has a rather nasal quality to our ears, proves to have a similar structure. Overtones visible after analysis at normal speed are resolved at quarter-speed in the wide band analysis as something like a rapid warble or fluctuating tone centred around 7 kHz. In neither case do the finely patterned overtones seem to be harmonics in the usual sense of the word, although at a quarter-speed we can see faint signs of second and third harmonics of the entire system centred around 14 and 21 kHz. How do such spectral systems relate to the harmonically organized overtones of musical sounds? To answer this, we must

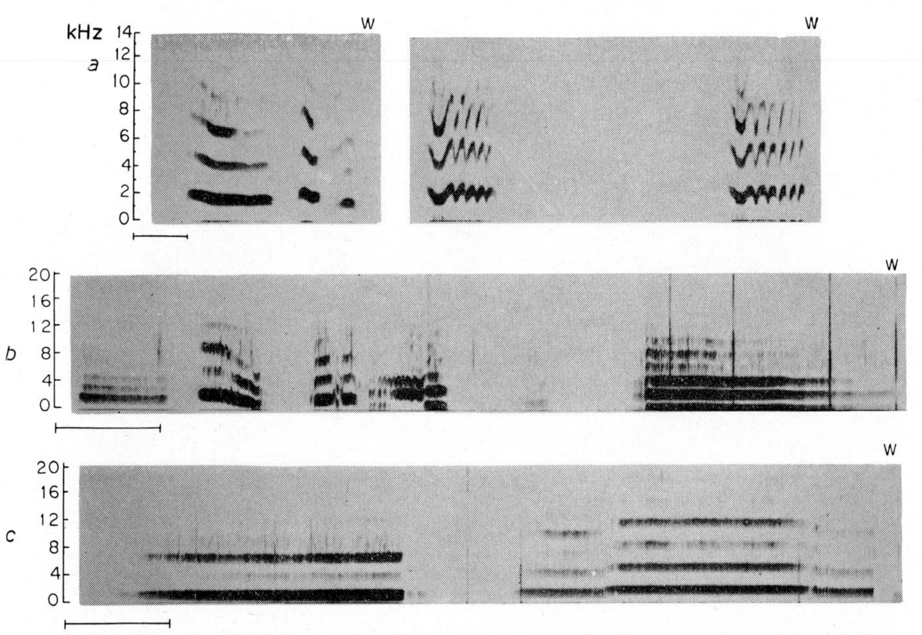

PLATE 1. Sound spectrograms of selected notes from three famous songsters, the mockingbird *a*, the Western meadowlark *b*, and the hermit thrush *c*. They all have strong harmonics and all show emphasis on odd harmonics. Compare with the clarinet in Figure 1. These are wide band analyses (marked W). The time marker is one-tenth of a second. The vertical scale is marked in kHz.

(*facing p. 8*)

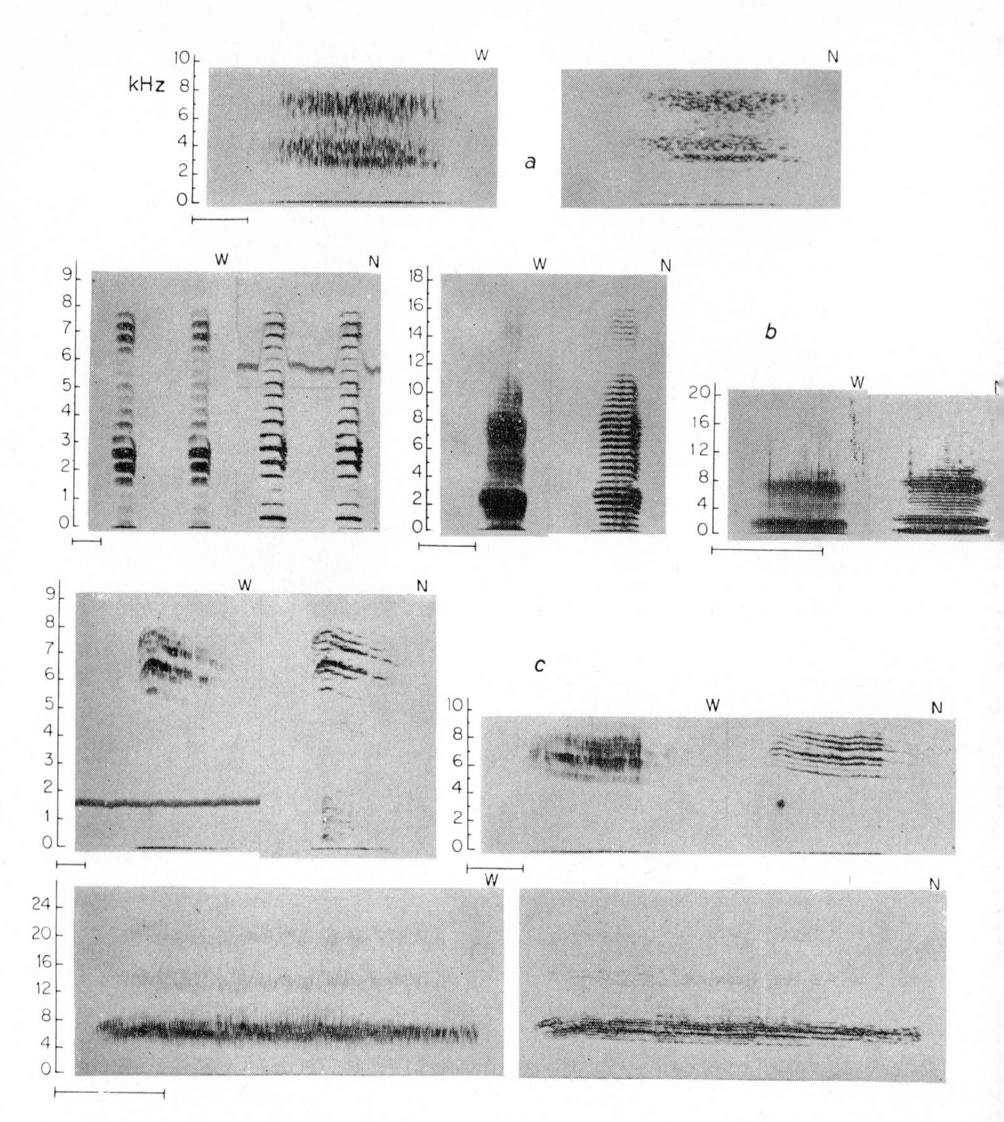

PLATE 2. Analyses of three nonmusical calls, the harsh 'wah' call of the Steller's jay *a*, the nasal 'ank' ca of the red-breasted nut-hatch *b*, and the 'thin peevish zpee' of the blue-gray gnat-catcher *c*—(see Peterso 1961; Brown, 1964). Both wide (W) and narrow band (N) analyses are shown. In *b* and *c* the tapes we played into the sound spectrograph at three different speeds, as indicated by the time markers, which a one-tenth of a second in duration. The vertical scale is marked in kHz. The two left figures in *b* each sho a pair of calls, the others each show a single call. The narrow band display on the left also includes a amplitude tracing which is irrelevant to the present discussion, as is the lower trace in the top left di play in *c*.

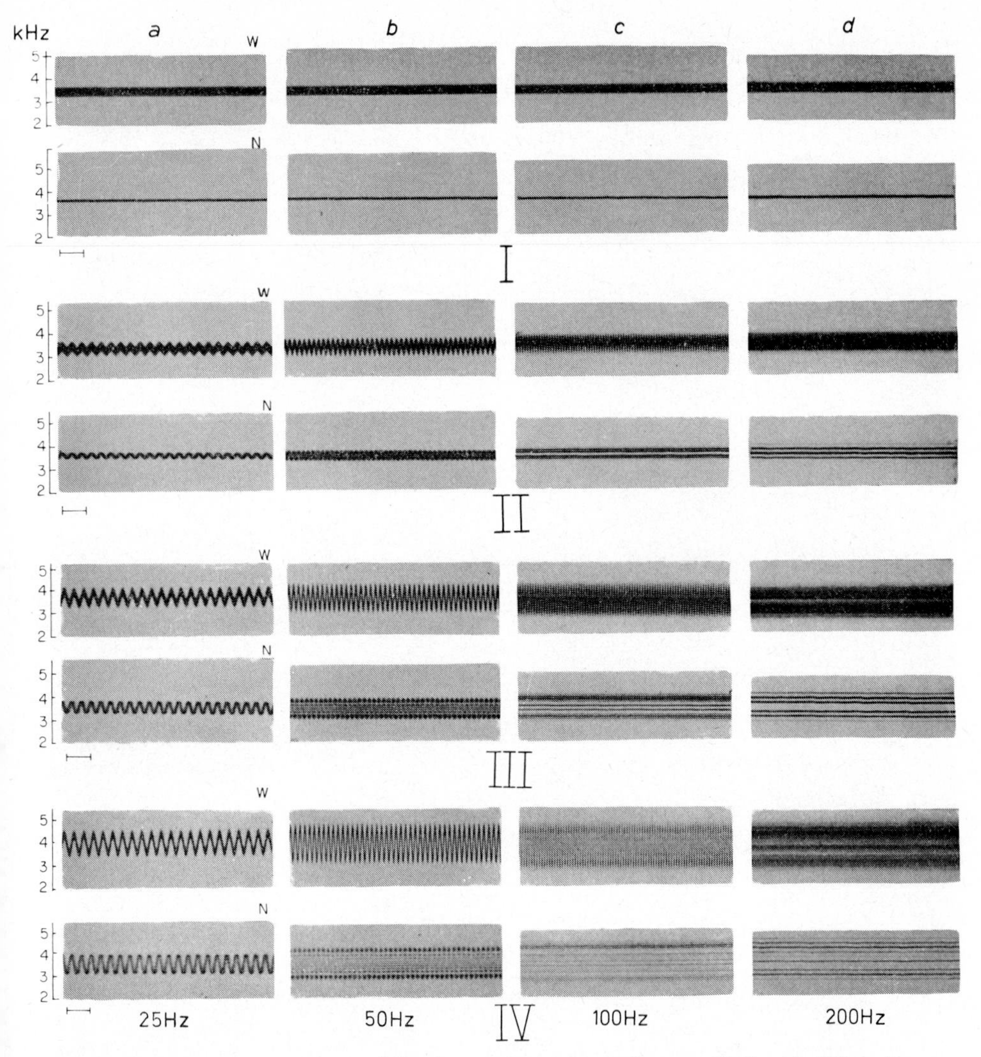

PLATE 3. A tone of about 3·5 kHz modulated at various rates and ranges. Both wide (W) and narrow band (N) analyses of each sound are shown, one above the other. The modulation rates are 25 Hz *a*, 50 Hz *b*, 100 Hz *c*, 200 Hz *d*. The range in I is zero. In II, III, and IV, the range is increased to approximately ±150 Hz in II, 250 Hz in III, and 450 Hz in IV. The time marker is one-tenth of a second.

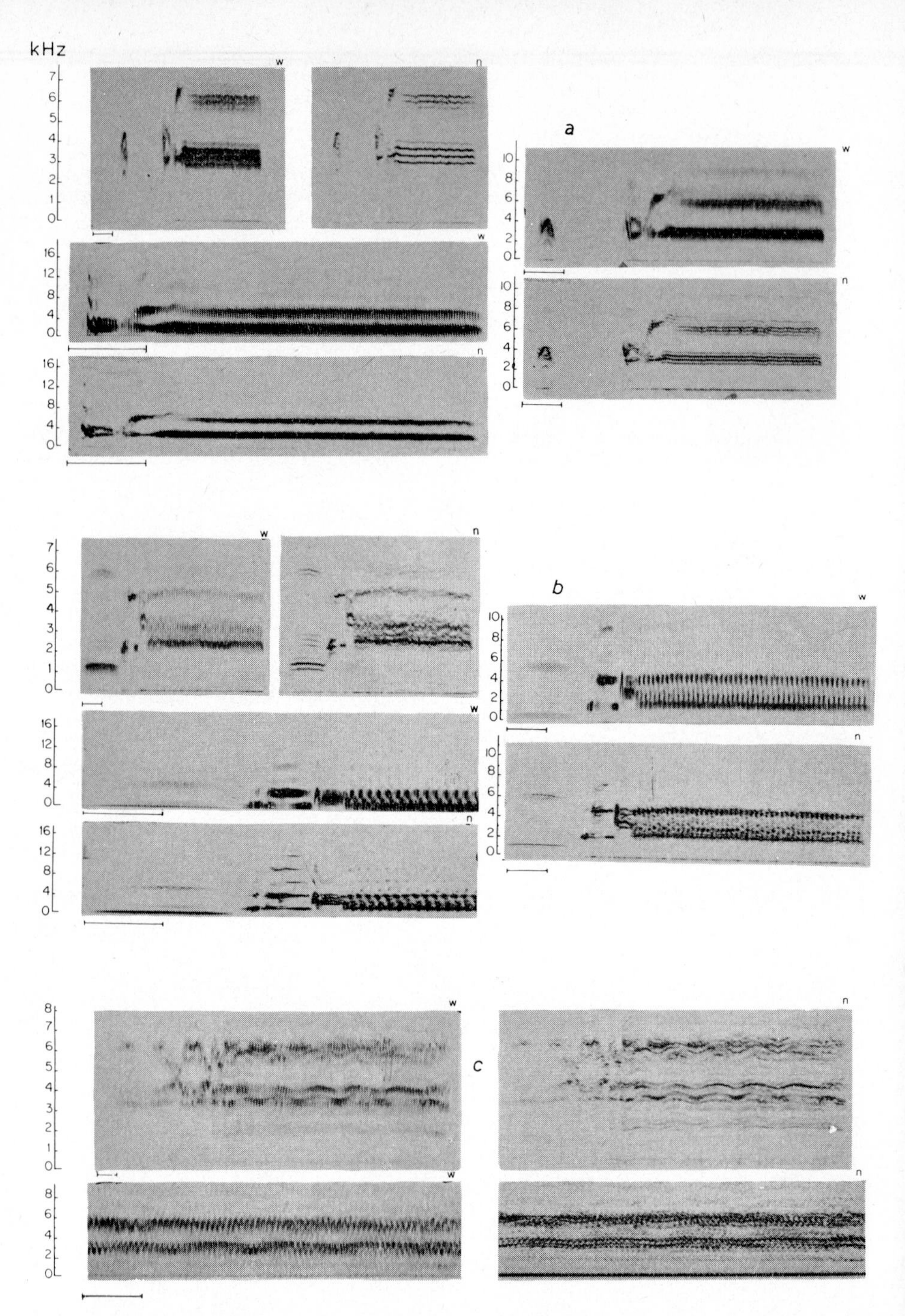

PLATE 4. Songs of three Icterids that include frequency-modulated tones. The second part of the Brewer's blackbird's song is not unmusical *a*. That of the red-winged blackbird *b* is more nasal, and that of the yellow-headed blackbird *c* has a harsh jangling tone to our ears. *a* and *b* are given at three playback rates, normal, half- and quarter-speed. *c* is given at two speeds, as can be seen from the time markers which are one-tenth of a second in each case. Both wide and narrow band analysis are shown of each sound.

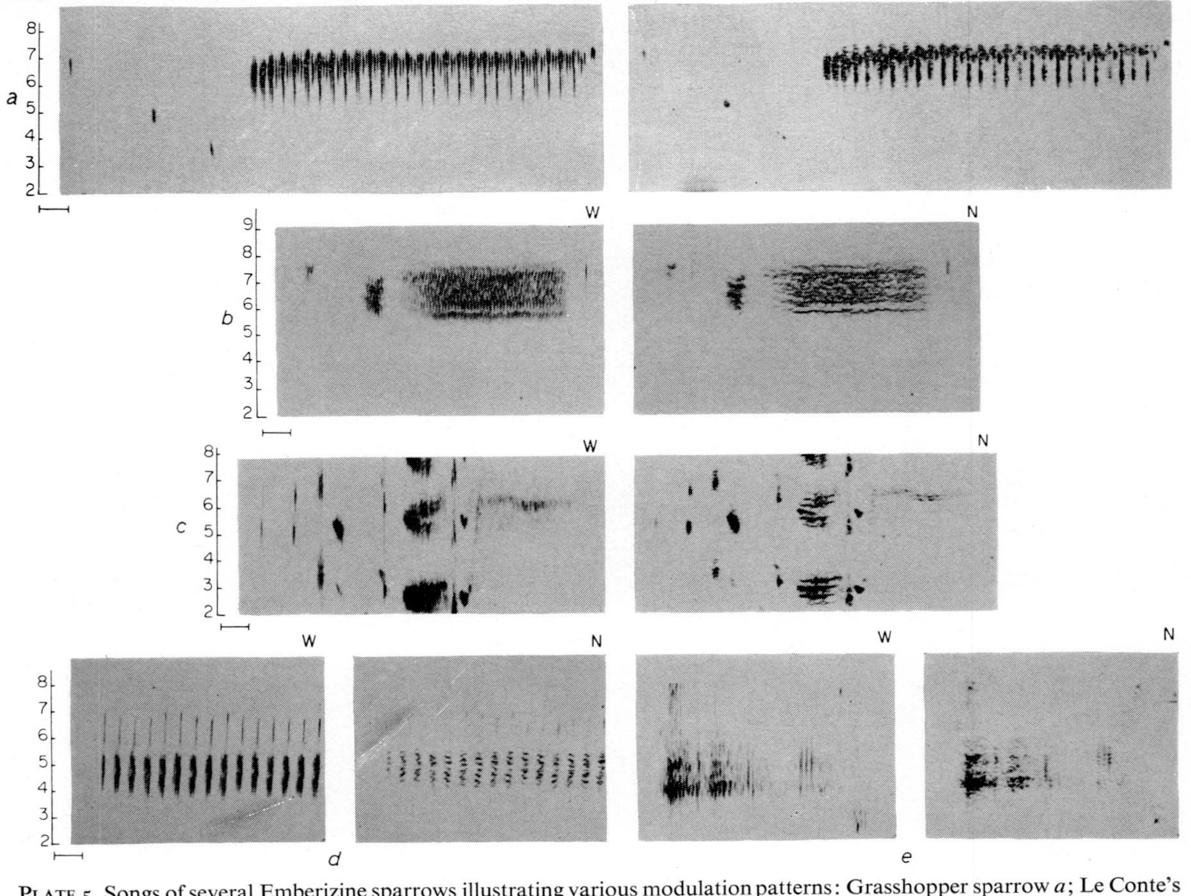

PLATE 5. Songs of several Emberizine sparrows illustrating various modulation patterns: Grasshopper sparrow *a*; Le Conte's sparrow *b*; seaside sparrow *c*; chipping sparrow (part of a song) *d*; sharp-tailed sparrow *e*.

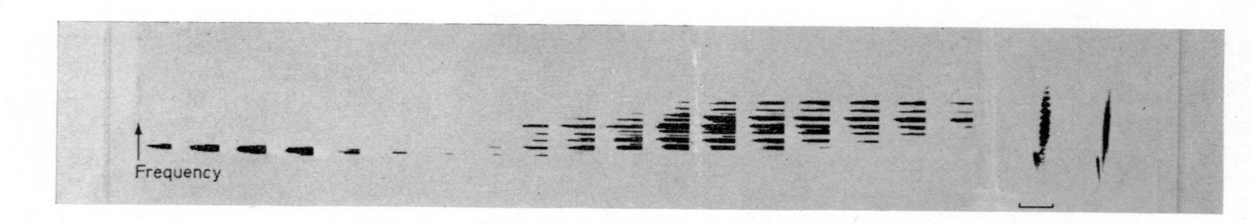

PLATE 6. A single syllable from a chipping sparrow song is shown in a narrow and wide band analysis on the right of the arrow. The rest of the figure shows a series of frequency-amplitude sections taken at about 2·5 msec intervals through the syllable from beginning to end. The length of the bars in each section show relative amplitude. The time marker at the right of the figure indicates one-tenth of a second.

consider another property of sounds which can contribute to tonal quality.

ANALYSIS OF THE VIBRATO

Suppose we play a pure tone on a signal generator and introduce a modulation into its frequency, sweeping it to and fro at a rate of about 1 Hz. We hear this as a warbling tone, the pitch of which glides rapidly up and down within the range we have set. Now we gradually increase the warble rate—the modulation frequency. At first, the only effect is an increase in the rate of fluctuation. But somewhere between six to eight modulations per second, our perception changes and we hear it not as a tone gliding up and down six times per second but as a single note with a definite pitch and intermittent variation of intensity. Together with this change, we also hear a modification of sound quality. It becomes richer in tone.

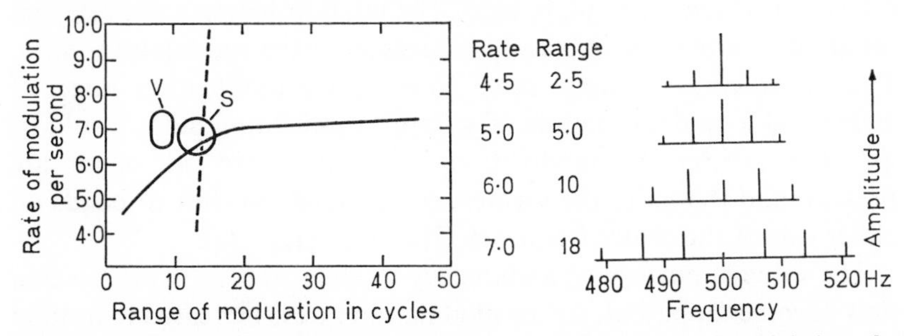

FIGURE 2. In the graph the solid line shows the rates and ranges of modulation of a 500 Hz tone giving an impression of singleness of pitch. The broken line shows the rates and ranges with the greatest richness of tone. The two enclosed areas define the vibrato used by accomplished singers (S) and violinists (V). The histograms on the right show the sound spectrum of a 500 Hz tone modulated at some of the rates and ranges illustrated. (After Stevens and Davis, 1938.)

Such frequency modulation is almost universal in human singing, and occurs in sounds of our instruments as well, particularly in those with strings. Skilled performers carefully control the rate with which tones are modulated, thus maximizing the richness of quality. The optimal rates again fall between 6 to 8 Hz. Both the modulation rates and the range of modulation used by artists coincide with the values found to be optimal in experiment (Figure 2).

What is responsible for the change in the impression we receive from a fluctuating tone as the modulation frequency is increased? If we ask an acoustician how the properties of such sound can be portrayed, we find that there are two alternatives. The sound can be viewed as a tone, the frequency of which is changing with time in a gliding fashion. It can also be viewed as a spectral system. For example, Stevens and Davis (1938) show how the spectrum of a 500-cycle tone varies with several combinations of modulation rate and range that give singleness of pitch (Figure 2). The spectrum consists of side-bands grouped around the original frequency, placed above and below at intervals which are multiples of the modulation rate. A modulation rate of 4·5 Hz yields a spectrum with peaks at 495·5 cycles, 500 cycles, and 504·5 cycles. With more rapid modulation rates, the side-bands are correspondingly apart. We also see that increase of the range of modulation results in additional side-bands, implying that a wider range of frequencies is represented. Here and elsewhere the term modulation 'range' refers to half the distance between the highest and lowest frequency reached during the warble (Stevens and Davis, 1938, p. 227). The relation between the rate and range of modulation is usually expressed as the modulation index. This is the ratio of range/rate. Thus with a modulation range of 20 Hz and a modulation rate of 5 Hz the modulation index would be 4·0. The greater the modulation index the more side-bands are present, and the more the greatest energy tends to shift to bands on either side of the centre frequency (Hund, 1942, p. 44).

The value of portraying a frequency-modulated tone in this fashion now becomes apparent. As the modulation rate is increased to around 6 or 8 Hz, we perceive it as a single tone with the complex spectrum that this analysis indicates. It consists, in fact, not of one but of several tones; and, if the modulation rate is still further increased to something above 12 Hz, an experienced listener can begin to detect several simultaneous tones rather than a single pitch (Stevens and Davis, 1938, p. 235). The richness of the tone is replaced by a rough quality that becomes more emphatic as the modulation rate is increased.

The choice between these alternative methods of viewing the structure of frequency modulated tones thus arises in our own perception, and seems to depend on the relationship between the rate of change of the sound patterning and the temporal characteristics of our ears. A critical modulation rate of about 7 Hz coincides well with von Békésy's estimate of the rate at which the sensation of a sound disappears after stimulation is terminated—about 0·14 sec. (Stevens and Davis, 1938, p. 238). If the notion of fusion is applicable to the perception of a modulated tone, this estimate would lead us to expect the modulations to fuse at about 7 Hz.

Although the richness of tone engendered by a vibrato of the correct rate seems to be the main explanation for exploitation of this device by musicians and singers, it may not be the whole story. The pitch of a vibrato is less certain than that of a pure tone, and Stevens and Davis (1938) suggest that a 'useful aspect of the vibrato from the musician's point of view is precisely this uncertainty of pitch, for it covers up slight errors in tuning' (Stevens and Davis, 1938, p. 240). Another possibility, suggested by Taylor (1965), is that frequency modulation minimizes the possibility of adaptation of receptors, by analogy with the role of eye movements in maintaining visual perception. It is also intriguing to note that vibrato in the human voice depends upon auditory feedback, as can be shown by playing masking sounds through earphones. 'The time delay in this feedback mechanism is about 0·15 sec. and this fits in very nicely with the 6·5 cycle modulation frequency that has already been mentioned' (Taylor, 1965, p. 111).

SOUND SPECTROGRAMS OF THE VIBRATO

Having considered how sounds with a vibrato are perceived by the human ear, we may ask how a machine designed for acoustical analysis will treat them—whether as spectral systems or as patterns changing with time. The answer again depends on the relationship between the rate of modulation and the temporal characteristics of the analyser. In a sound spectrograph the time constant of the filters relates in turn to their band-width.

As with many types of physical measurement, there is an uncertainty principle in the analysis of sounds, manifest as a conflict between requirements for precise temporal analysis of the sound and for precise determination of its frequency. For accurate temporal analysis we must be content with a wide band-pass filter and imprecise frequency analysis. Conversely, if we seek an exact picture of frequency, a narrow band-pass must be used, and the display of temporal properties is necessarily less precise (Joos, 1948; Cherry, 1966).

In the design of the Kay Electric Company sound spectrograph, this fundamental restriction on acoustical analysis is met by two band-widths, a narrow one of about 45 Hz and a wide one of 300 Hz. On the basis of the uncertainty principle, we may think of the resolution of the former along the time axis as of the order of 25 msec. and of the latter about 3 msec. (Cherry, 1966, p. 149).

We can now predict how such an instrument will treat frequency-modulated tones. As a single modulation cycle approaches the limit of temporal resolution of a filter, so the sound will be treated spectrally rather than as a frequency fluctuating from instant to instant. The limits will be different for the wide and narrow filters. The proper interpretation of the displays in Plate 2*b* and *c* now emerges. As a result of the difference in temporal resolution of the two band-widths, there is a class of frequency-modulated pattern that will be seen as a system changing with time in wide band analysis, and as spectral systems in narrow band analysis. The wide band analysis of the call of the red-breasted nuthatch at a quarter-speed shows a rapid train of separate clicks or pulses. The same sound is seen by the narrow band analyser as a system of side-bands placed at intervals of the pulse rate at 500 Hz (Plate 2*b*, right). If we compare the half-speed analysis, the doubled pulse rate now approaches the limit of temporal resolution of the wide band filter as well, although the pulse structure can still be discerned. At normal speed, both filters treat it as a spectral pattern. Three pairs of analyses of the call of the blue-gray gnatcatcher (Plate 2*c*) give a somewhat similar picture.

This contrast in the behaviour of wide and narrow band-pass filters is best displayed by presenting an artificial series of frequency-modulated tones to the sound spectrograph. Plate 3 shows a carrier frequency of about 3·5 kHz modulated at 25 Hz in *a*, 50 Hz in *b*, 100 Hz in *c*, and 200 Hz in *d*. The records appear in pairs with the narrow band analysis below the wide band analysis of the same sound. In series I, the range of the sinusoidal modulation is at zero. In series II, III, and IV, the range is increased to 150 Hz, 250 Hz and 450 Hz, progressively increasing the modulation index. Modulation at 25 Hz is clearly seen by both wide and narrow filters as a tone changing with time. With an increase to 50 Hz, approaching the band-width of the narrow filter, it begins changing to a spectral treatment, although the wide filter still resolves each modulation cycle. Even at 200 Hz, the wide band filter with a 300 Hz bandpass is still treating the modulation pattern as a sequence in time, although there is some indication of side-bands developing. We must, in fact, go above

300 Hz, to equal or exceed the bandwidth of the wide band filter before it changes over completely to spectral analysis. Plate 3 also illustrates the increasing numbers of side-bands that are generated as the modulation index is increased, and the complex changes in the relative distribution of amplitude among them. Thus in column d the modulation indices from top to bottom are 0, 0·75, 1·2 and 2·2. At 0·75 the centre frequency is still strong. At 1·2 the energy has shifted to both sides of the centre frequency. At 2·2 the centre frequencies are emphasized with side-bands of lower energy between them. In fact, a knowledge of the modulation index suffices to calculate the distribution of energy among the frequencies which are present (Hund, 1942, p. 352). Although these figures were obtained using a sinusoidal modulation, similar patterns are generated by trains of clicks or pulses, with side-bands spaced at intervals of the pulse frequency. Such patterns are known from the vocalizations of fish and cetaceans (Watkins, 1967, see below).

If we listen to such a range of sounds—and now we come to the point of this discussion—they have distinctive tonal qualities, quite unmusical to our ears; but nevertheless they offer a wide variety of alternative patterns. Are these exploited by birds? Having now some impression of the rough, twangy, nasal tone of rapidly modulated sounds, we know better where to seek for examples of their use.

ICTERIDS AND SPARROWS

One family in which the songs often have a nasal or twangy quality is the Icteridae, including the North American blackbirds. When recordings are analysed, it is found that many songs do, in fact, contain frequency-modulated tones. One species in which the modulated tones are smooth and almost musical to our ears is Brewer's blackbird, *Euphagus cyanocephalus*. Analysis at normal, half, and quarter speed (Plate 4*a*) shows that the second part of the song consists of a tone of about 3½ kHz, modulated at a rate of about 350 Hz through a range of approximately ±500 cycles. Both the carrier frequency and the modulation rate are stable through the duration of the note. The whole pattern is also represented as a strong second harmonic, and at a quarter-speed we can see traces of third and fourth harmonics. The modulation pattern serves to give the song of the Brewers' blackbird a distinctive and unmistakable tone.

Two other icterids make an interesting comparison. The red-winged blackbird's song, *Agelaius phoeniceus*, has a more squeaky and rasping tone. Analysis again reveals a frequency-modulated pattern. In the example shown in Plate 4*b*, the modulation rate is appreciably slower than in the Brewer's blackbird, resolvable as a warble pattern by wide band analysis even at a normal speed. The narrow band filter reveals a first part with clear side-bands, but a second part with side-bands that are blurred and fluctuating. The impression is of a much more noisy sound than that of the Brewer's blackbird. Slowing the recording reveals, in the second part of the song, a modulation pattern in which the structure of each cycle is complex, this complexity presumably being responsible for the lack of clarity in the spectral structure.

In the yellow-headed blackbird, *Xanthocephalus xanthocephalus*, this trend towards variability of the modulation pattern goes further. The terminal part of the song is a rasping buzz. Analysis at normal and half speed (Plate 4*c*) reveals the basis for this noisy quality. Both the carrier frequency, around approximately 3½ kHz, and the rate at which it is modulated, vary in the course of the note, generating a complex and variable pattern of side-bands and engendering a very distinctive tone. Other blackbirds and grackles illustrate further variations upon this theme.

North American sparrows (Plate 5) also exhibit a wide range of patterns in delivering sequences of syllables or pulses. In some species, such as the grasshopper sparrow, *Ammodramus savannarum*, the rate of delivery is slow (Plate 5*a*). Here the separate pulses are clearly audible, at a rate of about 20 per second. In Le Conte's sparrow, *Passerherbulus caudacutus* (Plate 5*b*), the main part of the song is modulated in a more rapid variable fashion at the rate of about 100 Hz. In the seaside sparrow, *Ammospiza maritima*, the sustained note in the middle of the song is modulated at a still higher rate, around 300 Hz (Plate 5*c*). The sharp-tailed sparrow, *Ammospiza caudacuta*, shows a trend in another direction, with rapid, irregular pulses spread over such a wide and varying frequency range that the effect is to generate a very noisy sound (Plate 5*e*). Each of these songs has a distinctive tonal quality.

Frequency modulation is not restricted to sustained notes. For example, Plate 5*d* shows notes from the song of the chipping sparrow, *Spizella passerina*, with an internal structure suggestive of a frequency-modulated tone. More detailed investigation of such notes is possible by frequency-amplitude sectioning. Plate 6 shows a series of 18 sections, taken from beginning to end through a single syllable, from the song of a chipping sparrow. This shows clearly that the first part of the syllable is a single tone deflected downwards and the second is a rising tone modulated at a rate of about 100 Hz. Again we may presume that such patterns contribute to the tonal quality of the song that is generated.

In addition to frequency modulation, patterns of amplitude modulation can also contribute to tonal quality and might be exploited by birds. The effects are again the result of side-bands, though in this case only two are produced, spaced above and below the centre frequency at intervals of the modulation frequency (Stevens and Davis, 1938). Amplitude and frequency modulations can be confused under some conditions. Thus with modulation indices of less than about 0·4 the spectrum of frequency modulated and amplitude modulated tones are identical (Hund, 1942) and it may be necessary to reduce the speed of playback into the sound spectrograph, or to use an oscillograph, to distinguish between them. As a still further complication, superposition of an amplitude modulation on a frequency modulation can cause asymmetrical suppression of side-bands below the centre frequency, which perhaps explains some of the irregularities in the distribution of side-band energy in certain bird sounds (Hund, 1942, p. 40).

According to Watkins (1967; personal communication) a train of distinct pulses or clicks is more properly treated as a pattern of amplitude modulation. Unlike amplitude modulation of a tone, which generates only two side-bands, trains of discrete pulses can generate a whole series of side-bands or overtones, beginning with the fundamental of the pulse rate. The principal difference between the spectra of a pulse train at a certain rate and a tone which is frequency-modulated at the same rate lies in the relative distribution of energy in the side-bands. We have seen the complex patterns which result with frequency modulation. With pulse trains the energy is more evenly distributed. Further examination of some of the bird sounds we have discussed suggests that they have this character. In Plate 2, *b* and *c* for example, the gnatcatcher call is clearly frequency-modulated, but the nuthatch call, even at the slowest speed, resolves not to a frequency-modulated tone but to a series of discrete pulses. The even distribution of energy in the sidebands across the spectrum tends to confirm this interpretation and the same may be true of the chipping sparrow song syllable analysed in Plate 6. Among the Icterid songs in Plate 4, *a* and *c* are clearly frequency-modulated, but the red-winged blackbird song in *b* is complex enough to suggest that some characteristics of a pulse-train are also present. All of these variations add to the types of spectra generated and therefore increase the variety of tonal qualities a sound can have.

Most of our discussion has concerned steady-state sounds. The range of possibilities is still further increased if we take account of variations in sonance. The manner of starting or finishing a sound can also vary greatly, as can the form in which frequency may change with time in a non-periodic fashion. For example, some notes in the song of the European robin, *Erithacus rubecula*, start very abruptly, endowing them with a distinctive tonal quality to our ears (Thorpe, 1959).

WHAT DOES TONAL QUALITY MEAN TO A BIRD?

In all of this discussion we have ignored the subjective nature of the concept of tonal quality. Are we justified in assuming that tonal quality has significance to birds and, if so, how are their impressions likely to compare with ours?

Since their ears function in essentially the same way as ours and respond to the same classes of sound properties, birds probably hear something analogous to both timbre and sonance. The range of frequencies they hear corresponds to our own, although the frequency they hear best varies widely, correlating with the size of the bird (Schwartzkopff, 1955). Their ability to discriminate sound frequencies also approaches our own. However, there are various reasons for thinking that birds are more responsive to the temporal properties of sounds than we are (Schwartzkopff and Winter, 1960; Schwartzkopff, 1963; Pumphrey, 1961; Schwartzkopff, 1957; Thorpe, 1963). This must influence the way in which frequency-modulated tones are heard.

Similar principles apply to the instrumental analysis of frequency-modulated sounds and to biological receptors. We may predict that the maximum modulation rate that is heard as a single complex tone rather than as a note whose pitch is gliding up and down, will be somewhat higher in birds than in humans. Nevertheless, a range of modulation rates above the transition will still be heard as complex tones. It is reasonable to suppose that the quality of these tones, as well as the patterns of sound in time, contribute something to species specificity. Proof of this must await playback experiments of the type Falls has conducted with the white-throated sparrow (Falls, 1963). Falls chose this species for his experiments partly because its songs were easy to synthesize. Now that the structure of some of the more complex songs is better understood, they can more readily be synthesized, thus opening up avenues for a new type of investigation.

ACKNOWLEDGEMENTS

Discussion with Drs Jack Bradbury, Donald Isaac and Kelvin Neil helped me to understand the problem of acoustical analysis considered here. I am indebted to Mrs Kathryn Bohning, Dr Jeremy Hatch, Mrs Helene Jordan, Mr William Martin, Dr William Watkins and Mr Richard Zigmond, for criticisms of the manuscript which is dedicated to Dr Thorpe in grateful acknowledgement of nearly twenty years of intellectual stimulation and friendship. The work was supported by an N.S.F. grant.

REFERENCES

BERGEIJK, W. A. VAN, PIERCE, J. R. and DAVID, Jr., E. E. (1960) *Waves and the Ear*. Doubleday, New York.

BROWN, J. L. (1964) The Integration of Agonistic Behaviour in the Steller's Jay, *Cyanocitta stelleri* Gmelin. University of California. *Publ. Zool.* **60** (4), 223–328.

CHERRY, COLIN (1966) *On Human Communication.* M.I.T. Press, Massachusetts.

DILGER, W. C. (1956) Hostile behavior and reproductive isolating mechanisms in the avian genera *Catharus* and *Hylocichla*, *Auk* **73**, 313–53.

FALLS, J. B. (1963) Properties of bird song eliciting responses from territorial males. *Proc. 13th Int. Orn. Congr.* **1**, 259–71.

HUND, A. (1942) *Frequency Modulation.* McGraw-Hill, New York.

JOOS, MARTIN (1948) Acoustic Phonetics. *Language Suppl.* **24** (2), 1–136.

PETERSON, R. T. (1961) *A Field Guide to Western Birds.* Houghton Mifflin, Boston.

PUMPHREY, R. J. (1961) Sensory organs: hearing. In *Biology and Comparative Physiology of Birds* (A. J. Marshall, ed.), **2**, 69–86. Academic Press, New York.

SCHWARTZKOPFF, J. (1955). On the hearing of birds. *Auk* **72**, 340–7.

(1957) Untersuchung der akustischen Kerne in der Medulla von Wellensittichen mittels Mikroelektroden. *Zool. Anz. Supply.* **21**, 374–9.

(1963) Morphological and physiological properties of the auditory system in birds. *Proc. 13th Int. Orn. Congr.* **2,** 1059–68.

SCHWARTZKOPFF, J. and WINTER, P. (1960) Zur Anatomie der Vogel-Cochlea unter naturlichen Bedingungen. *Biol. Zbl.* **79,** 607–25.

SEASHORE, E. C. (1938) *Psychology of Music.* McGraw-Hill, New York.

STEVENS, S. S. and DAVIS, H. (1938) *Hearing: Its Psychology and Physiology.* Wiley and Sons, New York.

TAYLOR, C. A. (1965) *The Physics of Musical Sounds.* Elsevier, New York.

THORPE, W. H. (1959) Talking birds and the mode of action of the vocal apparatus of birds. *Proc. Zool. Soc. Lond.* **132,** 441–5.

(1961) *Bird Song.* Cambridge University Press, Cambridge.

(1963) Antiphonal singing in birds as evidence for avian auditory reaction time. *Nature* **197,** 774–6.

THORPE, W. H. and LADE, B. I. (1961*a*) The songs of some families of the Passeriformes. I. Introduction: The analysis of bird songs and their expression in graphic notation. *Ibis* **103a,** 231–45.

(1961*b*) The songs of some families of the Passeriformes. II. The songs of the buntings (Emberizidae). *Ibis* **103a,** 246–59.

WATKINS, W. A. (1967) The harmonic interval; fact or artifact in spectral analysis of pulse trains. In *Marine Bioacoustics* (W. N. Tavolga, ed.), vol. 2. Pergamon Press, New York.

WOOD, A. (1962) *The Physics of Music.* Dover Publications, New York.

PART B
DEVELOPMENTAL ASPECTS

INTRODUCTION

It is a commonplace that many movement patterns develop in the normal manner in individuals reared in social isolation. To cite but one example, Burmese red junglefowl (*Gallus gallus*) reared in individual pens show many of the courtship and aggressive displays of the species, even though they have never witnessed the performance of those displays by other individuals (Kruijt, 1964). On the other hand some motor patterns develop only as a consequence of example from others, human speech being an extreme case. The peculiar interest attached to studies of the development of bird vocalizations arises not only from the extent to which these two extremes are represented, but also from the numerous cases which lie between them. On the one hand, many species-characteristic calls develop in the individual independently of experience of calls from other individuals, and many species seem incapable of vocal imitation. On the other, some species, like parrots and mynahs, seem able to imitate almost any kind of sound that they hear. Between these lie, for instance, many song-birds which produce aberrant songs if reared in social isolation, and sing the full species-characteristic song only if exposed to song from other individuals during a limited sensitive period, but are yet able to learn only a restricted range of sound patterns.

The first such species to be studied in detail was the chaffinch (*Fringilla coelebs*). Thorpe's (1958, 1961) principal findings can be summarized as follows:

(i) The normal adult song lasts about 2·5 sec., ranges from 2 to 6 kHz, and is divided into three phrases, each of several notes, followed by a terminal flourish. The three phrases are normally of successively lower frequency. Each individual has a repertoire of one to six songs of this type, differing in their details. Some of the notes in the song appear to develop from call-notes used in other contexts.

(ii) The song develops from a rambling subsong which has no definite duration, a wide range of notes, and no terminal flourish. The development seems to consists of the progressive omission of extreme frequencies and closer approximation towards the normal pattern. Although the subsong sometimes contains imitations of alien species, these are eliminated in the development to full song.

(iii) Chaffinches hand-reared in isolation from all other individuals from a few days of age develop only a simple song which, although recognizably a chaffinch song, is not divided into phrases and lacks the terminal flourish.

(iv) Chaffinches hand-reared in isolation from normal song from a few days of age, but kept in groups, develop songs which are more elaborate than those of individuals kept in total isolation, though they lack a terminal flourish and are usually quite unlike wild-type songs. The song patterns are usually similar in the members of each group, but differ markedly between groups: this indicates that learning by counter-singing is occurring within each group.

(v) Chaffinches caught in their first autumn and kept subsequently in isolation, develop songs more or less within the normal range. The song is divided into three sections and there is a terminal flourish. The difference between these birds and those reared in total isolation indicates that some learning must occur in the first few months of life, long before the birds themselves begin to sing.

(vi) Chaffinches caught in their first autumn and kept subsequently in groups, develop group songs which are just like those of wild birds. The difference between these birds and the autumn-caught isolates again indicates the importance of countersinging within the group both in elaboration of the song pattern and in mutual imitation.

(vii) Chaffinches reared in isolation and exposed to tape recordings of chaffinch song during their first autumn and winter subsequently produce nearly-normal songs. However, such birds will not learn anything which is played to them. For instance, an artificial chaffinch song in which the notes had a pure tonal quality and were thus unlike those of normal chaffinch song was not imitated, but chaffinch song played backwards, or with the ending placed in the middle, was. There is thus a restriction on what a chaffinch reared in isolation will learn. That this is not a matter of the sorts of sounds which the bird's own syrinx can produce is shown by the wider range of sounds in the subsong. Thorpe suggested that the criterion may be a resemblance in note structure to normal chaffinch song.

(viii) Chaffinches isolated only from their first autumn are even more restricted in what they will learn. If reversed or re-articulated song is played to them, they will not subsequently produce an imitation of it. They are, however, affected by hearing normal chaffinch song. Thus their early experience has had the effect of increasing the selectivity of their responsiveness to song.

(ix) Once a chaffinch has acquired its repertoire of songs in its first breeding season, little further change takes place.

Subsequent studies have shown that song-learning occurs in a basically similar way in a number of other species, though in each the process has its own peculiarities. For instance the length of the sensitive period in which song-learning is possible varies widely, and in some species the song of birds reared in isolation is simpler than the species-characteristic song, while in others it is more complex (e.g. Marler, 1967). Many of these studies are reviewed in the essays by Konishi and Nottebohm and by Immelmann in this volume (chapters 2 and 4).

Since the final song pattern develops gradually from the subsong, and since its form is influenced by experience of hearing song long before the bird itself starts to sing, it seems possible that the effect of experience is to increase the reinforcing effectiveness of song. On this view, the development of full song from subsong would be due to the reinforcing value to the bird of hearing itself sing the full song. Two predictions would follow from this:

(i) Deafening the bird after it had experienced song, but before it started to sing itself, would make it impossible for the bird to monitor its own output and thus would nullify the effects of the early experience (Konishi, 1964). Experiments on several species by Konishi and Nottebohm, summarized in the next chapter, show that deafening does eliminate the influence of early experience on song development, even though it does not markedly affect singing in individuals which had developed their songs before they were deafened.

(ii) Since reinforcers (defined as stimuli which increase the frequency of responses emitted immediately previously) are usually effective for a wide variety of responses, any reinforcing effect of hearing song on singing should also operate for other responses. Experiments demonstrating this are described later by Stevenson (chapter 3).

It is possible also to relate the influence of song on song development to its effect in eliciting song. Since chaffinch song-types occur as local dialects (see chapter 14), the particular characteristics of each song-type, as well as the more general species-characteristics of chaffinch song, must be influenced by hearing song from other individuals. If songs are played to adult chaffinches in the breeding season, the song-type which is most frequent in each individual's repertoire is also the most effective in eliciting song from that individual. It thus seems probable that earlier experience of a particular pattern both influences the subsequent production of that pattern (or one like it) and enhances the effectiveness of that pattern in eliciting song (Hinde, 1958).

One way of relating all these findings is to suppose that the naive bird has a crude 'model' or 'template' of the species-characteristic song, which is improved by experience of that song. On this view sounds which approximate to the model are reinforcing and are also effective in eliciting singing. It must be remembered, of course, that the postulation of a model is merely a device for tying together the results of several experiments: the corresponding physiological reality is quite another issue.

A further quite different type of observation is also compatible with this view. This concerns antiphonal singing—that is, cases in which two individuals sing different 'parts' in alternation. Usually, each individual sings its own particular note or song, and never that of the other bird. However there are a few records of one bird singing both parts in the absence of its partner. One possible hypothesis here is that both birds have a model of both parts against which their joint output (whether self-produced or from the partner) must be matched. Although normally singing only its own part, each individual can produce the whole when necessary. This interesting form of singing, described earlier by Thorpe and North (1965), is discussed by Hooker and Hooker in chapter 9 of this volume.

Studies of song-learning have implications for a number of more general problems, and it is worthwhile mentioning a few of these briefly.

(i) The complex interweaving of the effects of genetic, environmental and response-produced factors in song development demonstrates the impossibility of dichotomizing behaviour into categories such as 'innate' or 'instinctive' and 'learned' behaviour. Development involves a constant interaction between organism and environment at all stages, and its understanding requires the teazing apart of this interaction—a task which over-simple dichotomies can only hinder.

(ii) The initial phase of song-learning occurs before the bird itself begins to sing. This first phase therefore fits neither the paradigm of classical conditioning, which involves the attachment of a response to a new stimulus, nor that of operant conditioning, which involves a change in the frequency of a response as a consequence of reinforcement. Thus those who believe all learning can be described as either classical or operant conditioning must consider either the possibility that song-learning does not fit easily into their categories, or that additional assumptions are necessary to make it do so.

(iii) In that the initial learning does not involve the response by which its occurrence is subsequently revealed, parallels to song-learning can be found in a number of other contexts—for instance:

(a) *Perceptual learning*. Rats exposed to cut-out figures on the side of their cages subsequently learn a discrimination involving those figures more readily than rats not so exposed (Gibson and Walk, 1956; Gibson *et al.*, 1958; Gibson, Walk and Tighe, 1959). The importance of perceptual learning in a wide variety of contexts has been emphasized by Thorpe (1956, 1963), though the category is a difficult one to define (Bevan, 1961).

(b) *Discrimination learning*. Lawrence (1949, 1950) showed that an acquired ability to discriminate between two stimuli had some measure of independence from the particular response the animal was required to make to them. For instance rats which had learnt to discriminate between two cues in one situation subsequently learnt to discriminate between them in a different situation more rapidly than rats not so trained, even when the significance of the cues was reversed. More recently Sutherland (1959) has suggested that discrimination learning involves two stages: first a learning of the qualities by which the two situations are differentiated, and then a learning to attach particular responses to the characters in question. An hypothesis of this sort accounts more satisfactorily for various phenomena of 'overlearning' than do rival theories.

(c) *Latent learning*. If rats are allowed daily access to a maze which has no reward in the goal-box, the number of 'errors' remains constant or falls only slowly in comparison with rats which are rewarded. If a reward is subsequently introduced, performance improves rapidly to a level similar to that of animals rewarded throughout. Thus the initially rewarded animals must have learnt something in the early trials, though this was utilized in reaching the goal-box only subsequently (e.g. MacCorquodale and Meehl, 1954).

(iv) Current theories of learning presume that what an animal learns is primarily determined by the stimuli to which it is exposed and/or the consequences of its behaviour—in either case, the restrictions are environmentally imposed. It is however becoming increasingly apparent that there are species-characteristic limitations on what an animal learns, and that these are not limitations of capacity (Lorenz, 1965): song-learning is an interesting example. As noted above, in the chaffinch the limitation seems to be set by the resemblance in note structure to normal song, but in other species different mechanisms operate. Nicolai's (1956) observations on bullfinches indicate that in that species each individual learns only from the male which reared him. Extensive observations by Immelmann showing a similar mechanism to operate in Estrildid finches are reviewed in chapter 4.

(v) The category of reinforcing stimuli can be defined in terms of the consequences of stimuli upon the probability that the response(s) which preceded their appearance will be repeated. If the factors influencing the occurrence of learning are to be understood, however, it becomes important to find independent criteria by which the consequences of stimuli on learning can be predicted. One route to this end is to describe reinforcing stimuli according to immediately preceding motivational conditions, as with food and water reinforcement. Such descriptions, however, are not invariably useful, as the influence of some stimuli on an operant response can be predicted more easily from their relevance to the behaviour of the species concerned in other contexts than from a preceding history of deprivation. As discussed by Stevenson in chapter 3, song is an addition to this previously little studied category.

(vi) Another issue raised by song-learning concerns the nature of sensitive periods. It is well-known that learning in many contexts is more likely to occur at some phases of the life cycle than others. For instance, the 'imprinting' of the following response of nidifugous birds is limited to a brief period in the first few days of life (Lorenz, 1935; Bateson, 1966). But it is important to remember that the factors limiting such 'sensitive periods' may differ from case to case, and generalizations about sensitive periods can be at best superficial. The factors limiting the sensitive periods for song-learning not only differ from those of imprinting, but also differ between species. Thus in the chaffinch it comes to an end during the individual's first spring, but in the red-backed shrike (*Lanius collurio*) and canary it recurs seasonally (Blase, 1960; Poulson, 1959). The matter is considered further by Konishi and Nottebohm in chapter 2, and by Immelmann in chapter 4.

(vii) If the hypothesis discussed above is at all close to the mark, and the second stage of song-learning involves an adjustment of the motor output to conform to the 'model', there are again parallels to studies in a number of other contexts. Thus in man the acquisition of many skilled movements can best be described in terms of comparison with a model of the required movement. Furthermore, the carrying out of skilled movements depends on comparison between the feedback expected as a consequence of the movement on the basis of previous experience, and that actually received: with experience the necessity for feedback is reduced (Held, 1961; Held and Bossom, 1961; Held and Freedman, 1963).

(viii) Stereotyped movements vary greatly in the extent to which their patterning depends on feedback. In mammals even movements like walking may be disrupted by deafferentation of the limb (e.g. Lassek and Mayer, 1953), though it has recently been demonstrated that the learning of simple avoidance responses can occur in the absence of sensory feedback (Taub, Bacon and Berman, 1965). In insects, fishes and amphibia, by contrast, there is considerable evidence that the coordination of many species-characteristic movements can be independent of feedback. Nearly all the experimental studies, however, have been concerned with locomotor movements, and there is no evidence about species-characteristic signals. Although Konishi and Nottebohm showed that deafening prevented normal song development in the species they worked with, it had little effect once song had been acquired. Thus the stereotyped movements of singing are independent of at least that source of feedback which seems most likely to be important. Song thus seems to contrast with speech, which is seriously disrupted if the normal time relations between speech and the auditory feedback are changed, though the consequences of the total removal of auditory feedback are less dramatic.

It is thus apparent that studies of the development of bird song have more than parochial interest, and link with studies of the ontogeny of behaviour drawn from widely different contexts.

REFERENCES

BATESON, P. P. G. (1966) The characteristics and context of imprinting. *Biol. Rev.* **41**, 177–220.

BEVAN, W. (1961) Perceptual learning: an overview. *J. Gen. Psychol.* **64**, 69–99.

BLASE, B. (1960) Die Lautäusserungen des Neuntöters (*Lanius c. collurio* L.), Freilandbeobachtungen und Kasper-Hauser-Versuche. *Z. Tierpsychol.* **17**, 293–344.

GIBSON, E. J. and WALK, R. D. (1956) The effect of prolonged exposure to visually presented patterns on learning to discriminate them. *J. comp. physiol. Psychol.* **49**, 239–42.

GIBSON, E. J., WALK, R. D., PICK, H. L. and TIGHE, T. J. (1958) The effect of prolonged exposure to visual patterns on learning to discriminate similar and different patterns. *J. comp. physiol. Psychol.* **51**, 584–7.

GIBSON, E. J., WALK, R. D. and TIGHE, T. J. (1959) Enhancement and deprivation of visual stimulation during rearing as factors in visual discrimination learning. *J. comp. physiol. Psychol.* **52**, 74–81.

HELD, R. (1961) Exposure-history as a factor in maintaining stability of perception and coordination. *J. nerv. ment. Dis.* **132**, 26–32.

HELD, R. and BOSSOM, J. (1961) Neonatal deprivation and adult rearrangement: complementary techniques for analyzing plastic sensory-motor coordinations. *J. comp. physiol. Psychol.* **54**, 33–37.

HELD, R. and FREEDMAN, S. J. (1963) Plasticity in human sensorimotor control. *Science* **142**, 455–62.

HINDE, R. A. (1958) Alternative motor patterns in chaffinch song. *Anim. Behav.* **6**, 211–18.

KONISHI, M. (1964) Effects of deafening on song development in two species of Juncos. *Condor* **66**, 85–102.

KRUIJT, J. P. (1964) Ontogeny of social behaviour in Burmese red junglefowl (*Gallus gallus spadiceus*) Bonnaterre. *Behaviour Suppl. No.* 12.

LASSEK, A. M. and MAYER, E. K. (1953) An ontogenetic study of motor deficits following dorsal brachial rhizotomy. *J. Neurophysiol.* **16**, 247–51.

LAWRENCE, D. H. (1949) Acquired distinctiveness of cues: I Transfer between discriminations on the basis of familiarity with the stimulus. *J. exp. Psychol.* **39**, 770–84.

(1950) Acquired distinctiveness of cues: II Selective association in a constant stimulus situation. *J. exp. Psychol.* **40**, 175–88.

LORENZ, K. (1935) Der Kumpan in der Umwelt des Vogels. *J. Ornith.* **83**, 137–213, 289–413.

(1965) *Evolution and modification of behavior.* University of Chicago.

MACCORQUODALE, K. and MEEHL, P. E. (1954) Edward C. Tolman. In *Modern Learning Theory* (Eds. Estes *et al.*). Appleton-Century-Crofts, New York.

MARLER, P. (1967) Comparative study of song development in sparrows. In *Proceedings 14th Orn. Congr.* (D. Snow, ed). Blackwell's Oxford.

NICOLAI, J. (1956) Zur Biologie und Ethologie des Gimpels (*Pyrrhula pyrrhula* L.) *Z. Tierpsychol.* **13,** 93–132.

POULSON, H. (1959) Song learning in the domestic canary. *Z. Tierpsychol.* **16,** 173–8.

SUTHERLAND, N. S. (1959) Stimulus analyzing mechanisms. In *Proc. sym. mechanization of thought processes.* H.M. Stationery Office, London.

TAUB, E., BACON, R. C. and BERMAN, A. J. (1965) Acquisition of a trace-conditioned avoidance response after deafferentation of the responding limb. *J. comp. physiol. Psychol.* **59,** 275–9.

THORPE, W. H. (1956, 1963) *Learning and Instinct in Animals.* Methuen, London.

(1958) The learning of song patterns by birds, with especial reference to the song of the chaffinch, *Fringilla coelebs. Ibis* **100,** 535–70.

(1961) *Bird Song.* Cambridge University Press.

THORPE, W. H. and NORTH, MYLES E. W. (1965) Origin and significance of the power of vocal imitation: with special reference to the antiphonal singing of birds. *Nature* **208,** 219–22.

2. EXPERIMENTAL STUDIES IN THE ONTOGENY OF AVIAN VOCALIZATIONS

by **MASAKAZU KONISHI and FERNANDO NOTTEBOHM**
Princeton University, New Jersey, and The Rockefeller University & New York Zoological Society, New York

INTRODUCTION

The study of behavioural development has been handicapped by the weight of common views. The conceptual and terminological laxity of the old nature-nurture controversy affected scientific discussion of the issue, and the polarization of opposing views enhanced existing biases still further. Experimental approaches to the nature-nurture problem can take two different courses, genetical or developmental. The cross-breeding of animals differing in behaviour reveals the relative importance of genetic and environmental factors in causing their differences. The study of ontogeny, on the other hand, involves the manipulation of environmental factors: genetic variance is either kept small or is assumed not to affect the results, so that any non-random variation observed is considered due to the environmental factors under test. However, neither genetical nor developmental study alone suffices to evaluate the relative weight of hereditary and environmental determinants of behaviour. Recent studies control both genetical and environmental variances (Parsons, 1967).

A conceptual dichotomy between inheritance and acquisition of behaviour tends to obscure the importance of developmental processes which involve varying degrees and modes of interplay between genetical and environmental variables. Analysis of this interplay is often difficult: some of the environmental factors are hard to define, some cannot be controlled independently, and some are indispensable for the development of the normal anatomical and physiological prerequisites for behaviour. Stimuli from the environment are not the only variables that affect behavioural ontogeny. The developing animal creates stimuli which may in turn influence the course of its ontogeny. This type of interaction cannot be easily fitted into genetical or environmental categories. Analysis of the interaction is often made difficult because the motor and sensory functions cannot be studied separately.

Vocal behaviour offers unique opportunities to study behavioural ontogeny. Vocal patterns can be tape-recorded and visually displayed with the audiospectrograph. The auditory environment can be easily manipulated and defined. Most important, the control of the auditory environment by acoustic isolation, by addition of external stimuli, or even by deafening does not directly interfere with the functioning of the vocal motor system. Audition stands in contrast with other types of perception such as proprioception, experimentation with which may interfere directly with the motor performance itself (e.g. Grohman, 1938).

Several reviews have been written on the ontogeny of avian vocalizations (Lanyon, 1960; Thorpe, 1961; Marler, 1964; Marler and Hamilton, 1966). In this chapter we will not only review recent findings but also attempt a reappraisal of some of the older work. We will also try to discuss avian vocal development within a broad perspective of behavioural ontogeny.

ACOUSTIC ISOLATION

Sensory deprivation experiments are designed to test whether particular behavioural traits and differences develop in the absence or presence of certain environmental information. As early as the 1920s, Schjelderup-Ebbe (1923) conducted acoustic isolation experiments with domestic chickens (*Gallus domesticus*). Birds were hatched and raised without exposure to sounds of adult chickens. Most of the normal vocalizations developed, although they first appeared slightly later than in normal chickens. No more systematic studies ensued until the 1950s, when members of the German school under the direction of Koehler (1954) began to publish their work. Sauer (1954) successfully raised two male whitethroats (*Sylvia communis*) from the egg in soundproof chambers. These birds are said to have produced all of the normal calls and subsong before they died prior to the onset of full song. Sauer believes that they would have sung normal full songs, since all the elements of normal full song were present in their subsong. Messmer and Messmer (1956) reared two blackbirds (*Turdus merula*) from the egg in acoustic isolation, and concluded that they developed normal songs. However, these early observations and conclusions have been accepted rather uncritically. The musical notations, oscillographic display, and spectral band analysis used by these early workers were not adequate to reveal the subtler frequency or temporal characteristics of vocal signals. These are the very features which later turned out to be most susceptible to modification by isolation or deafening.

Advent of the audiospectrograph introduced greater objectivity and reliability into the study of avian vocal development. For example when Thielcke-Poltz and Thielcke (1960) made audiospectrograms of the so-called normal songs of the same birds Messmer and Messmer had raised, they were found to lack some features characteristic of the songs of wild blackbirds. Blase (1960) was the first to follow the entire course of avian vocal development from the egg in acoustic isolation, using the audiospectrograph. He raised one male and two female red-backed shrikes (*Lanius collurio*) from the egg in individual isolation. No systematic abnormalities were observed in their calls. However, the isolates maintained the so-called juvenile song, which the wild red-backed shrike usually transforms to the 'call song' by the addition of sounds copied from other species.

Mulligan (1966) had three song sparrows (*Melospiza melodia*) foster-reared by canaries from the egg in a soundproof room. These birds produced songs indistinguishable from the average song of wild song sparrows. It is also interesting to note that they did not copy any of the canary sounds. This is, perhaps, the first and only reliable study of passerine birds in which the audiospectrograms did not show any qualitative differences between the isolate and the wild-type songs. However, in all these cases one can speak only about qualitative differences because of the small numbers of isolates raised to maturity.

The great difficulties encountered in hand-rearing newly hatched passerines forced most workers in this field to be content with isolating nestlings. Thorpe (1954, 1958) reported conspicuous qualitative differences between the songs of individual chaffinches (*Fringilla coelebs*) isolated as nestlings and that of wild conspecifics, an observation made earlier by Poulsen (1951). In the blackbird, contrary to the previous conclusion of Messmer and Messmer (1956), audiospectrographic analysis showed that nestling-isolates could not arrange the component sounds to produce normal wild-type songs without exposure to wild blackbirds (Thielcke-Poltz and Thielcke, 1960).

Some more recent studies of song development have used New World species. Among them the white-crowned sparrow (*Zonotrichia leucophrys*) was found to be an excellent subject for this purpose (Marler and Tamura, 1962, 1964). Unlike other species, each white-crown sings only one type of song which is shared by all the members of a given population, giving the impression of 'dialects'. The songs of nestling-isolates not only lack the dialect characteristics of their birthplace but also differ from any of the known song types of wild conspecifics, although they contain some of the basic properties shared by all wild populations (Plate 1). Furthermore, in contrast with the chaffinch, in which group isolation leads to a community song pattern, no systematic differences can be detected between the songs of group and individual isolates in the white-crowned sparrow. The isolation of fledged white-crowns, on the other hand, produces entirely different results; they can reproduce the dialect of their birthplace, indicating that exposure to conspecific song during the fledgling period determines the course of song development. Other experiments with the white-crowns will be mentioned later.

So far the specific effects of deprivation have been discussed in terms of the presence or absence of normal features in the resulting song. Marler, Kreith and Tamura (1962) discovered an additional effect of modifying the auditory environment on song development. The nestling-isolates of the Oregon junco (*Junco oreganus*) produced wild-type songs in individual isolation with a wild-type overall pattern but with a simpler syllabic structure. However those birds, which were raised in group isolation or in general laboratory rooms where they could hear other species, produced more complex syllables, and larger repertoires of songs which included sound patterns that were neither present in any of the known wild junco populations nor copied from other species. This phenomenon was attributed to improvisation. The junco work pointed out another interesting fact: the ability to develop normal song in isolation does not exclude the faculty to copy and to improvise new sounds. Mulligan (1966) made a similar observation in the song sparrow.

While it may be generally true that the effects of isolation of nestlings approximate those of isolation of eggs, the early exposure of the young to the natural auditory environment might have some effects upon their vocal development. Judging from the audiospectrograms prepared by Thielcke-Poltz and Thielcke (1960) the songs of their nestling-isolates and those of Messmers' egg-isolates are different. Since the song of the blackbird is very variable, it is difficult to say how many of the differences are due to individual variation and how many to the differences in their auditory experience. Blase (1960) presented rather convincing measurements indicating similarities between the song of an egg-isolate and that of a nestling-isolate in the red-backed shrike. Marler and Tamura (1964) reported that exposure to tape-recorded songs during the nestling period did not have any specific effects on song development in the white-crowned sparrow.

THE NATURE OF SONG LEARNING

What sounds are imitated

In the preceding section it was implied that acoustic isolation caused abnormal song development because of lack of exposure to the wild-type song. This assumption must be tested by exposing isolated birds to the wild-type song: this constitutes the tutoring experiment.

If hand-reared chaffinches are exposed to a tape-recorded wild chaffinch song during part of their first ten months of life, they will reproduce this song the following spring (Thorpe, 1958). Similar results were obtained with the white-crowned sparrow. Birds which heard a wild conspecific song on tape during part of their first three months of life develop a good copy of the model song (Marler and Tamura, 1964). In both of these species young birds showed a predisposition to imitate conspecific song. In the chaffinch exposure to alien songs had no noticeable effects except that one hand-reared bird imitated the song of a tree pipit (*Anthus trivialis*) with which it had been tutored. In frequency range and overall pattern, the song of the tree pipit is remarkably reminiscent of that of the chaffinch. Some of Thorpe's hand-reared chaffinches learned wild-type songs in which the usual order of the component syllables and phrases had been rearranged. He concluded that the chaffinch shows a well-defined tendency to imitate 'chaffinch-like' sounds. Individuals caught as juveniles, and which, therefore, had heard conspecific song in the field during their first spring and summer, did not respond to training with 'rearranged' songs, nor to songs of tree pipits. Exposure to conspecific song, therefore, seems to enhance their predisposition to copy chaffinch song.

Hand-reared white-crowned sparrows tutored with wild-type

conspecific song and the songs of two other sympatric sparrow species, the song sparrow and Harris's sparrow (*Zonotrichia querula*), did not imitate the song of the two latter species but copied that of their own. When the same alien songs were presented alone, hand-reared white-crowns did not copy them but developed songs resembling those of nestling-isolates.

The songs that chaffinches and white-crowned sparrows acquire during their first year of life remain virtually unaltered in future years. Further training with other wild-type songs of their own species or with songs of other species will not enlarge the song repertoire or alter the established patterns, whether they be normal or abnormal.

Not all species are as selective in their song learning as chaffinches and white-crowned sparrows. Some birds which in nature do not imitate other species may copy alien songs and other sounds when they are raised away from their natural environment. According to Lanyon (1957), eastern and western meadowlarks (*Sturnella magna* and *S. neglecta*) isolated from their own species at the time of fledging failed to develop the species-specific song but acquired songs from other species they heard. A *neglecta* male, for example, learned songs of a wood pewee (*Contopus virens*) and of a yellowthroat (*Geothlypis trichas*); and males of both species copied the song of the red-winged blackbird (*Agelaius phoeniceus*). A similar observation has been described for the linnet (*Carduelis cannabina*) (Poulsen, 1954). In the bullfinch (*Pyrrhula pyrrhula*) juvenile birds will learn whatever songs their father sings. If bullfinches are foster-reared by canaries, they will adopt canary song and ignore the normal song of conspecifics kept in the same room (Nicolai, 1959). Immelmann (1965, 1966 and this volume) describes a similar condition in the zebra finch (*Taeniopygia guttata*). Nicolai (1964) found a most interesting case of vocal mimicry in the parasitic weaver-finches (*Viduinae*). These birds learn not only the songs but also the calls of their host species.

There are, of course, well-known cases of species which under natural circumstances are 'open-minded' as to the kinds of environmental sounds they will include in their vocal repertoire. Popular examples are the mockingbird (*Mimus polyglottus*; Borror and Reese, 1959), the starling (*Sturnus vulgaris*) and some of the bower birds (Marshall, 1954). It is not known whether these species are as indiscriminate in their borrowing from the auditory environment as popular accounts indicate, and experimental studies with these species have yet to be conducted.

The factors determining the range of vocal imitation are largely unknown. There must be constraints on the auditory and motor systems. A bird cannot learn what it is incapable of hearing. Even when a sound is heard, its vocal reproduction may not be possible. Besides these determinants there may be other conditions that bias the selection of the song model. There may be a built-in neuronal mechanism for recognition of the conspecific song. The selectivity may also result from the individual hearing its own vocalization during ontogeny; the sounds of young birds may contain some of the qualities present in the model song. Cues other than the qualities of song may determine the acquisition of song; for example, filial bonds play a decisive role in the selection of the song model in bullfinches and zebra finches.

The 'critical period' for song learning

Restriction of the incorporation of environmental sounds into song to a certain 'sensitive' or 'critical' period in the bird's life has been described for several species: chaffinches (Poulsen, 1951; Thorpe, 1958), eastern and western meadowlarks (Lanyon, 1957), white-crowned sparrows (Marler and Tamura, 1964) and zebra finches (Immelmann, 1965, 1966). The precise relationship between learning and the 'critical periods' is not always clear. Song learning includes both hearing a sound and reproducing that sound, and the role of these two processes may be hard to separate in some cases. Thus, under natural conditions, a male chaffinch is able to learn song during the first ten months of its life. At the end of this period, the bird establishes its song themes in their final stereotyped patterns in a process known as 'crystallization' of song. Once this comes to an end the chaffinch will not alter its song themes nor enlarge its song repertoire, which remains virtually unchanged in future years (Thorpe, 1958). The coincidence of crystallization of the motor pattern and loss of the capacity to imitate makes it difficult to separate the roles of sensory and motor processes.

What are the determinants of the onset and cessation of the critical period? Is it strictly an age-dependent phenomenon or does it depend on hormonal and experiential factors? In the chaffinch, singing behaviour is under the control of rising testosterone levels (Collard and Grevendal, 1946; Poulsen, 1951). Song makes its first appearance in early spring when birds born the previous season are some nine months old.

In order to answer the question of age dependency, the following experiment was undertaken. A male chaffinch was prevented from coming into song by castration at approximately seven months of age, and induced to sing at two years of age by implantation of testosterone. Before its singing reached the final stereotyped form, this bird was exposed to recordings of two wild-type themes which clearly differed from the bird's own developing song. The results of this experiment are shown in Plate 2. After the bird crystallized its two song patterns, both showed unmistakably many details of one of the tutor themes. The same bird was exposed to a different wild-type theme the following year. This exposure did not add to the song repertoire nor change the two song themes of the previous year (Nottebohm, 1967 and manuscript in preparation).

Thus a chaffinch can develop a normal *motor* pattern of song well after the usual 'critical period' for song learning is over, an observation also made by Thorpe over a shorter time period (Thorpe, 1958). Furthermore, at two years of age a chaffinch that has not yet sung retains the ability to incorporate new auditory stimuli into its song. The end of the critical period for song learning is not strictly an age-dependent phenomenon. Rather, it follows the crystallization of song, at whichever age this occurs. It is not possible at this time, however, to decide whether this termination results from the crystallization of song as a motor pattern *per se*, or whether it is determined by some other correlate of the high testosterone levels that accompany song crystallization.

In species such as the white-crowned sparrow and the zebra finch, the critical period ends before the onset of full song and corresponds to the end of the birds' ability to establish a template of the song model. In this case motor fixation cannot be invoked as the immediate cause for the termination of the critical period. It would be interesting to investigate whether postponement of vocal reproduction of the model would result in the recurrence of the critical period in later seasons.

It seems likely that the critical period for song learning in birds is not a unitary phenomenon. It may result from the interaction of two critical periods: one for the establishment of the auditory template, and another for its vocal reproduction. The two periods may overlap to varying extents in different species. In the chaffinch both periods seem to terminate simultaneously, even though the acquisition of the song template may precede the onset of singing (Thorpe, 1958). In this case a temporally restricted motor fixation may impose an end to the manifestation of further auditory learning. Independent manipulation of the two phenomena, if this can be accomplished, should shed considerable light on our understanding of the 'critical period for song learning' and the physiological events underlying it.

THE ROLE OF AUDITORY FEEDBACK

The common assertion by some extreme environmentalists that since an isolated bird can hear itself, it can 'practice' or 'learn' to sing, obscures an important distinction between modes of acquiring information. First it assumes that hearing is involved in vocal development. This must be tested. Second, if hearing is essential for vocal development, exogenous as well as self-generated sounds must be considered. A bird reared in auditory contact with wild conspecifics can acquire from the latter an auditory model against which to match its own vocalizations. Contrariwise a bird reared in isolation can only develop its vocal repertoire by reference to itself. In either case, auditory feedback seems the most likely method of controlling vocal development. This was the theoretical expectation with which the following deafening experiments were conducted. Birds were deafened by extirpation of the cochlea as described by Schwartzkopff (1949) and Konishi (1963, 1964).

Normal vocal development in deaf birds

Domestic chickens (*Gallus domesticus*, Konishi, 1963) and ring doves (*Streptopelia risoria*, Nottebohm, in preparation), deafened soon after hatching developed all of the normal species vocal signals under study (Plates 3 and 4). Because of variation in these calls both within and between individuals, especially in the chicken, it is difficult to evaluate how much of the slight deviation occurring was attributable to deafness. Nevertheless, the deaf birds could certainly produce a complete repertoire of vocalizations, the general structure of which was normal.

Moreover, deaf chickens and ring doves uttered their calls under the normal stimulus situations: roosters produced aerial alarm calls upon seeing flying objects; their food calls were good enough to attract normal hens to them. Some of the chicken vocalizations were graded according to the stimulus situation and the internal state of the bird: this grading of signals occurred also in the deaf chickens.

Invalidation of auditory experience by deafening

Prenatal deafening has not yet been attempted. Since precocial birds are known to be capable of hearing several days before hatching, this is considered a serious limitation on the interpretation of deafening experiments. Therefore, it is important to study whether or not prenatal or at least pre-deafening auditory experience can contribute to vocal development.

In the white-crowned sparrow song learning consists of two stages separated in time, namely hearing and vocal reproduction of the model song. The young bird does not immediately imitate what it hears, but rather remembers the model until several weeks or months later, when it reproduces it vocally. Furthermore, during its vocal reproduction the bird does not have to hear the model from any external source. After the 'critical period' additional auditory exposure of any sort does not affect song development (Marler and Tamura, 1964). The following experiments were conducted with this background information.

When white-crowned fledglings which had been exposed to the song of their birthplace in the field during the critical period were deafened before the onset of singing, they failed to reproduce the model song. These birds would have reproduced the model song had they been left intact. When birds collected as nestlings and hand-reared in auditory isolation were deafened, they produced songs resembling those of the first group. Both groups of deaf birds produced songs which were radically different both from the wild-type and from isolate songs (Plate 5; see also Plate 1). Evidently deafening renders any auditory experience acquired previously unusable for further vocal development. These results are expected if auditory feedback is used to match vocal output with an auditory model in the form of a memory trace (Konishi, 1965*a* and *b*).

Abnormalities in the songs of deaf birds

So far as is known, no passerine bird produces perfectly normal song, if deafened before the completion of song crystallization. In all of the species carefully studied, early deprivation of auditory feedback results in some consistent abnormalities in their songs. These abnormalities can be summarized as follows:

I. Absence of unit organization: the most extreme effect of deafening is the disappearance of all the recognizable structural entities of song; notes, syllables, and phrases may be lacking in the songs of deaf birds.[1]

II. Abnormal notes and syllables: if deaf birds produce distinct notes and syllables at all, these tend to contain abnormal patterns of frequency modulation. Notes and syllables produced by deaf birds usually appear irregular and fuzzy on the audiospectrogram.

III. Instability in the forms of notes and syllables: in sharp contrast with the songs of intact birds, the notes and syllables of deaf birds are not repeated in exactly the same form either from song to song or within a song, even though their general patterns are maintained (Plate 6). This type of fluctuation is seldom observed among intact birds. Despite this short-term instability, deaf birds can maintain to a considerable extent the individual characteristics of their songs in successive years (Konishi 1964, 1965*a* and *b*).

Effects of deafening in relation to stages of vocal development

Quite apart from effects of previous auditory experience of environmental sounds, pre-deafening sensory-motor experience of vocalizations might influence song development in a bird that was subsequently deafened. This seems to be implied by experiments in which birds were deafened after their songs had stabilized. White-crowned sparrows and chaffinches with songs crystallized before deafening maintained them after the operation, relatively unaltered, for a long time, up to two or three years (Plate 7). These results led to more systematic studies in which chaffinches were deafened at different stages of vocal development (Nottebohm, 1966, 1967, 1968).

Song development in the chaffinch is characterized by a series of

[1] *Terminology of bird song:* Any continuous marking on the time-frequency audiospectrogram will be named a note. If there is any discontinuity in the distribution of intervals separating notes, the new temporal unit is called a syllable. Such units may contain one or more notes. Several syllables may group together to form a phrase. A simple song consists of repetitions of the same note or syllable. The most complex type of song contains phrases composed of different notes and syllables.

progressive changes consisting of the early nestling stage, juvenile and adult subsong, 'plastic' song and final full song stages. The transition from one stage to the next is gradual: elements of nestling calls may be found in early subsongs, which gradually become more stereotyped, louder, and recurrent during transformation to 'plastic' songs. These are unstable but contain some of the basic features of chaffinch song. They finally develop into highly stereotyped and discrete full songs, which appear normally during the bird's first spring at about ten months of age (Plate 8).

First-year male chaffinches were deafened at various stages of their vocal development. One bird was deafened before its two song themes had reached their final stereotyped forms, after twelve days of plastic song experience. In the two months following the operation the songs of this chaffinch underwent considerable deterioration: syllables and phrases gradually changed and lost their identity (Plate 9, also Plate 8). There was a regression from the level of motor skill in singing established before the operation. Since continued singing was ensured in this case by testosterone therapy, the regression cannot be attributed to hormone deficiency.

Similarly, other first-year male chaffinches were deafened at various times before the onset of full song, which normally occurs during the bird's first spring, when it is ten months old. Some were deafened after they had spent one or two days in plastic song; others were operated upon in the middle of their first winter at about seven months of age after subsong had started but before the 'plastic' song stage. Still others lost hearing in the middle of their first summer at 88 and 107 days of age, probably before adult subsong had begun. All these birds sang during their first spring. The quality of their songs varied over an enormous range (Plate 10).

The birds in the last group, operated upon around 100 days of age, had pre-deafening experience with their nestling and fledgling vocalizations, and presumably were starting to develop some of their juvenile calls. The songs they produced as adults were very rudimentary and unstructured. These defects were more marked in the bird deafened at the youngest age. The individuals deafened in the middle of their first winter, with some previous experience of subsong, produced songs with syllables coming closer to the wild-type than the previous group, though still of considerable simplicity. The bird deafened in early spring after two days of plastic song

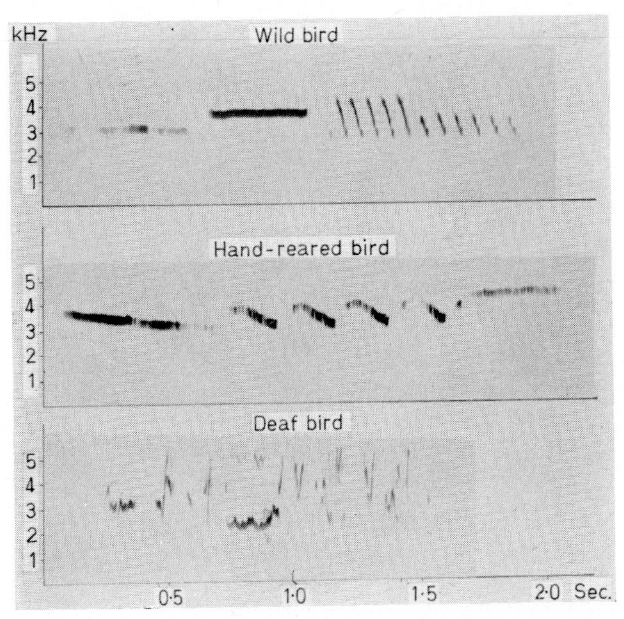

PLATE I. Songs of three white-crowned sparrows raised in different auditory environments: a wild bird, a hand-reared individual and a bird deafened before the onset of singing.

(*facing p. 40*)

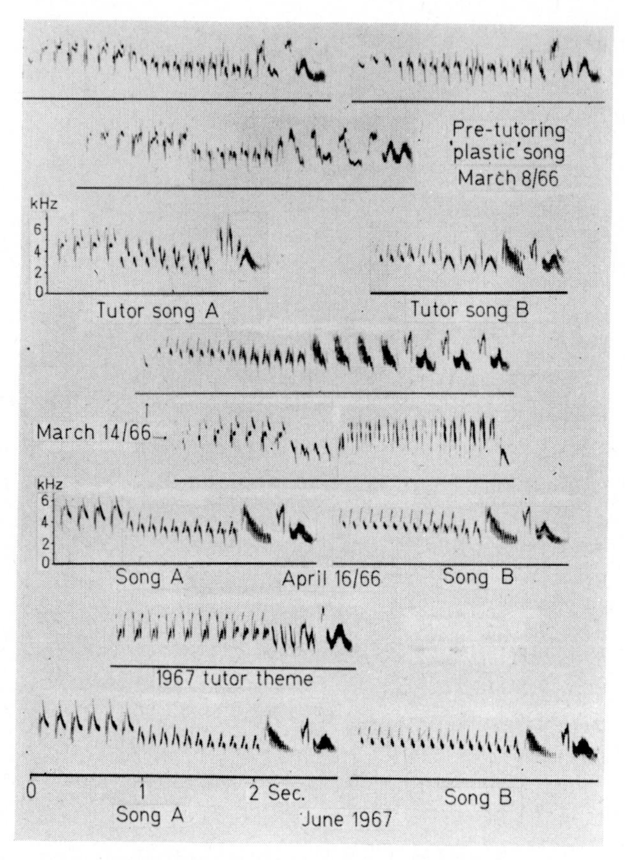

PLATE 2. Pre- and post-tutoring song of a castrated male chaffinch. Elements from tutor songs A and B were imitated during 'plastic' song (14 March 1966). Components of tutor song B were retained in the final version of songs 'A' and 'B' (16 April 1966). These two songs were repeated with small changes in the following year (June 1967). The tutor theme used in 1967 had no effect on the song structure of the experimental bird.

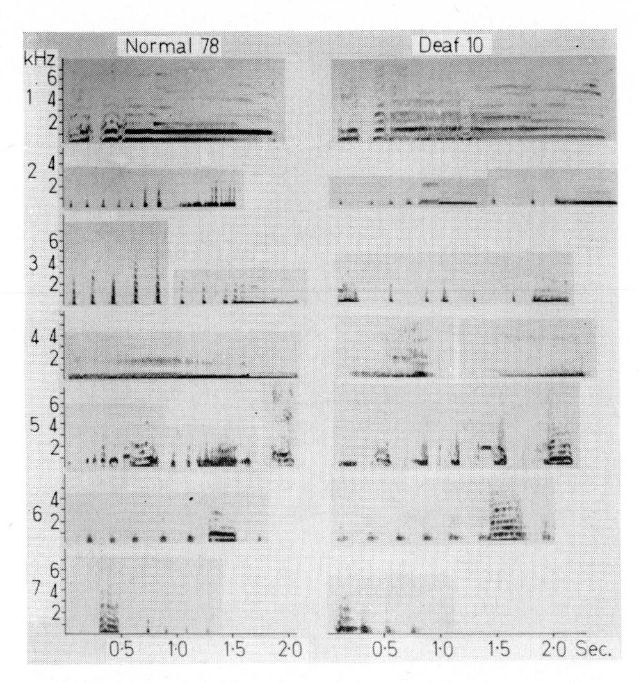

PLATE 3. Seven different calls of a normal and a deaf domestic chicken: 1. Crowing; 2. Aggressive call; 3. Food call; 4. Aerial alarm call; 5. Distress calls and cries; 6. Ground enemy alarm; and 7. Alerting call.

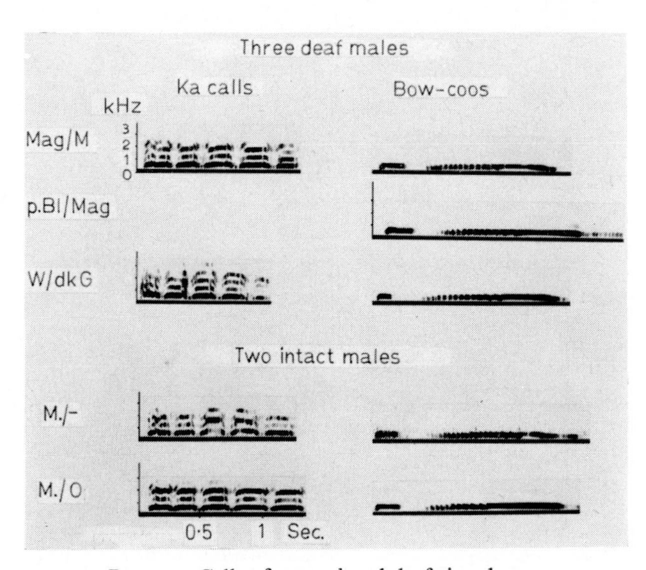

PLATE 4. Calls of normal and deaf ring-doves.

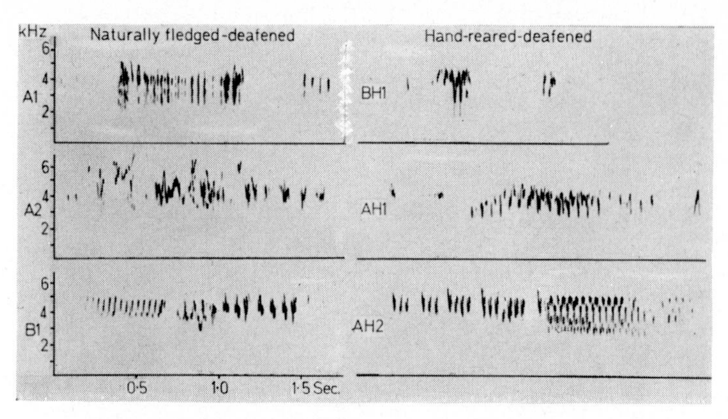

PLATE. 5. Songs of six white-crowned sparrows deafened before the onset of singing. Birds on the left column had been exposed to a wild white-crown song during the critical period and then lost their hearing before being able to reproduce the model song. Those in the right column had been acoustically isolated before deafening.

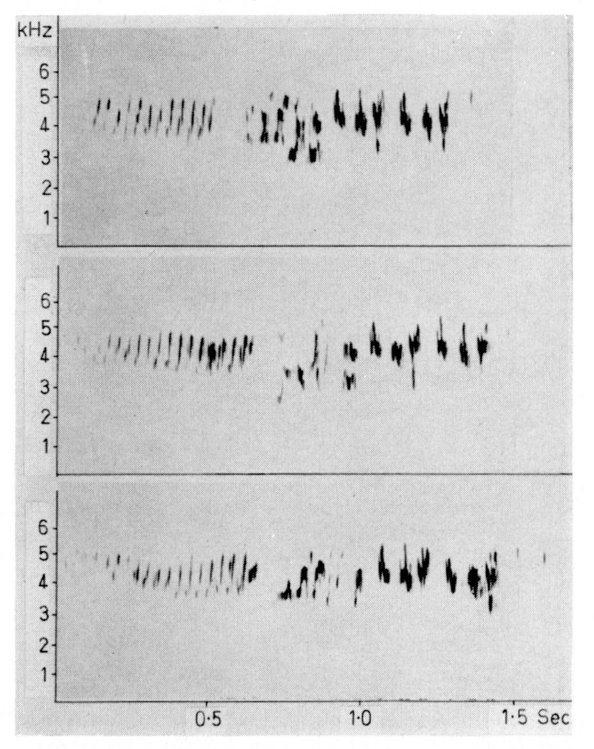

PLATE 6. Unstable song characteristic of deaf passerines: three consecutive songs of a deaf white-crowned sparrow. While the overall pattern of the song is stable, its fine details fluctuate.

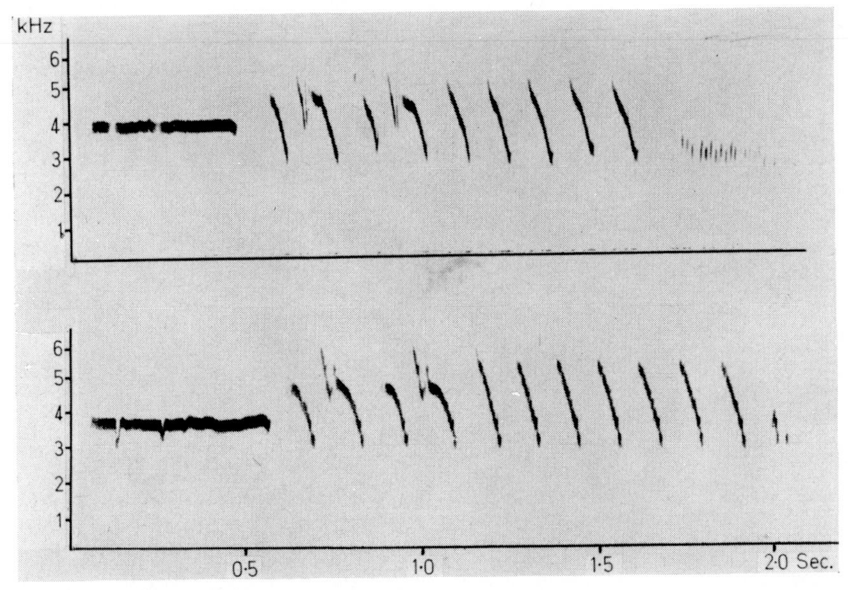

PLATE 7. Maintenance of song without auditory feedback: a wild white-crowned sparrow was recorded before (*top*) and fourteen months after deafening (*bottom*). Note that most of the fine details of song remained relatively unchanged.

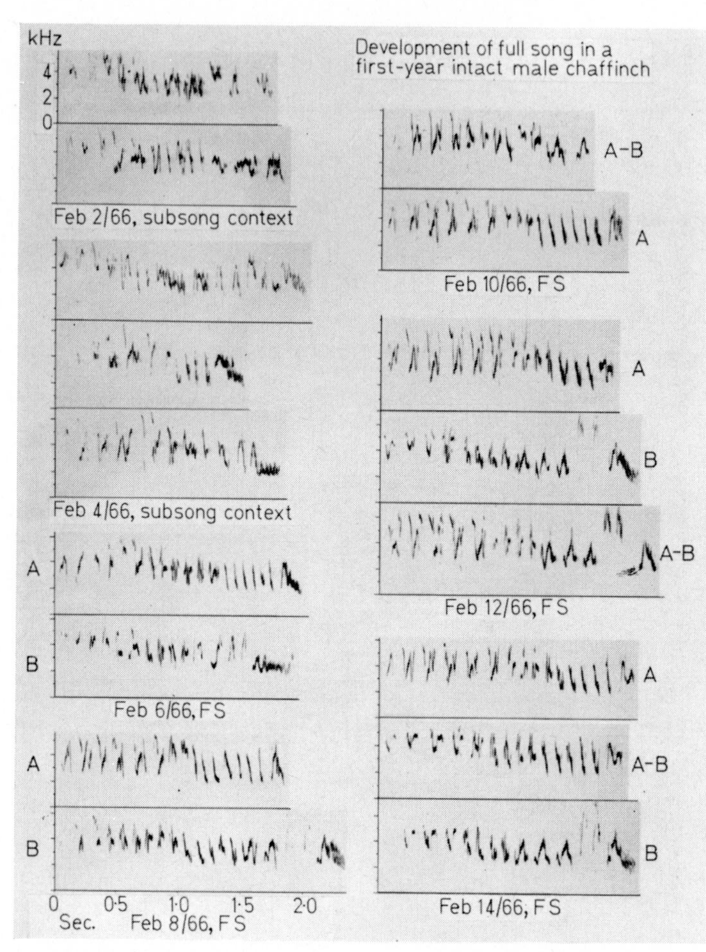

PLATE 8. Development of two song themes 'A' and 'B' in a first-year male chaffinch, indicating gradual progress from subsong to full song. Note sound patterns intermediate between A and B themes.

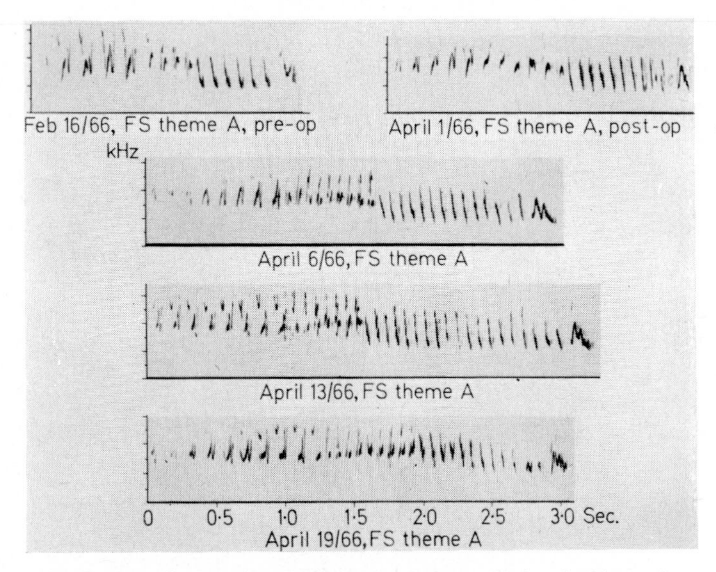

PLATE 9. Regression of song after deafening in the same bird as in Figure 8: note the gradual deterioration of theme 'A' during the course of two months. This bird was deafened immediately after the recording of 16 February.

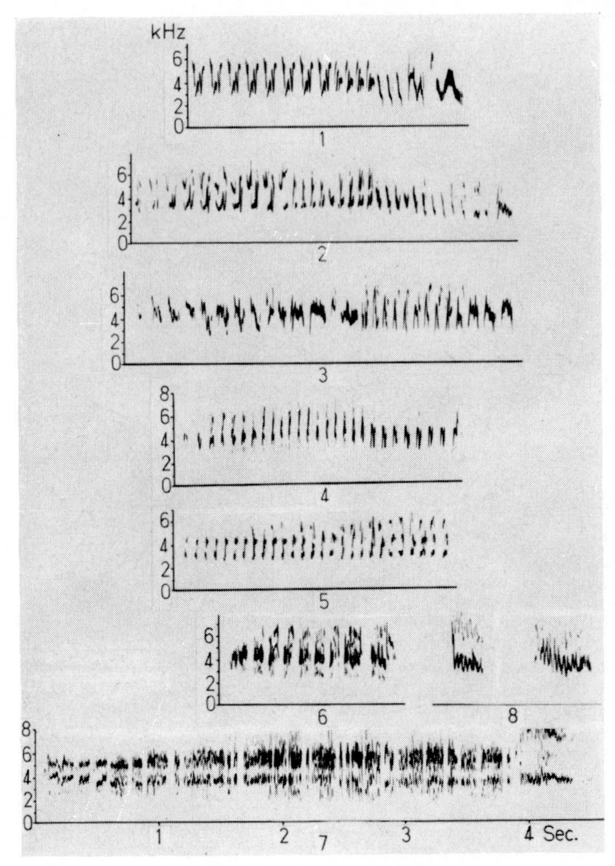

PLATE 10. Songs of male chaffinches deafened at different stages of vocal development: 1. Adult, with one or more seasons of singing experience; 2. Birds deafened in their first spring after they had been in 'plastic' song for 12 days; 3. For 2 days; 4. For 1 day; 5. Bird deafened in the middle of its first winter; 6. Bird operated at 107 days of age; 7. Individual operated at 88 days of age; 8. Two subsong 'chirrups' by an intact first-year male.

experience did better than the one that had only engaged in one day of plastic song before it was operated upon. Both these birds produced songs closer to the normal wild-type than the birds deafened in mid-winter, although they were less accomplished than the bird that had sung plastic song for twelve days before deafening.

There is thus a correlation between the stage of vocal development at which hearing is lost and the quality of the song that develops after deafening. The earlier the stage, the greater the effects of deafening on the resulting song. The sound patterns established before the loss of hearing are probably incorporated, with some loss or modification, into full song. Records suggestive of this process are found in the song of the bird deafened at 88 days of age. It included a sound highly reminiscent of the juvenile 'chirrup', which also occurs as the food-begging call of fledglings (see 7 and 8 in Plate 10).

These results cast light on several problems of song development. First, they indicate that song development in the chaffinch is a gradual integrative process which cannot be considered separately from the development of other vocalizations. Second, they show that the effects of deafening cannot be evaluated without reference to the stage of vocal development at which the operation occurs, making it necessary to reexamine earlier work that was thought to show relatively normal song development of deaf birds. In Oregon juncos and black-headed grosbeaks (*Pheuticus melanocephalus*) birds deafened before the onset of singing apparently developed more normal songs than in other species (Konishi, 1964, 1965*a*). But the onset and duration of song development differ from species to species, and if all were to be deafened at comparable stages in their vocal development, more uniform results might be obtained.

Even the development of normal vocalizations in deaf chickens and ring doves may need some reinterpretation. Chicks of nidifugous species are known to produce several different calls even before hatching. Experience of these sounds may contribute in some way to the development of normal vocalization after postnatal deafening.

Interspecific variation in the effects of deafening may result from differences in the developmental strategies followed by different species (Marler, 1967). Deafening experiments can discriminate only between two types of strategy; those that require auditory feedback and those that do not. The song sparrow does not copy sounds extensively in normal song development, whereas white-crowned sparrows shape their songs after external auditory models. Notwithstanding these two different modes of vocal development, deafening has equally drastic effects on song development in these species. Auditory monitoring of vocalization is essential for normal song development in both of them.

In this connection it should be noted that deafening does not seem to affect the different properties of song uniformly: some features such as frequency range and duration develop more normally than others. The development of these normal features may not require auditory feedback. Another line of evidence in support of the above conjecture can be derived from the fact that deaf birds show gradual progress from subsong on to full song, as in normal individuals, even though the course and end product of development may be extremely abnormal. This suggests the presence of developmental processes, operating independently of auditory feedback. It is thus conceivable that species differ from one another in the extent to which such processes dictate the normal course of song development.

Effects of deafening on the temporal pattern of singing

Most passerine birds deliver their songs in bouts and the intervals between songs within a bout tend to be rather constant. When a bird has more than one song type, different themes are delivered in separate bouts, such as AAAA BBBB, where A and B are song themes (Hinde, 1958; Isaac and Marler, 1963).

Deaf birds in some of the species studied exhibited rather normal temporal patterns of singing, even when their songs were abnormal. Other species showed some abnormalities: for example, deaf American robins (*Turdus migratorius*) delivered syllables at intervals shorter than those found among wild robins (Konishi, 1965*a*). However, rotation of song themes was observed in all of the species studied. For instance, one of the deaf Oregon juncos sang 1,274 songs comprising nine different themes in the following sequence and frequencies during a period of three and one-half hours: E(43) – D(16) – A(154) – C(120) – A(2) – C(11) – B(1) – C(1) – B(1) – C(1) – B(1) – C(1) – B(65) – I(47) – F(11) – G(33) – H(33) – B(1) – G(35) – D(509) – G(107) – E(81). (The letters indicate the song themes and the numbers are frequencies of their occurrence).

In ring doves perch-cooing and bow-cooing develop normally in the absence of auditory feedback (see Plate 4). These vocalizations are delivered in bouts. The average intra-bout interval between cooings is not significantly different between deaf and intact birds. Furthermore, intact as well as deaf ring doves show a tendency to deliver their perch-cooing bouts in clusters, although both perch- and bow-cooings tend to occur in longer bouts among deaf birds than in normal individuals. The frequency of occurrence of perch- and bow-cooings during each phase of the breeding cycle is also normal in deaf doves (Nottebohm, in preparation).

So far the methods used for analysing the temporal distribution of songs have not been adequate for detecting the characteristics of distribution other than averages and variances. Schleidt (1964), using an advanced method, compared the temporal distribution of gobbling calls between normal and deaf turkeys (*Meleagris gallopavo*). Although all deaf birds produced normal gobbling calls, the short-term intervals between successive calls were less constant in four out of five deaf turkeys than in normal birds recorded in a soundproof room. One deaf bird exhibited a normal interval distribution. We have, therefore, at least one reliable case in which the normal temporal distribution of vocalizations can occur in the absence of auditory environmental and feedback stimuli.

It is of course well known that external stimuli can elicit sound and calls in birds, and can even affect their frequency and distribution (Marler, 1956; Hinde, 1958; Kneutgen, 1964).

THE ROLE OF PROPRIOCEPTIVE FEEDBACK

Song is produced by the expiratory flow of air which presumably sets the membranes of the syrinx into vibration. Two motor components are involved. The expiratory musculature exerts force on the air sacs; the syringeal musculature controls the tension of its membranes and the bore of the air passage at that level. The syrinx acts as a valve, and thus determines the impedance against which the expiratory musculature is working. The activity of the expiratory and syringeal muscles might be monitored by proprioception. The control of vocalization may therefore depend not only on auditory feedback but also on proprioceptive feedback.

As mentioned previously, deafening does not alter song patterns after their crystallization even if they are copies of external models. Songs developed in the absence of auditory feedback remain unstable. In intact birds, stable vocal patterns are established gradually, and auditory feedback is essential if full stability is to be achieved. At the same time some other mechanism must gradually take over so that stability will persist even in the absence of auditory feedback. Such a mechanism may involve proprioceptive feedback. Suppose that as a given vocal pattern is gradually stabilized, its auditory and proprioceptive feedback are compared so as to establish a reference through which the vocal output can be evaluated by either feedback. Maintenance of the pattern should then be possible by proprioception alone in the absence of auditory feedback.

While the role of auditory feedback is easily studied, that of proprioceptive feedback is hard to analyse, since no method has yet been developed to sever selectively all the proprioceptive feedback pathways involved in song. One example will suffice to illustrate the ambiguities that arise. In the chaffinch, the hypoglossal nerve probably provides both sensory and motor innervation to the syrinx. Cutting this nerve in intact or deaf birds alters the syllable structure and tonality of songs but does not affect their patterns of phrasing, duration, and stability on successive renderings. A bird operated upon in this way also delivers different song themes in alternating sequences (Nottebohm, 1967). The abnormal syllable structure and tonality evident after the operation may be due to the loss of either motor or sensory fibres controlling these properties of song, or both. The maintenance of other features might result from contributions by motor and sensory fibres other than those in the hypoglossus. Nevertheless the above experiment shows clearly that proprioceptive matching of vocal output, if operative at all, is not a unitary process. Different properties of song can be independently controlled, perhaps through different proprioceptive pathways. Equally open is the possibility that the motor pattern of song, once stabilized, can be maintained without any sensory monitoring, auditory or proprioceptive. Whichever is the case, chaffinches retain many temporal characteristics of normal singing behaviour in the presence of altered auditory and proprioceptive feedback.

DISCUSSION

The effects of sensory deprivation are usually assessed by the presence or absence of particular behaviour patterns. Although the aim of such experiments is to explore the developmental origin of behaviour, they rarely elucidate the details of the ontogenetic process. Rather, they evaluate the effects of a given set of environmental conditions on the end products of ontogeny. A motor pattern may be generated in the central nervous system without any patterned peripheral sensory input being necessary for its control (e.g. Wilson, 1961). Although the ontogeny of such a central mechanism has not been studied, it is conceivable that its development does not require patterned sensory inputs. In domestic chickens and ring doves, neither external sound models nor auditory feedback are needed for normal vocal development insofar as is known. However, this does not necessarily prove central patterning, since proprioceptive feedback can in theory supply the central nervous system with patterned inputs resulting from vocal motor coordination.

Among passerine birds we know at least one species that can develop normal song without previous exposure to it. Song sparrow egg-isolates develop normal song, whereas song sparrows deafened at an early age produce extremely abnormal song. This implies that the song sparrow must use some internal reference to steer its vocal development by auditory feedback. By analogy to the acquired song template one might infer that intact song sparrows already possess at the very start of vocal development an auditory template sufficient to ensure normal song patterns. It is interesting to note that the performance of such a system is indistinguishable from the developmental process steered by an external song model which must first be incorporated in the central nervous system as a reference memory.

Normal song development in the absence of an external song model could also be explained without postulating either central motor coordination or a built-in song template. A vocal control system consisting of the central nervous system, vocal motor system and sensory feedback may undergo progressive changes in the mode of its operation. The progressive changes leading to song crystallization may involve temporal integration of successive stages so that the nature of the preceding event determines that of the following one. However, insofar as song development progresses and ends in predictable patterns, there must be genetically determined criteria by which such experiential processes are integrated. These criteria, for example, could be set relationships between auditory feedback and motor commands at successive stages of vocal ontogeny.

In the white-crowned sparrow and in the chaffinch the song of nestling-isolates also differs from that of birds deafened as juveniles.

However, these two examples require some reservation. Song development in nestling-isolates may be influenced by auditory experience before isolation, so that the resulting song pattern may not have been derived by reference to a pre-existing template.

Removal of auditory feedback during ontogeny would interfere with any of the alternative developmental strategies mentioned with the exception of built-in central motor patterning. To this extent, then, deafening studies fail to distinguish between the different programmes which might steer the vocal development.

We have discussed different strategies of vocal development. At one extreme there are birds which can develop species-specific vocalizations without hearing. At the other end of the spectrum we find species which normally copy song from their conspecifics. Tempting as it may be to theorize on the evolutionary significance of these differences, this is probably premature. Ideally we would like to relate the modes of song development to social, population, and ecological variables, but the number of species sampled is still far too small for this. Further studies of this kind will broaden our understanding of the evolution of communication in animals, as well as elucidating the physiological mechanisms underlying the ontogeny of vocal and other patterns of motor coordination.

ACKNOWLEDGEMENTS

Our own studies reviewed here were conducted under the guidance of Dr Peter Marler to whom we are most grateful, both for discussion and correction of the manuscript. Thanks are due to Dr W. H. Thorpe for his sponsorship of part of the work by one of us (Nottebohm) at Cambridge. Preparation of this article was made possible by the National Science Foundation grants BG 5274, GB 6979, and BG 5697.

REFERENCES

BLASE, B. (1960) Die Lautäusserungen des Neuntöters (*Lanius c. collurio* L.), Freilandbeobachtungen und Kaspar-Hauser-Versuche. *Z. Tierpsychol.* **17**, 293–344.

BORROR, D. J. and REESE, C. R. (1959) Mocking bird imitations of Carolina wren. *Bull. Mass. Audubon Soc.* 16 pp.

COLLARD, J. and GREVENDAL, L. (1946) Etudes sur les caractères sexuels des Pinsons, *Fringilla coelebs et F. montifringilla. Gerfaut* **2**, 89–105.

GROHMAN, J. (1938) Modification oder Funktionsregung? Ein Beitrag zur Klärung der wechselseitigen Beziehungen zwischen Instinkthandlung und Erfahrung. *Z. Tierpsychol.* **2,** 132–44.

HINDE, R. A. (1958) Alternative motor patterns in Chaffinch song. *Anim. Behav.* **6,** 211–18.

IMMELMANN, K. (1965) Prägungserscheinungen in der Gesangsentwicklung junger Zebrafinken. *Naturwiss.* **52,** 169–70.

(1966) Zur ontogenetischen Gesangsentwicklung bei Prachtfinken. *Verh. dt. zool. Ges.* (*Göttingen*), 1966, Suppl. 320–32.

ISAAC, D. and MARLER, P. (1963) Ordering of sequences of singing behavior of mistle thrushes in relation to timing. *Anim. Behav.* **11,** 179–88.

KNEUTGEN, J. (1964) Beobachtungen über die Anpassung von Verhaltensweisen an gleichförmige akustische Reize. *Z. Tierpsychol.* **21,** 763–79.

KOHLER, O. (1954) Vorbedingungen und Vorstufen unserer Sprache bei Tieren. *Verh. dt. zool. Ges.* (*Tübingen*), 327–41.

KONISHI, M. (1963) The role of auditory feedback in the vocal behaviour of the domestic fowl. *Z. Tierpsychol.* **20,** 349–67.

(1964) Effects of deafening on song development in two species of juncos. *Condor* **66,** 85–102.

(1965*a*) Effects of deafening on song development in American robins and black-headed grosbeaks. *Z. Tierpsychol,* **22,** 584–99.

(1965*b*) The role of auditory feedback in the control of vocalization in the white-crowned sparrow. *Z. Tierpsychol.* **22,** 770–83.

LANYON, W. E. (1957) The comparative biology of the meadowlarks (*Sturnella*) in Wisconsin. *Pub. Nuttal Ornith. Club*, No. 1, Cambridge, Mass.

(1960) The ontogeny of vocalizations in birds. In *Animal Sounds and Communication* (W. E. Lanyon and W. W. Tavolga, eds), AIBS, Washington, D.C.

MARLER, P. (1956) Behavior of the Chaffinch, *Fringilla coelebs. Behaviour Suppl. V.*

(1964) Inheritance and learning in the development of animal vocalizations. In *Acoustic Behaviour of Animals* (R. G. Busnel, ed). Elsevier, Amsterdam.

(1967) Comparative study of song development in emberizine finches. *Proc. 14th Int. Orn. Congr.* (1966), 231–44.

MARLER, P. and HAMILTON III, W. J. (1966) *Mechanisms of Animal Behavior.* John Wiley and Sons, New York.

MARLER, P. and TAMURA, M. (1962) Song 'dialects' in three populations of white-crowned sparrows. *Condor* **64,** 368–77.

(1964) Culturally transmitted patterns of vocal behavior in sparrows. *Science* **146,** 1483–6.

MARLER, P., KREITH, M. and TAMURA, M. (1962) Song development in hand-raised Oregon juncos. *Auk* **79,** 12–30.

MARSHALL, A. J. (1954) Bower birds. *Biol. Rev.* **29,** 1–45.

MESSMER, E. and MESSMER, I. (1956) Die Entwicklung der Lautäusserungen und einiger Verhaltensweisen der Amsel (*Turdus merula merula* L.) unter natürlichen Bedingungen und nach Einzelaufzucht in schalldichten Räumen. *Z. Tierpsychol.* **13,** 341–441.

MULLIGAN, J. A. (1966) Singing behaviour and its development in the song sparrow, *Melospiza melodia. University of California Publ. Zool.* **81,** 1–76.

NICOLAI, J. (1959) Familientradition in der Gesangsentwicklung des Gimpels (*Pyrrhula pyrrhula* L.). *J. Ornith.* **100,** 39–46.

(1964) Der Brutparasitismus der Viduinae als ethologisches Problem. Prägungsphänomene als Faktoren der Rassen- und Artbildung. *Z. Tierpsychol.* **21,** 129–204.

NOTTEBOHM, F. (1966) The role of sensory feedback in the development of avian vocalizations. Ph.D. dissertation, University of California, Berkeley.

(1967) The role of sensory feedback in the development of avian vocalizations. *Proc. 14th Int. Orn. Congr.* (1966), 265–80.

(1968) Auditory experience and song development in the chaffinch (*Fringilla coelebs*): ontogeny of a complex motor pattern. *Ibis.* **110,** 549–568.

in preparation. Vocalizations and breeding behavior of surgically deafened ring doves.

in preparation. The 'critical period' for song learning in birds.

PARSONS, P. A. (1967) *The Genetic Analysis of Behaviour.* Methuen & Co. London.

POULSEN, H. (1951) Inheritance and learning in the song of the chaffinch, *Fringilla coelebs. Behaviour* **3,** 216–28.

(1954) On the song of the linnet (*Carduelis cannabina* L.) *Dansk. orn. Foren. Tidsskr.* **48,** 32–7.

SAUER, F. (1954) Die Entwicklung der Lautäusserungen vom Ei abschalldicht gehaltener Dorngrasmücken (*Sylvia c. communis* Latham) im Vergleich mit später isolierten und mit wildlebenden Artgenossen. *Z. Tierpsychol.* **11,** 10–93.

SCHJELDERUP-EBBE, T. (1923) Weitere Beiträge zur Social- und Individual-Psychologie des Haushühns. *Z. Psychol.* **132,** 289–303.

SCHLEIDT, W. M. (1964) Über die Spontaneität von Erbkoordinationen. *Z. Tierpsychol.* **21,** 235–56.

SCHWARTZKOPFF, J. (1949) Über Sitz und Leistung von Gehör und Vibrationssinn bei Vögeln. *Z. vergl. Physiol.* **31,** 527–608.

THIELCKE-POLTZ, H. and THIELCKE, G. (1960) Akustisches Lernen verschieden alter Schallisolierter Amseln (*Turdus merula* L.) und die Entwicklung erlernter Motive ohne und mit künstlichen Einfluss von Testosteron. *Z. Tierpsychol.* **17,** 211–44.

THORPE, W. H. (1954) The process of song-learning in the chaffinch as studied by means of the sound spectrograph. *Nature* **173,** 465.

(1958) The learning of song patterns by birds, with especial reference to the song of the chaffinch, *Fringilla coelebs. Ibis* **100,** 535–70.

(1961) *Bird Song.* Cambridge University Press.

WILSON, D. M. (1961) The central nervous control of flight in a locust. *J. exp. Biol.* **38,** 471–90.

3. SONG AS A REINFORCER

by **JOAN G. STEVENSON**
sub-Dept. of Animal Behaviour, Madingley Cambridge

INTRODUCTION

A chaffinch reared in isolation from adult song produces a song which is abnormal in that it is not broken up into phrases and lacks a terminal flourish. Thorpe (1958, 1961) has shown that normal song is learned in part during the bird's first summer, before it has started to sing, and in part during its first spring, as subsong develops into full song. Chaffinches will not, however, learn just any song during these periods. Even if tutored from a loudspeaker when beginning to sing, birds reared in isolation from song learn only songs similar to normal song. Learning a song rearticulated with the end in the middle, or the song of a tree pipit (*Anthus trivialis*) which has note structure similar to that of chaffinch song, is possible; but learning a chaffinch-like song played on an organ with rather pure tones is not. That the restriction is not due to limitations in effectors is implied by the much wider range and quality of notes found in early subsong. Birds caught in the autumn are even more restricted in what they will learn, presumably because they have already learned some aspects of normal song in the wild before they themselves were singing (see also Introduction to this section and Chapter 2).

Since more than physical factors are involved, the restriction must be related to the stimulus properties of normal song. That notes in subsong approximating those of normal song increase in probability suggests that normal song has reinforcing properties. The greater restriction to normal song with autumn-caught as compared with isolate birds might be correlated with a greater reinforcing value of normal song for the former. The work with autumn-caught and hand-reared birds discussed in this chapter made use of operant techniques to investigate this possibility. Birds were permitted to make an operant response which turned on a playback of an adult song. Song was called a reinforcer if it increased the subsequent probability of the operant response. Little is known about auditory reinforcers and, unlike song, most auditory stimuli studied from this point of view have no particular relevance to the animal in its natural environment. It is worthwhile, however, to survey some of these studies on other species.

AUDITORY REINFORCERS FOR OTHER SPECIES

Most experiments with auditory reinforcers have involved rather simple stimuli such as white noise or pure tones. As might be predicted, high-intensity stimuli have negative rather than positive reinforcing properties for mice. For example, when intense white noise (98 ± 1 db) was presented to mice (C57 Black 6 strain) in a situation where depression of a platform turned off the noise, the amount of time spent on the platform increased. With another group of mice, for which platform depression turned on the noise, the amount of time on the platform decreased (Barnes and Kish, 1957). However, at low intensities, white noise is a weak reinforcer for mice. Barnes and Kish (1961) tested 1,000 male and female mice with either white noise or any one of nine tones, ranging from 700 to 16,000 Hz, at any one of five intensities, from 45 to 85 db. Results showed that 'response-contingent low-intensity sounds, toward the lower end of the examined frequency spectrum, generally maintain behavior at levels slightly but significantly above control group levels' (p. 168). The authors suggest that reinforcing properties are exhibited at frequency values toward the limits of the mouse's hearing range. The suggestion is supported at the lower limit from this study, and evidence remains to be collected at the upper limit.

That mice can hear extremely high frequencies is suggested by the sounds emitted by rodent pups in conditions of distress. For the mother, the sounds are probably a guide towards lost pups, and they affect retrieving and nest-building positively. Mice and vole pups emit ultrasounds at 60 to 80 kHz, and rat pups at 35 to 55 kHz (Noirot, 1968). If reinforcing effects of such sounds could be demonstrated in an operant situation it not only would agree with the prediction of Barnes and Kish (1961), but also might relate to the function of such stimuli in a natural context.

Furthermore, stimuli relevant in a normal context are usually not as simple as pure tones, and this suggests that increasing complexity might increase reinforcing value. From a study with visual stimuli, of mice bar-pressing to produce a filled circle, an open square, or a cross, Barnes and Baron (1961) concluded 'that the magnitude of the sensory reinforcing effect is related to the amount of information presented to the subject by the sensory reinforcer, decreasing redundancy leading to increasing magnitudes of reinforcement' (p. 468). To see if increased complexity of auditory stimuli would increase the reinforcing effect, eight groups of mice were tested: for one group, touching a lever produced no sound; for another, contact produced a continuous, intense (70 db), 3,000 Hz tone for the duration of contact; for the third group, contact produced the tone at regular intervals; and for the fourth, contact produced 2,000, 3,000, and 4,000 Hz tones in sequence. Another four groups received the same conditions, except that each contact produced a light as well. Whereas light presentation increased response rates, sound decreased response rates. The decrease was greatest in the patterned groups, regardless of whether light was produced as well or not (Baron and Kish, 1962). Thus only low intensity sounds have been shown to be reinforcing for mice, and this was a very weak effect, not increased by slightly increasing complexity.

Although weak reinforcing properties of a 6,000 Hz tone at 86 to 98 db have been reported for rats (Andronico and Forgays, 1962), other results suggest that pure tones are not reinforcing. Symmes and Leaton (1962) failed to observe reinforcing effects of white noise, a 2,000 Hz tone, or a warbling tone, presented at 70 db. Other studies have shown that rats will work to turn off white noise. When a lever press turned off noise for fifteen seconds, the higher the intensity (74, 86, 98, or 110 db), the higher the rate of responding (Lavery and Foley, 1965). Furthermore, rate increased when each response terminated noise for only three seconds as compared with sixteen seconds, clearly indicating a negative reinforcing effect (Harrison and Tracy, 1955).

Thus, with mice and rats, auditory stimuli have been shown to have either a negative reinforcing effect, or at low intensities, a very weak positive one. This could mean either that auditory stimuli in general are not very strong positive reinforcers for the animals (as suggested by Kish, 1966), or that more effective auditory stimuli could be chosen. Baron and Kish (1962) did suggest that increasing complexity might increase reinforcing properties, but their results demonstrated the reverse. However, too often complexity is attained in a rather timid way, such as presenting three pure tones instead of one; and 'several sense modalities compounded in some complex fashion' (Kish, 1966, p. 116) is confused with lack of experimental control. Presenting a truly complex stimulus, which from what one knows of the animal is also relevant in other contexts, might lead to more meaningful results than using large numbers of animals to find a weak reinforcing effect of white noise or pure tones.

More complex sounds were used by Butler (1957), who showed that rhesus monkeys would learn to press whichever of two levers produced sounds of the colony for fifteen seconds. Butler (1958) went on to find that auditory stimuli differed in the following order, from most to least reinforcing: monkey feeding sounds, single monkey sounds, white noise, monkey sounds of rage, and dog sounds. From this, one could go on to investigate just what factors are responsible for the reinforcing properties of these stimuli.

With children, Thompson, Heistad and Palermo (1963) have used a taped story with music and short children's songs as reinforcers, each operant response playing ten seconds of tape. One method has involved a group situation, where a lever is simply left in a room of nursery school children. Frey (1960) has found the intriguing result that in such a situation, children press a lever as much to hear 'That's bad', as to hear 'That's good'.

Thus with monkeys and children, a reinforcing effect of auditory stimuli has been found. Since the stimuli used were relevant to behaviour in other contexts, significant stimuli might be reinforcing for mice and rats as well. To apply this argument to chaffinches, white noise, which is not relevant to the normal life of a chaffinch, should not be a reinforcer; but adult song, which is relevant, as discussed above, should be.

SONG *VS.* WHITE NOISE AS A REINFORCER

To test the prediction that adult song but not white noise would be reinforcing for chaffinches, conditions in which song-learning normally occurs were used. That is, since Thorpe has shown that some learning occurs in a bird's first summer, the birds were autumn-caught; since song learning also occurs when a bird begins to sing for the first time, they were in their first year and testosterone-injected. For the duration of testing, each bird lived alone in a sound-proof chamber with three parallel perches: the end perches operated counters when perched on, and the middle perch prevented hopping directly from one end perch to the other. During an initial control period, the frequency of perching to each end perch was simply recorded. Then the end perch which had received the fewer responses became the 'active' perch throughout testing. For eight birds, responding to the 'active' perch switched on a tape loop which played one normal, adult song of 2·5 seconds in duration, varying around

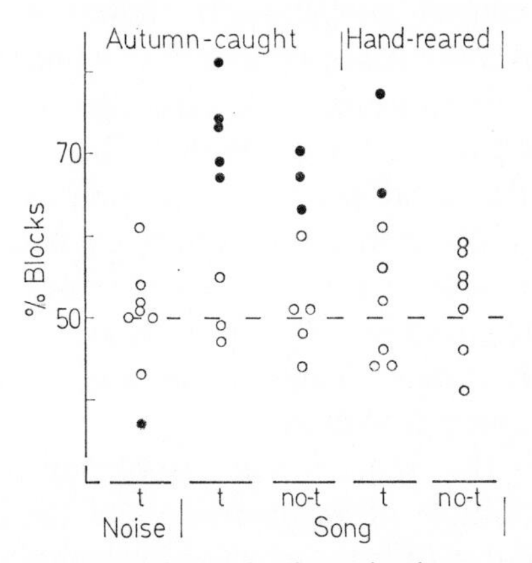

FIGURE 1. Percentage of blocks of no stimulus, stimulus, no stimulus, in which the number of responses to the active perch, divided by responses to both perches, was greater when the active perch turned on the stimulus than when it did not. Each point represents an individual bird in one of the five groups.

64 db. For another eight birds, all conditions were the same, except that responding to the active perch produced a burst of white noise (74 db) of the same duration and loudness as the song. For each bird, testing lasted for 72 blocks, each of four ten-minute periods. Within each block the periods consisted of:

Ten minutes of control conditions in which responding to each end perch was simply recorded.

Two ten-minute periods of stimulus conditions in which cue lights came on and every fifth or sixth response (alternately) to the 'active' perch played the tape loop.

Ten minutes in which responding to each end perch was again simply recorded.

(For further details, see Stevenson, 1967).

In Figure 1, each point provides a measure of the reinforcing effect for an individual. It indicates whether or not perching on the two end perches was differentially affected, and in what direction, by producing either noise or song. If the number of responses to the *active* perch divided by responses to both perches was greater during the two ten-minute stimulus conditions than during the surrounding two control conditions, a 'plus' was assigned to that block. The y-axis is the percent blocks assigned a plus rather than a minus, with equals disregarded. A filled circle indicates that the absolute number of plusses was significantly different from the absolute number of minusses (Sign test, two-tailed, $P<0.05$). The two left columns represent the sixteen birds being considered: injected autumn-caught birds, turning on either song or noise. No bird in the noise group gave significantly more plusses than minusses, and one gave significantly less. By contrast, five birds in the song group gave significantly more plusses, indicating that a higher proportion of the perching was to the active perch during song than during no-song conditions. This relative measure, which controls for increases in perching in general during stimulus conditions, was significantly higher for the song group than for the noise group (Mann-Whitney U test, one-tailed, $P<0.025$). Thus, a positive reinforcing effect was found with song, but not with noise.

Figure 2 shows the frequency of responding to each end perch under the two conditions of this procedure: to the *active* perch (solid lines) during stimulus (filled circles) and no-stimulus (open circles) conditions; and to the *inactive* perch (dashed lines) during stimulus (filled circles) and no-stimulus (open circles) conditions. For each group, the median frequency is plotted against successive sets of six blocks, each point representing the median for a group over two hours of responding. As seen from the upper left graph, responding by the noise group to the *active* perch (solid lines) was higher only 3 times out of 12 when this perch produced noise (filled circles) than when it did not (open circles). In the song group, responding to the *active* perch was higher when it produced song than when it did not for all 12 sets of blocks. Furthermore, there was no consistent tendency for the noise group to go less to the inactive perch during noise than no noise, but the song group went to the inactive perch less during song than no song on 9 out of 12 sets.

Although a reinforcing effect of song has thus been demonstrated, it did not involve all the changes in responding which might be expected to occur if food had been the reinforcer. For example, in Figure 2 median responding to the inactive and initially preferred perch remained higher than to the active perch which turned on the song. Furthermore, although the four median curves for the song

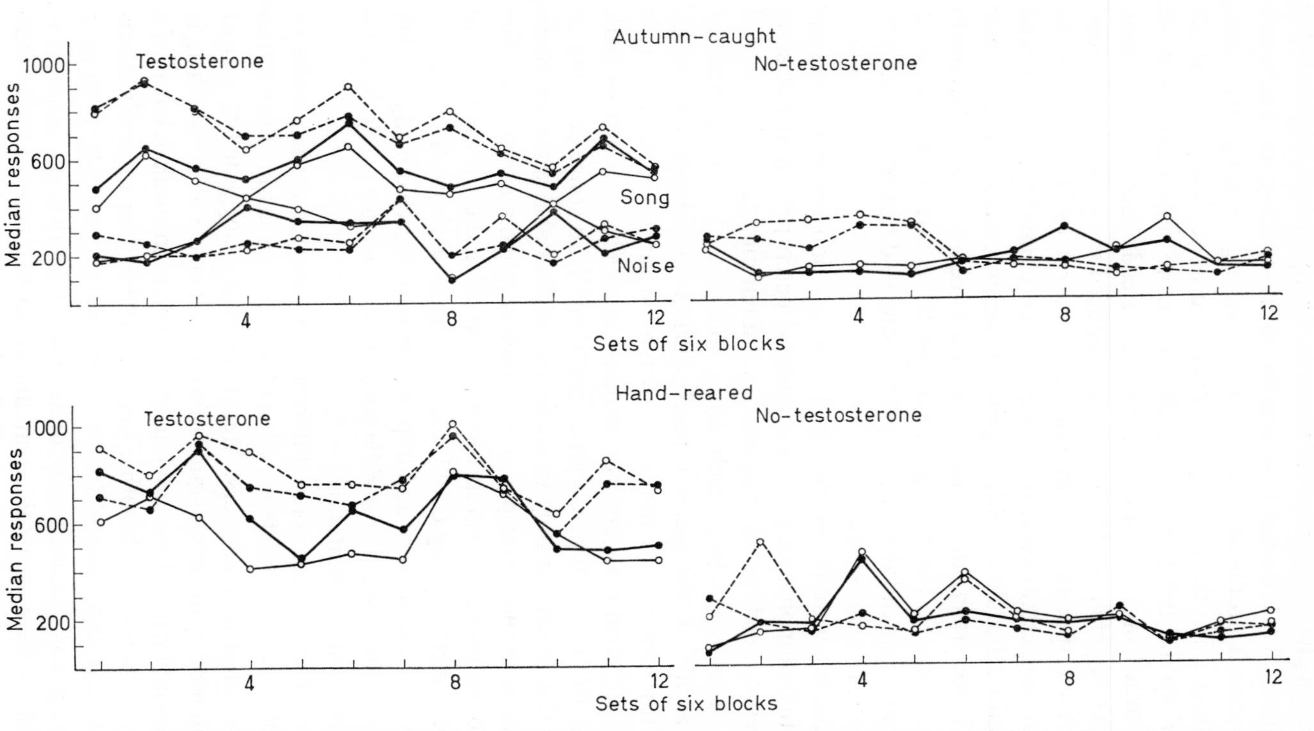

FIGURE 2. Median number of responses over sets of six blocks: to the active perch when it turned on the stimulus (—●—); to the active perch when it did not turn on the stimulus (—○—); to the inactive perch when the active perch turned on the stimulus (--●--); to the inactive perch when the active perch did not turn on the stimulus (--○--).

group are higher than those for the noise group, this did not hold for individuals. That is, total frequency of responding to the active perch when it produced song did not differ significantly from total responding when the active perch produced noise, implying that the song group did not turn on significantly more song than the noise group turned on noise. Finally, no high, steady rate of responding to the active perch occurred when it produced song, as with 'typical' fixed-ratio responding. Thus although control of this operant by song has been established, this control differs from that usually found on a ratio schedule of either 'primary reinforcement' (e.g. Ferster and Skinner, 1957) or 'conditioned reinforcement' (e.g. Stevenson and Reese, 1962). The difference between the control of the response in this situation and the others could be due to differences in subject, procedure, or reinforcer.

Chaffinches were therefore tested in a situation which was similar except that hopping to the less-preferred perch produced not song but access to seed for six seconds. Within ten minutes, birds learned to hop from the active perch down to the feeder. They perched hardly at all on the inactive perch during food conditions, when every fifth response to the active perch operated the feeder. This was very different from song conditions, in which every fifth or sixth response to the active perch produced song. With song, perching to the inactive perch not only was maintained, but also was sometimes higher than to the active perch. So much responding to the inactive perch could not have been due to birds perching on it while listening to song, something they could not have done while eating food, for birds did not usually turn and hop off the active perch quickly enough to get from there over the middle perch to the inactive perch during a 2·5-second song presentation.

A more likely factor is the difference between the consequences of deprivation of food and deprivation of song. In discussing the relation between different reinforcers, Meehl (1950) has suggested 'a sort of continuum of reinforcing states of affairs at one end of which it is most easy and natural and obviously very useful to speak in terms of a need . . . whereas at the other end . . . the notion of needs seems relatively less appropriate' (p. 71). The greater control by food over responding was obtained only under severe deprivation conditions. If the birds were maintained at about 90 per cent of their free-feeding weight and not given any food for six hours before testing, they would not always perform the above sequence, which involved operating and quickly approaching the feeder in spite of its normally aversive bang. Often such birds would show only intention movements toward the active perch, but not actually hop on it. Only if deprivation was increased to near-starvation for a chaffinch (i.e. 80 per cent free-feeding weight and no seed for six hours) did perching to the active perch always occur. Thus with food, but not song, it was possible to manipulate the condition of even a wild bird so that responses involved in normal feeding behaviour dropped out. Only perching and approaching the feeder in a rather automatic way remained.

EFFECTS OF EARLY EXPERIENCE AND TESTOSTERONE

To return to song, which as a reinforcer neither permits such drastic deprivation conditions nor produces such drastic behavioural changes, what conditions are responsible for its reinforcing effect? It was argued from Thorpe's experiments that both early experience of song and testosterone injections to induce singing should produce the maximum reinforcing effect of adult song. The next experiments were designed to examine the effectiveness of song for first-year chaffinches not treated in one or both of these ways. Thus the above autumn-caught, testosterone-injected birds were compared with autumn-caught birds not injected with testosterone and not singing. To see if isolation from adult song decreased the reinforcing effect, hand-reared birds also were tested, both with and without testosterone treatment.

To consider first the median responding for each group, the trend mentioned in Figure 2 for the autumn-caught, testosterone-injected birds also held for the hand-reared, injected birds. For this group, responding to the *active* perch when it did turn on song was higher than when it did not for 10 out of 12 points. That this was not simply an increase in total activity due to song presentations is shown by responding to the *inactive* perch, which was higher when the active perch did turn on song than when it did not only 3 out of 12 times. No such trend held for the no-testosterone groups, whose median responding was also lower than for the injected groups.

A single measure of the reinforcing effect of song for individuals in each of these four groups is presented in Figure 1. Whereas 5 out out of 8 autumn-caught, injected birds showed a significant number of blocks in which the number of responses to the active perch divided by responses to both was greater when the active perch did turn on song than when it did not, only 3 out of 8 non-injected birds did this. For the hand-reared birds, only 2 out of 8 did if injected, and none of the non-injected birds did. However, the only significant difference between the four groups was between the autumn-caught, testosterone group and the hand-reared, no-testosterone group (Mann-Whitney U test, one-tailed, $P<0.05$).

Thus the reinforcing value of song is related to the two conditions for song-learning in that birds both exposed to song and testosterone-treated showed a significantly higher effect than birds not exposed to song and not testosterone-treated. More birds are currently being tested to see if the groups satisfying only one of these conditions (i.e. autumn-caught, no testosterone; and hand-reared, testosterone) do fall in between the two extreme groups, as Figure 1 suggests. However, the difference in early experience between the autumn-caught and hand-reared birds clearly involves more than experience of song. Therefore, a fifth group which was hand-reared, tutored throughout its first summer, and injected is currently being tested.

If the tutored, injected group showed less of a reinforcing effect than the autumn-caught, injected group, then aspects of early experience other than just hearing adult song must be responsible for the difference. For example, experience in the wild might determine whether particular stimuli elicited particular responses. With a territory-owning male, song elicits behaviour involved in territorial defence (Marler, 1956), and this might increase the reinforcing effect of song. It has been shown that some visual stimuli which elicit aggressive behaviour are also reinforcers. Siamese fighting fish (*Betta splendens*) will swim through a loop more often if that response produces a mirror image or a model fish than if it does not; and fighting cocks will peck a key more often if key-pecking produces a mirror image or a view of another cock than if it does not (Thompson, 1963, 1964). Although the autumn-caught, injected chaffinches in the sound-proof chamber did not show any observable display to the adult song, a measure from another situation is now being correlated with how much each bird worked for song. That is, immediately after testing, each bird is placed in an owl-mobbing situation (Hinde, 1960), and the number of chink calls to a stuffed owl recorded. The above possibility of a weaker reinforcing effect with tutored birds than with autumn-caught birds and its relation to the development of behaviour other than song-learning would suggest that the tutored birds might show a weaker owl-mobbing response than the autumn-caught birds.

However, if the reinforcing effect of song is as high for birds tutored in their first summer as it is for autumn-caught birds, then hearing song, regardless of context, will have been sufficient to establish reinforcing properties once the birds begin to sing. This reinforcing effect of song would help explain why those notes in subsong which approximate the adult song heard early on increase in probability until finally a pattern of full, normal song results.

ACKNOWLEDGEMENTS

The experiment with food as a reinforcer for chaffinches was done in collaboration with B. Darlow and E. Williams. It is a pleasure to thank Professor W. H. Thorpe for his support and encouragement, Professor R. A. Hinde for his advice on all aspects of this work, and Professor T. Thompson for his comments on the manuscript.

REFERENCES

ANDRONICO, M. P. and FORGAYS, D. G. (1962) Sensory stimulation and secondary reinforcement. *J. Psychol.* **54**, 209–19.

BARNES, G. W. and BARON, A. (1961) Stimulus complexity and sensory reinforcement. *J. comp. physiol. Psychol.* **54**, 466–9.

BARNES, G. W. and KISH, G. B. (1957) Reinforcing properties of the termination of intense auditory stimulation. *J. comp. physiol. Psychol.* **50**, 40–3.

(1961) Reinforcing properties of the onset of auditory stimulation. *J. exp. Psychol.* **62**, 164–70.

BARON, A. and KISH, G. B. (1962) Low-intensity auditory and visual stimuli as reinforcers for the mouse. *J. comp. physiol. Psychol.* **55**, 1011–13.

BUTLER, R. A. (1957) Discrimination learning by rhesus monkeys to auditory incentives. *J. comp. physiol. Psychol.* **50**, 239–41.

(1958) The differential effect of visual and auditory incentives on the performance of monkeys. *Am. J. Psychol.* **71**, 591–3.

FERSTER, C. B. and SKINNER, B. F. (1957) *Schedules of Reinforcement.* Appleton-Century-Crofts, New York.

FREY, R. B. (1960) The effects of verbal reinforcers on group operant behavior. Unpublished M.A. thesis, University of Maine.

HARRISON, J. H. and TRACY, W. N. (1955) Use of auditory stimuli to maintain lever pressing behavior. *Science* **121**, 373–4.

HINDE, R. A. (1960) Factors governing the changes in strength of a partially inborn response, as shown by the mobbing behaviour of the chaffinch (*Fringilla coelebs*): III. The interaction of short-term and long-term incremental and decremental effects. *Proc. R. Soc., B,* **153**, 398–420.

KISH, G. B. (1966) Studies of sensory reinforcement. In *Operant Behavior: Areas of Research and Application* (W. K. Honig ed.), pp. 109–59. Appleton-Century-Crofts, New York.

LAVERY, J. J. and FOLEY, P. J. (1965) Bar pressing by the rat as a function of auditory stimulation. *Psychon. Sci.* **3,** 199–200.

MARLER, P. (1956) Territory and individual distance in the chaffinch *Fringilla coelebs. Ibis* **98,** 496–501.

MEEHL, P. E. (1950) On the circularity of the law of effect. *Psychol. Bull.* **47,** 52–75.

NOIROT, E. (1968) Ultrasounds in young rodents. II. Changes with age in albino rats. *Anim. Behav.* **16,** 129–34.

STEVENSON, J. G. (1967) Reinforcing effects of chaffinch song. *Anim. Behav.* **15,** 427–32.

STEVENSON, J. G. and REESE, T. W. (1962) The effect of two schedules of primary and conditioned reinforcement. *J. exp. Analysis Behav.* **5,** 505–10.

SYMMES, D. and LEATON, R. N. (1962) Failure to observe reinforcing properties of sound onset in rats. *Psychol. Rep.* **10,** 458.

THOMPSON, T. I. (1963) Visual reinforcement in Siamese fighting fish. *Science* **141,** 55–57.

(1964) Visual reinforcement in fighting cocks. *J. exp. Analysis Behav.* **7,** 45–49.

THOMPSON, T., HEISTAD, G. T. and PALERMO, D. S. (1963) Effect of amount of training on rate and duration of responding during extinction. *J. exp. Analysis Behav.* **6,** 155–61.

THORPE, W. H. (1958) The learning of song patterns by birds, with especial reference to the song of the chaffinch *Fringilla coelebs. Ibis* **100,** 535–70.

(1961) *Bird Song.* Cambridge University Press, London.

4. SONG DEVELOPMENT IN THE ZEBRA FINCH AND OTHER ESTRILDID FINCHES

by **KLAUS IMMELMANN**
Zoologisches Institut der Technischen Universität, Braunschweig, Germany

INTRODUCTION

The grass finches (Estrildidae)[1] have recently been studied by several authors: the calls and song of numerous species have been described and their biological significance has been discussed (reviews by Moynihan and Hall, 1954; Morris, 1958; Kunkel, 1959; Hall, 1962; Harrison, 1962; Immelmann, 1962*a*; Goodwin, 1965).

The male (in some species also the female) possesses a soft and sometimes almost toneless song phrase which is characterized by a wide frequency range (Plate 1). Unlike that of other passerines it is never aggressive in motivation and is thus never used in fighting or territorial behaviour. Its primary function is sexual, and it forms part of the highly elaborated courtship display characteristic of most of the species. Outside courtship, an undirected song is performed by many species during the breeding season or even throughout the year. Its exact significance has not yet been determined, though it may help to strengthen the pair bond and to promote flock cohesion. Hitherto the brief and stereotyped song phrase of estrildid finches was believed to be largely independent of experience. In recent years, however, it has been reported for several species—mainly by aviculturists—that individuals raised in captivity in a mixed collection of birds, or by strange foster parents, may develop an abnormal song and may even imitate the songs of other species (Goodwin, 1960; Nicolai, *in litt.*; F. Karl, *in litt.*). This suggests that some learning may be involved in song development—a possibility also supported

[1] The term grass finches is applied here to the whole family Estrildidae. Hitherto, the name has been reserved for the Australian species, while the family as such was called weaver finches and some African and Australian species were known as waxbills. Reasons for the renaming are given elsewhere (Immelmann, 1965*b*).

by the great individual variability which is to be observed in the songs of wild-caught Australian zebra finches (see page 63).

METHODS

Our studies of song development have been part of a research project on sexual imprinting in grass finches. So far, experiments have been carried out with three species, the Australian zebra finch (*Taeniopygia guttata castanotis*), the African silverbill (*Euodice cantans*), and the Bengalese or society finch, a domesticated form of the Asian striated finch (*Lonchura striata*). In the silverbill, only descendants of wild-caught stock have been used. In the zebra finch, descendants of wild-caught birds as well as domesticated stock were studied. In the Bengalese finch, only domesticated stock was available, but some wild-caught striated finches are at present under investigation.

The basic procedure was as follows: single eggs of one of these species were placed into a fresh clutch of another species. Only the domesticated Bengalese and zebra finches could be used for fostering purposes, however: they readily accept the hatching nestling of another species and usually brood and feed it even in the presence of nestlings of their own kind. Unless mentioned otherwise, the young were reared by their foster parents in sound-proof chambers. When independent, the young were separated from the foster parents and isolated on their own in sound-proof cages.

The songs were recorded with Telefunken–AEG M 5, M 5 A and M 95 tape recorders and a Brüel and Kjaer half-inch condenser microphone (type 4133), at tape speeds of 15 or 7·5 inches per second. The sound spectrograms were prepared with a Kay Electric Co. Sonograph, using narrow band-pass filter (45 Hz) in order to record the high frequencies as completely as possible, and wide band-pass filter (200 Hz) in order to record the overall pattern of the song phrase.

DEVELOPMENT OF YOUNG GRASS FINCHES

Since the development of grass finches has been described previously (Immelmann, 1962*a*, 1956*b*), only facts relevant to song development will be mentioned here. Both parents participate in caring for the nestlings. The young first leave the nest at three weeks of age and are fed by the parents for another seven to twelve days.

Young estrildid finches are characterized by a very early onset of sexual activity. In the species studied there seems to be a complete lack of a juvenile refractory period. In the zebra finch, for example, the first sexual behaviour patterns occur at six to nine weeks of age, i.e. at the beginning of juvenile moult. A colour-banded female in Australia laid her first egg on her 86th day of life, and other females started breeding at thirteen and fourteen weeks of age. Two pairs of captive zebra finches, whose parents were captured in the interior of Northwestern Australia, started their first clutch at eleven weeks of age. The Australian Gouldian finch (*Chloebia gouldiae*) may even start to breed in juvenile plumage if conditions are favourable (Immelmann, 1963).

Singing also may start very early. In zebra and Bengalese finches it is first heard at about 35 to 40 days of age. It develops continuously from the begging calls of the young: the first indications usually consist of a repetitive series of begging-like sounds which are put together to a very slow and uniform 'juvenile song'. From this source the final song develops during the following weeks by emancipation of characteristic syllables, increase of sound intensity, and the formation of brief song phrases. This process is completed by nine to eleven weeks of age. Juvenile moult starts at about eight weeks of age and is completed at about twelve to fourteen weeks.

THE SONG OF THE ADULT

The physical characteristics of estrildid song have been described in detail by Hall (1962) for a considerable number of species, including zebra finch, silverbill and striated finch. In the three species under investigation numerous call-notes are incorporated into the song phrase. Except for the calls there is a limited number of fairly simple other notes (Plate 1). Courtship and undirected song (see page 61) are identical in all physical characteristics, except that in the former the song phrases may be uttered in quicker succession (Plate 2).

A striking feature of the song of thirteen wild-caught male zebra finches was the individual variability: though ten males were caught at exactly the same locality near Katherine (Northern Territory) and the remaining three males at the same place near Coolawanya (Western Australia), their song phrases varied in length of strophes, and in number, sequence and nature of motifs, as much within these groups as between them. This is in contrast to several Asian species of the genus *Lonchura* for which Hall (1962) reported individual variation to be present but comparatively slight.

In the silverbill, the number of individuals hitherto tested is too small to draw any conclusions about song variation. For the Bengalese finch no data about the wild form (striated finch) are yet available. A great deal of individual variation is also known from other groups of passerine birds, e.g. the Emberizidae (see Marler, 1960; Thorpe and Lade, 1961; Konishi, 1964; Mulligan, 1966), where it is supposed to serve individual recognition. Apparently the same applies to the zebra finch.

Our first preliminary results showed that a zebra finch raised by Bengalese finches as described above developed a song almost identical to that of the foster father (Plates 3 and 4) (Immelmann, 1965*a*, 1967*b*). The same applied to male silverbills fostered by Bengalese finches, and to Bengalese finches fostered by zebra finches. This means that the song of experimental birds must have been acquired during early juvenile life, i.e. before separation from the foster parents. It does not mean, however, that there is no innate basis for the song at all. We, therefore, wanted to study the nature of the acquisition process and the possible occurrence of an innate song template.

THE INNATE SONG TEMPLATE

Fifteen male zebra finches and four male Bengalese finches were raised without any song tutor. They were reared in sound proof cages either by hand (two zebra finches) or by females of their own (three zebra finches, three Bengalese finches) or of another species (ten zebra finches, one Bengalese finch). The song of the birds was recorded when three to six months old. The following differences from the song of normally raised males were noted:

(i) *Number of elements.* In both species the song phrase was composed of a smaller number of varied syllable types than in normal birds (average for *Taeniopygia*: 5·3 syllables per song phrase without a tutor, 9·4 elements when raised with a tutor; average for *Lonchura*: 6·6 : 10·6). Furthermore, the syllables tended to be rather simple and uniform, and considerably longer in duration than in the wild-type song (cf. Plates 5 and 6). The more complicated increasing or decreasing elements were almost completely absent. Males raised by females of another species regularly included some of the calls of the foster mother in their song.

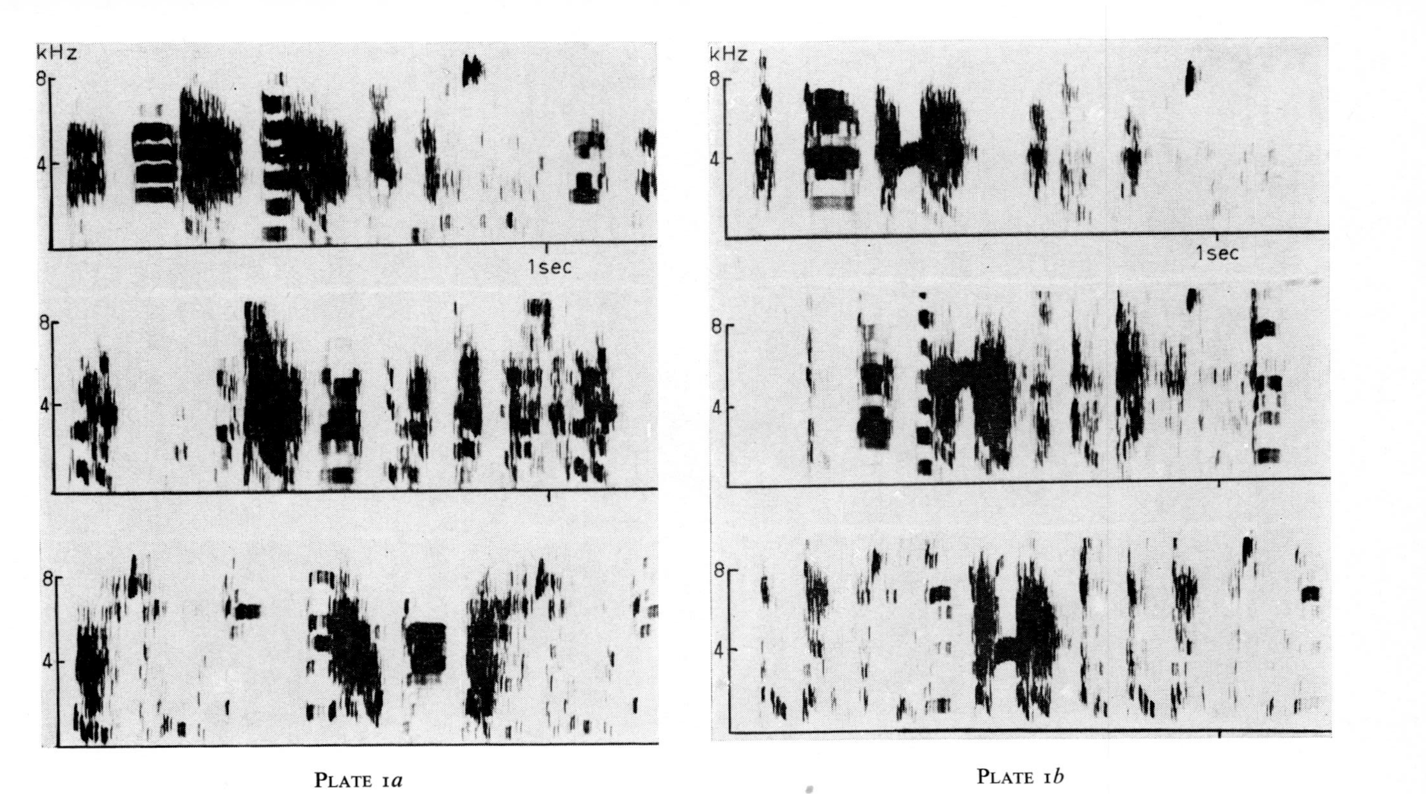

PLATE 1*a*

PLATE 1*b*

PLATE 1*a* and *b*. Song phrase of six male zebra finches caught near Katherine, Northern Territory, Australia, demonstrating wide frequency span and great individual variability.

(*facing p. 64*)

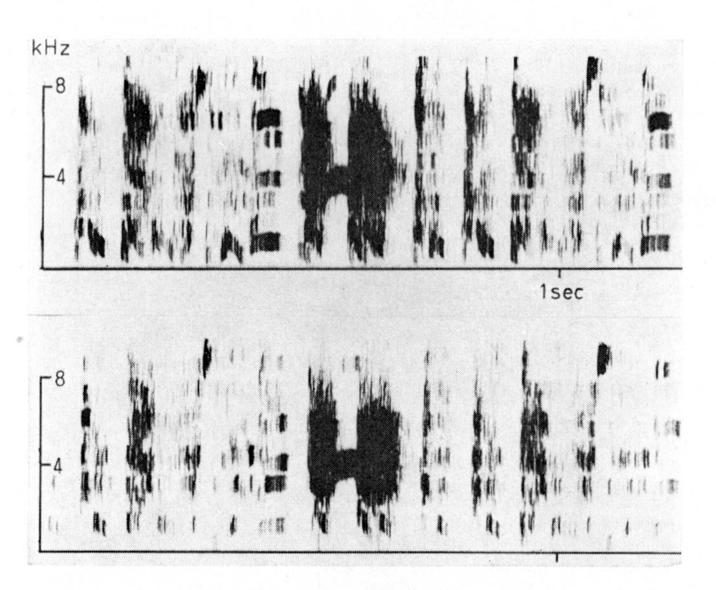

PLATE 2. Courtship song (*above*) and undirected song (*below*) of a wild-caught zebra finch.

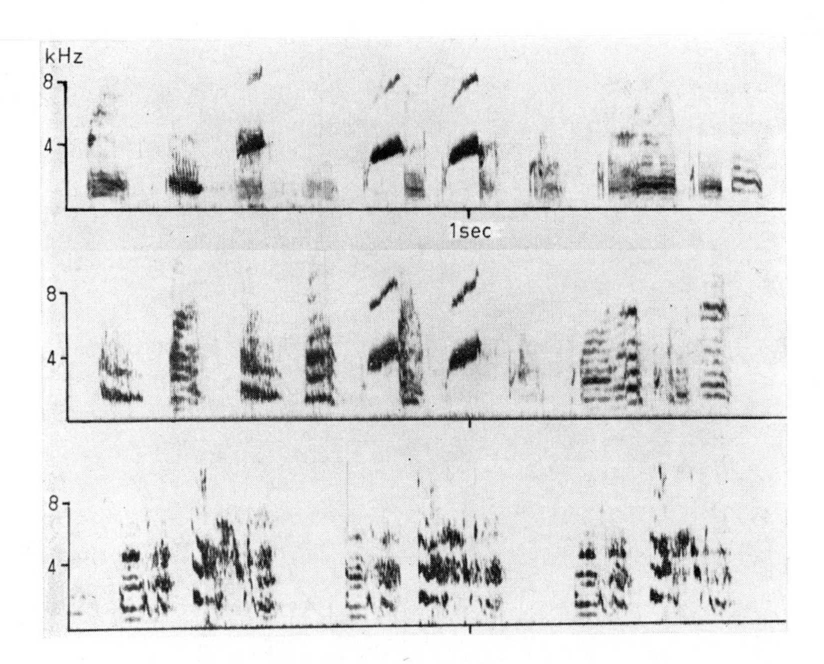

PLATE 3. Song phrase of a Bengalese finch (*above*), of a zebra finch raised by the Bengalese finch (*middle*) and of a zebra finch raised by its own parents (*below*). The latter is the natural father of the second individual.

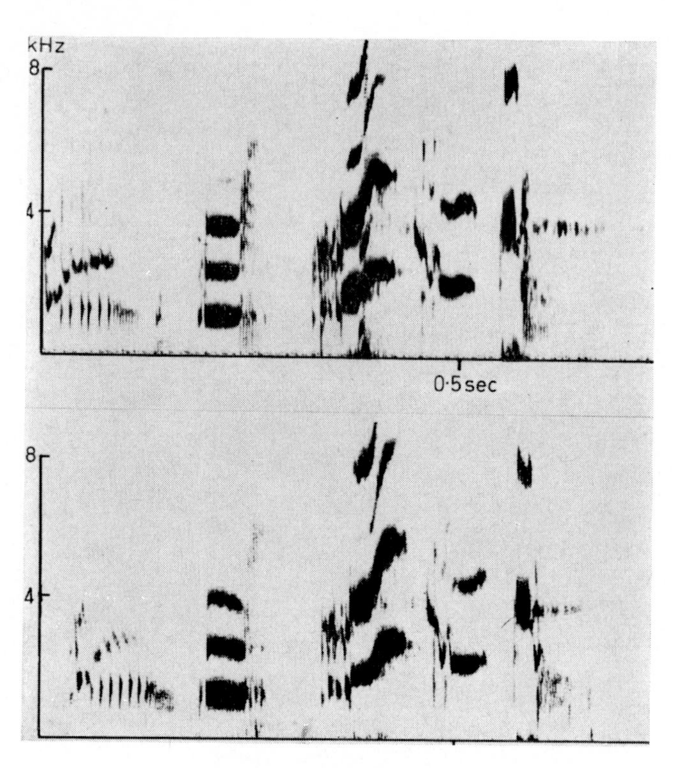

PLATE 4. Song phrase of a Bengalese finch (*above*) and of a zebra finch raised by the Bengalese finch.

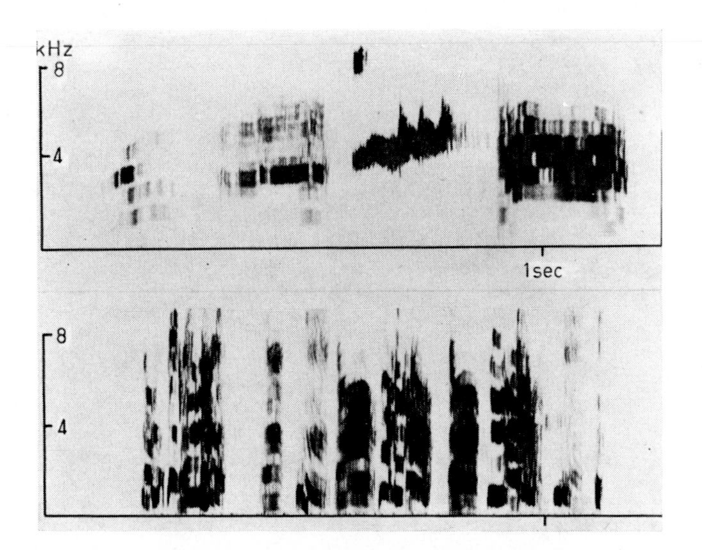

PLATE 5. Song of a zebra finch raised without tutor (by two Bengalese finch females) (*above*) and song of a zebra finch raised by its own parents (*below*).

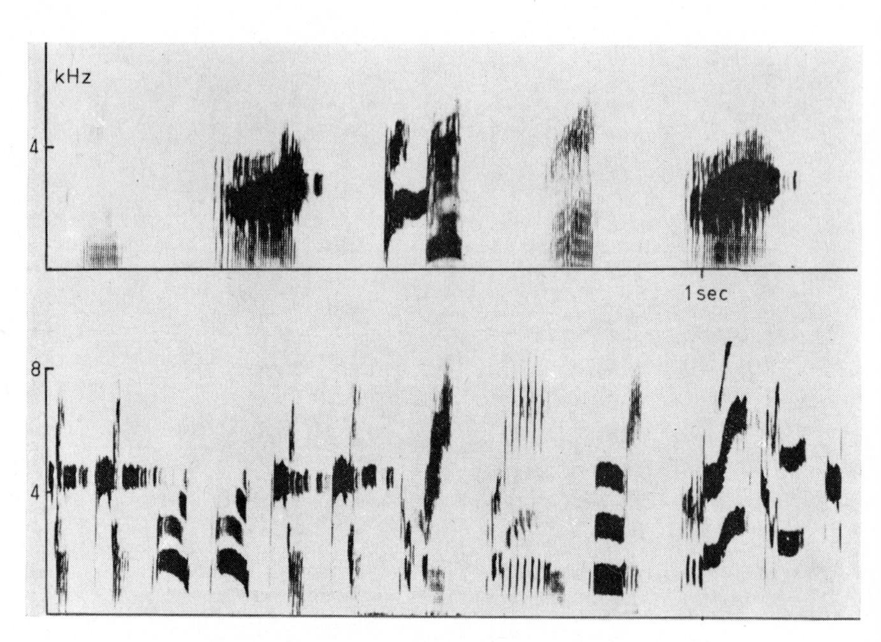

PLATE 6. Song of a Bengalese finch raised without tutor (by two females) (*above*) and song of a Bengalese finch raised by its own parents (*below*).

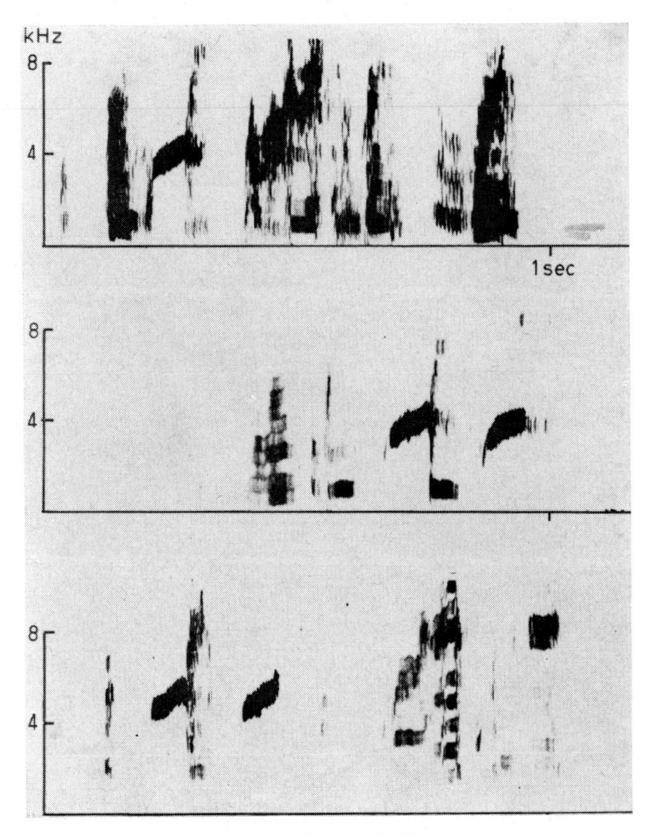

PLATE. 7. Song of zebra finches isolated from the Bengalese foster father at the 94th (*above*), 40th (*middle*) and 63rd (*below*) day of life.

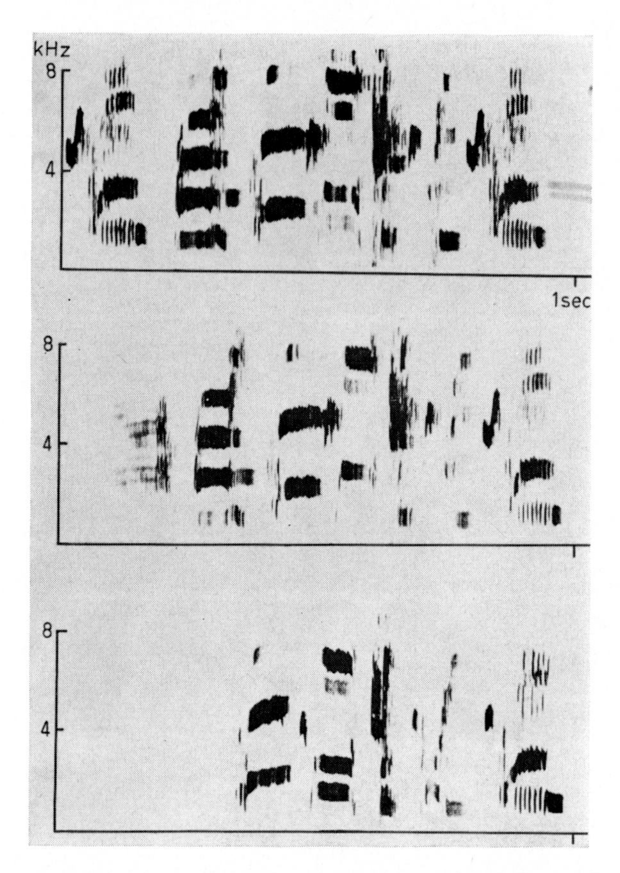

Plate 8. Song of zebra finch male raised by Bengalese finches recorded before listening to species-specific song (*above*), after fourteen months of acoustic contact with conspecific males (*middle*), and after five years of such contact (*below*).

A remarkable fact was the great individual variability in the song of isolated birds which was to be observed even in sibling males. As the number of F_1 descendants from wild-caught stock is still small, it is not yet known if the great variability is due to the influence of domestication.

(ii) *Rhythm*. Because the syllables are longer the song phrase of males raised without a tutor tends to be slower in overall tempo than the song of normal birds (number of song elements per second in isolated and in normal birds respectively: *Taeniopygia* 6·1 : 14·8; *Lonchura* 8·3 : 14·4) (see Plates 5 and 6). This means that the slow rhythm characteristic of the first period of juvenile song is permanently retained by the isolated males.

Similar results have been obtained in other species of birds. The song elements of blackbirds reared in isolation are longer and more uniform than those of wild birds (Thielcke and Thielcke, 1960). Isolated chaffinches develop a song 'of abnormally uniform tonal quality' (Thorpe, 1958), while hand-raised linnets produce a song which is slower and consists of fewer elements than in the normal male (Poulsen, 1954). Song sparrows isolated at the egg stage also differ from wild-type individuals in having a smaller repertoire (Mulligan, 1966).

THE LEARNING PROCESS

In studying the acquisition process in more detail we concentrated on two main topics: the possible occurrence of a sensitive period for song-learning, and the possible selectivity of imitation for certain song patterns.

Sensitive period

In order to learn more about the life span during which the bird is able to acquire its definite song pattern, we varied the age at which the young male was isolated from its foster parents. So far the number of birds used in this series of experiments is sufficient only in the zebra finch and the following results refer only to this species:

(i) All males isolated after the 80th day of life developed a song which in all recognizable physical characteristics was identical with the song of the foster father (six males).

(ii) Males which had been isolated between the 38th and 66th day of life (i.e. before their own song phrase was completed) developed a song which consisted of song elements only of the foster father, but which was *not* identical in the sequence of elements or in the total length of the song phrase (eleven males).

(iii) Males which had been isolated before the 40th day of life developed a song containing *some* very marked song elements of the foster father but differing in almost all other characters (length, number and sequence of elements etc.). Their song tended to be slow and uniform and thus resembled the song of males raised without a tutor (eleven males) (Plate 7).

It can be concluded, therefore, that the single characters of the zebra finch song are acquired at different ages. As a first step the song elements of the tutor are accepted. This can be done *before* the young male begins with its juvenile song. Male 4, for example, reproduced two song elements of its foster father though it was isolated on its 25th day of life and did not show the slightest trace of its juvenile song before the 35th day of life. More than thirty days after it had last had the opportunity to listen to its tutor, these elements became recognizable in the juvenile song. Another male (male 13) which was isolated on its 38th day of life, developed a song including *all* elements of the foster father's song phrase. This means that the acquisition of song elements must already have been *completed* at a time when juvenile song was *just* beginning.

The length and sequence of song elements as well as the rhythm, on the other hand, are fixed between the 40th and 80th day of life, i.e. *during* song development. At this time, the fairly slow innate rhythm (see above) may be accelerated and the comparatively long elements may be abbreviated through listening to the song of a normally raised tutor.

As the above-mentioned groups overlapped slightly, there seems to be some individual variation in the sensitive periods for song acquisition. Again it is not yet known if this variability is due to the influence of domestication.

Finally, we tried to find out if listening to the tutors *after* the 80th day of life influences song development. For this purpose three males which had been raised by Bengalese finches and consequently had accepted the foster father's song were introduced into an aviary with 10–20 conspecific males when 80 to 90 days old. They were kept in the aviary for from four to fifteen months. Afterwards, their song was recorded again and the spectrograms compared with those taken at the 80th day of life. Despite the constant acoustic contact with conspecific males possessing a normal song, the experimental birds did *not* supplement their song phrases further to any great extent. Only a few very minor alterations could be detected in some of the birds (Plate 8). This means that once the song phrase of a bird is fully developed, basic alterations are apparently no longer possible. This applied even to birds which, because they had been raised by another species, did not yet possess the species-specific song, and which afterwards had opportunity to listen to conspecific males. The song phrase seems to be irreversibly fixed once it has been fully developed.

Selectivity of song acquisition

In species in which the song is largely learnt it is necessary that the imitative ability should be restricted to elements of the species-specific song. This is especially important for the Australian zebra finch which frequently breeds in mixed colonies (see page 71). In the literature, two methods for this restriction have been described:

(i) a predisposition to learn species-specific song patterns in preference to those of other species (e.g. *Fringilla coelebs*, Thorpe, 1958; *Zonotrichia leucophrys*, Marler and Tamura, 1964);

(ii) a predisposition to learn the song of birds to which an 'emotional' relationship of some kind exists (father: *Pyrrhula pyrrhula*, Nicolai, 1959; territorial neighbour: *Galerida cristata*, Tretzel, 1965).

In order to determine whether grass finches do focus their attention on a certain limited range of song patterns, we have carried out the following experiments:

(i) Seven male zebra finches were raised by Bengalese finches in our bird rooms where they had the opportunity to hear and see other zebra finches in neighbouring breeding cages. Five of these birds developed a song identical to their Bengalese finch foster-father and did *not* include any species-specific elements into their song phrase. Only in two of the males did the song later contain a few song elements of the neighbouring conspecific males. Interestingly enough the Bengalese foster-father of these males had a song of rather poor and uniform quality.

(ii) Eight male zebra finches were raised by Bengalese foster parents in an aviary containing not only the foster parents but also two pairs of zebra finches, so that the experimental birds could come into close contact with members of their own species after having left the nest.

The birds of this group showed interesting differences in song development. All males which had been fed exclusively by their foster parents imitated the song of their foster father (five males). Different results were obtained from those males which, after fledging, had been fed not only by the foster parents but also by a pair of zebra finches which happened to have youngsters of approximately the same age and had probably been attracted by the begging calls of the strange zebra finch. The song of these males contained elements of their foster father and of the zebra finch male, clearly preferring the latter (three males).

(iii) Three male zebra finches were raised only by females, though not in soundproof chambers, but in our general bird rooms. They could hear and see conspecific males as well as males of several other species of grass finches in neighbouring cages. These males developed a song consisting only of species-specific elements. In contrast to normally raised males, however, they did not imitate the song of a certain male but rather accepted single elements from different neighbours.

The results lead to the following conclusions: All males of group (i), as well as the first mentioned males of group (ii), did acquire the strange song of the foster father in spite of the presence of conspecific males. This means that in the young zebra finch there is a tendency to restrict the imitative ability to those tutors with which a strong personal bond exists. Their song is preferred even if they do not sing in the species-specific manner and even if the song of conspecific males can be heard in close proximity. The same has been reported in other songbirds raised by strange species (bullfinch, Nicolai, 1959; Viduinae, Nicolai, 1964; pied flycatcher, Vilka, 1966).

The males of group (iii), as well as the latter category in group (ii), clearly show, however, that there is also a selective responsiveness to the species-specific song, possibly for the right tonal quality. If there is no personal bond with a singing male (group (iii)), the youngster will select the species-specific song elements from the many different songs to be heard in the vicinity. And if there is a personal bond with two males, one of which is singing in the species-specific manner (group (ii)), the latter is preferred.

It is to be concluded, therefore, that in the zebra finch a double security exists: there are innate preferences both for the father's song and for the right tonal quality. If the two do not coincide, the former is likely to be the stronger.

Under natural conditions both preferences do always refer to the same individual, the natural father, and a maximum security for the acquisition of species-specific song patterns seems thus to be achieved. Possibly the preference for the species-specific tonal quality may be especially advantageous if the natural father happens to have a rather poor and incomplete song phrase, so that the auditory template of the youngster is completed only by species-specific song elements.

DISCUSSION

Our results can be summarized as follows. In the three species of estrildid finches under investigation the basic outline of the song as a sequence of syllables and with a certain approximate duration is independent of experience of the species song. This template, however, is characterized by a high degree of individual variation. It is completed, or—if the young male is raised by members of another species—covered over, by a song template which is acquired during early juvenile life. The acquisition process differs slightly from that found in other groups of song birds.

The most striking feature is the fairly early sensitive period for the acquisition of song elements, and especially the early end of that period. While the length and sequence of song elements are acquired from other birds *during* the phase of juvenile song, the acquisition of song elements as such may be considerably earlier; it starts well before the young male is able to produce any kind of song and closes during the early phases of juvenile song, though the acquired song patterns become recognizable only several weeks later. Even if the young bird has no further opportunity to listen to its father or foster father during its own song development it will later produce a song almost identical to the model. It follows that the young bird is able to produce a copy of the song heard during the first weeks of life without listening to the original model again.

Similar results have been obtained by Marler and Tamura (1964) and Konishi (1965) for the white-crowned sparrow (*Zonotrichia leucophrys*). M. Heinroth (1911) mentioned a nightingale (*Luscinia megarhynchos*) which in June at the age of six weeks was given the opportunity to listen to a male blackcap (*Sylvia atricapilla*) for ten days. When this bird started to sing at the beginning of the following year it developed a perfect imitation of the blackcap's song. Under natural conditions young nightingales also acquire some details from their father's song immediately after leaving the nest, though they do not reproduce them until about seven months later (Stresemann, 1948). Finally young chaffinches having been isolated in September, i.e. before the onset of their own juvenile song, will develop a more complete song pattern than individuals isolated immediately after birth (Thorpe, 1958) (for further examples, see Armstrong, 1963).

These examples indicate that several species of songbirds can acquire a song template before the birds themselves can sing, and keep it for months without practising it. What seems to be exceptional in the song development of grass finches, however, is the early *end* of the acquisition period. As a rule, the young bird acquires some song elements in its first summer and autumn but completes the song in the next spring, partly under the influence of neighbouring males (cf. *Melospiza melodia*, Nice, 1943; *Fringilla coelebs*, Marler, 1956; *Lanius collurio*, Blase, 1960; *Turdus merula*, Thielcke, 1961; *Richmondena cardinalis*, Lemon and Scott, 1966). This points to the existence of another sensitive period for the acquisition of song elements in the second year of life. Some species may acquire new song elements even in later years (*Mimus polyglottus*, Laskey, 1944; *Turdus merula*, Messmer and Messmer, 1956; Thielcke and Thielcke, 1960; *Junco oreganus*, Marler, Kreith and Tamura, 1962). In the zebra finch, in contrast, the sensitive period ends between the 80th and 90th day of life and is not repeated any more. A similar case has been reported by Mulligan (1966) for the song sparrow, in which sensitivity is high at five to ten weeks and diminishes at fourteen weeks of age.

The early end of the sensitive period in estrildid finches seems to be due to peculiarities in their annual periodicity: birds of higher geographical latitudes usually undergo a complete regression of the gonads outside the breeding season (for survey of the literature see Farner and Follett, 1966; Immelmann, 1967*a*). During this period, song and all other sexual activities are extinguished more or less completely. At the beginning of the next breeding season song matures again, and there may be a new sensitive period during which the song may be further supplemented. The zebra finch, on the other hand—and probably many of the tropical and subtropical grass finches—displays an almost constant activity of the gonads caused by a tonic gonadotrophic activity of the hypothalamo-hypophyseal system (Farner and Serventy, 1960). It expresses itself through an almost equally constant courtship and song activity that persists throughout the year and probably is an adaptation to a fairly irregular breeding season in the birds' natural environment (for discussion see Immelmann, 1963).

It seems possible, therefore, that the lack of a second sensitive period for song development is due to the absence of an annual regression and re-activation of song-activity. As the three species of grass finches are the first tropical birds used in an experimental study of song development it seems possible that a similar situation may be found in other song-birds of equatorial regions.

The biological significance of the very early song acquisition in the zebra finch may be explained in connection with its breeding biology. The species inhabits the arid regions of northern and central Australia. It usually breeds in loose breeding colonies ('neighbourhoods', cf. Immelmann, 1967*c*), very often in the immediate vicinity of nests or breeding colonies of other species of grass finches. The breeding individuals as well as the fledged young may thus frequently come into contact with other species during feeding, bathing, drinking and other activities. After the end of the breeding season several species may even form mixed parties (Immelmann, 1962*b*). Due to the very early sexual maturity of the zebra finch (Immelmann, 1963) the young birds may soon form pairs within those flocks.

As a consequence of the frequent and fairly close contact with members of closely related species it seems to be advantageous, therefore, if the species-specific song can be acquired immediately after fledging—i.e. when the birds are still fed by the parents and stay in the immediate vicinity of the nest—and can afterwards be fully developed without further exposure to the original model. This can be achieved by the early onset and end of the sensitive period. The above mentioned twofold selectivity in song acquisition may also serve the same purpose. The same may apply to *Lonchura* and *Euodice* which also display an early sexual maturity, though the data about their breeding biology in the wild are not yet sufficient to draw any definite conclusions. In this respect a study of song development in the solitary-living forest grass finches would seem highly desirable.

The biologically advantageous early song acquisition in the zebra finch is facilitated by the fact that the birds usually have several broods, and the male thus increases its courtship and song activity again as soon as the young have left the nest. It also seems to be possible that the individual differences in the songs of wild zebra finches depend partly on the amount and duration of their father's song activity, i.e. on whether their parents produced another brood or not.

ACKNOWLEDGEMENTS

This work was supported by the Deutsche Forschungsgemeinschaft. I would also like to thank the staff of the laboratory 5K of the Physikalisch-Technische Bundesanstalt, Braunschweig; especially Dr W. Kallenbach, Dr H. J. Schroeder, and Miss G. Noch for kindly preparing the sound spectrograms and for their most useful help in all technical problems. I am also deeply grateful to Dr M. Konishi for his valuable criticism and many most interesting discussions.

REFERENCES

ARMSTRONG, E. A. (1963) *A Study of Bird Song*. London.

BLASE, B. (1960) Die Lautäusserungen des Neuntöters (*Lanius c. collurio* L.). *Z. Tierpsychol.* **17**, 293–344.

FARNER, D. S. and FOLLETT, B. K. (1966) Light and other environmental factors affecting avian reproduction. *J. Anim. Sci., Suppl.*, **25**, 90–118.

FARNER, D. S. and SERVENTY, D. L. (1960) The timing of reproduction in the arid regions of Australia. *Anat. Rec.* **137**, 354.

GOODWIN, D. (1960) Observations on Avadavats and Golden-breasted Waxbills. *Avicult. Mag.* **66**, 174–99.

(1965) A comparative study of captive Blue Waxbills (Estrildidae). *Ibis* **107** 285–315.

HALL, M. F. (1962) Evolutionary aspects of estrildid song. *Symp. Zool. Soc. Lond.* **8**, 37–55.

HARRISON, C. J. O. (1962) An ethological comparison of some waxbills (Estrildini), and its relevance to their taxonomy. *Proc. Zool. Soc. Lond.* 139, 261–82.

HEINROTH, M. (1911) Zimmerbeobachtungen an seltener gehaltenen europäischen Vögeln. *Verh. V. Int. Orn. Kongr.*, Berlin 1910, 703–64.

IMMELMANN, K. (1962*a*) Beiträge zu einer vergleichenden Biologie australischer Prachtfinken (Spermestidae). *Zool. Jb. Syst.* **90**, 1–196.

(1962*b*) Biologische Bedeutung optischer und akustischer Merkmale bei Prachtfinken (Aves: Spermestidae). *Verh. dt. zool. Ges., Saarbrücken*, 1961. 369–74.

(1963) Drought adaptations in Australian desert birds. *Proc. 13th Int. Orn, Congr.*, Ithaca, 1962, 649–57.

(1965*a*) Prägungserscheinungen in der Gesangsentwicklung junger Zebrafinken. *Naturwissenschaften* **52**, 169–70.

(1965*b*) *Australian Finches*. Sydney.
(1967*a*) Periodische Vorgänge in der Fortpflanzung tierischer Organismen. *Studium generale*, **20**, 15–33.
(1967*b*) Zur ontogenetischen Gesangsentwicklung bei Prachtfinken. *Verh. dt. zool. Ges.*, Göttingen, 1966, 320–32.
(1967*c*) Verhaltensökologische Studien an afrikanischen und australischen Estrildiden. *Zool. Jb. Syst.* **94**, 609–86.
KONISHI, M. (1964) Song variation in a population of Oregon Juncos. *Condor* **66**, 423–36.
(1965) The role of auditory feedback in the control of vocalization in the White-crowned Sparrow. *Z. Tierpsychol.* **22**, 770–83.
KUNKEL, P. (1959) Zum Verhalten einiger Prachtfinken (Estrildinae). *Z. Tierpsyschol.* **16**, 302–50.
LASKEY, A. R. (1944) A mockingbird acquires his song repertoire. *Auk* **61**, 211–19.
LEMON, R. E. and SCOTT, D. M. (1966) On the development of song in young cardinals. *Canad. J. Zool.* **44**, 191–7.
MARLER, P. (1956) The voice of the Chaffinch and its function as a language. *Ibis* **98**, 231–61.
(1960) Bird songs and mate selection. In: Animal Sounds and Communication. *Am. Inst. Biol. Sci.*, Publ. No. 7, 348–67.
(1963) Inheritance and learning in the development of animal vocalizations. In *Acoustic Behaviour of Animals* (R. G. Busnel, ed.). Amsterdam, 228–43.
MARLER, P. and TAMURA, M. (1964) Culturally transmitted patterns of vocal behaviour in sparrows. *Science* **146**, 1483–6.
MARLER, P., KREITH, M. and TAMURA, M. (1962) Song development in hand-raised Oregon Juncos. *Auk* **79**, 12–30.
MESSMER, F. and MESSMER, E. (1956) Die Entwicklung der Lautäusserungen und einiger Verhaltensweisen der Amsel (*Turdus merula merula* L.) unter natürlichen Bedingungen und nach Einzelaufzucht in schalldichten Räumen. *Z. Tierpsychol.* **13**, 341–441.
MORRIS, D. (1958) The comparative ethology of grassfinches (Erythrurae) and mannikins (Amadinae). *Proc. Zool. Soc. Lond.* **131**, 389–439.
MOYNIHAN, M. and HALL, M. F. (1954) Hostile, sexual, and other social behaviour patterns of the Spice Finch (*Lonchura punctulata*) in captivity. *Behaviour* **7**, 33–76.
MULLIGAN, J. A. (1966) Singing behaviour and its development in the Song Sparrow *Melospiza melodia. University of California Publ. Zool.* No. 81.
NICE, M. M. (1943) Studies in the life history of the Song Sparrow. Part II. *Trans. Linn. Soc. N.Y.* **6**, 1–328.
NICOLAI, J. (1959) Familientradition in der Gesangsentwicklung des Gimpels (*Pyrrhula pyrrhula* L.) *J. Ornith.* **100**, 39–46.
(1964) Der Brutparasitismus der Viduinae als ethologisches Problem. *Z. Tierpsychol.* **21**, 129–204.
POULSON, H. (1954) On the song of the Linnet (*Carduelis cannabina* L.). *Dansk. orn. Foren. Tidsskr.* **48**, 32–7.
STRESEMANN, E. (1948) Nachtigall und Sprosser: ihre Verbreitung und Ökologie. *Orn. Ber.* **1**, 193–222.

THIELCKE, G. (1961) Ergebnisse der Vogelstimmen-Analyse. *J. Ornith.* **102**, 285–300.

THIELCKE, H. and THIELCKE, G. (1960) Akustisches Lernen verschieden alter schallisolierter Amseln (*Turdus merula* L.) und die Entwicklung erlernter Motive ohne und mit künstlichem Einfluss von Testosteron. *Z. Tierpsychol.* **17**, 211–44.

THORPE, W. H. (1958) The learning of song patterns by birds, with especial reference to the song of the Chaffinch *Fringilla coelebs*. *Ibis* **100**, 535–70.

(1959) Learning. *Ibis* **101**, 337–53.

(1961) *Bird Song*. Cambridge University Press, Cambridge.

THORPE, W. H. and LADE, B. I. (1961) The songs of some families of the Passeriformes. II. The songs of the buntings (Emberizidae). *Ibis* **103a**, 246–59.

TRETZEL, E. (1965) Über das Spotten der Singvögel, insbesondere ihre Fähigkeit zu spontaner Nachahmung. *Verh. dt. zool. Ges.*, Kiel, 1964, 556–65.

VILKA, I. (1966) The significance of heredity and imitation in the formation of bird song. *Atlantic Naturalist*, **21**, 78–87 (translation from the Latvian).

PART C
PHYSIOLOGICAL ASPECTS

INTRODUCTION

The causal basis of behaviour in the individual involves two overlapping but distinguishable problems. The first, considered in the previous section, concerns the long-term development of the potentiality for showing particular types of behaviour. The second, to which we shall now turn, concerns the more immediate factors controlling whether the behaviour occurs at this moment or at that. Analyses of the immediate bases of behaviour are often handicapped by difficulties of measurement, and this is especially the case in studies of the relations between neural functioning and behaviour. The experimental methods available for dissecting these relations—ablation, intracranial recording and stimulation—are crude, even when micro-electrodes are used, in comparison with the delicacy of functioning of the central nervous system: if they are to be effective, clearly recognizable behavioural end-points are essential.

In the study of avian neural functioning, vocalizations are particularly useful. In each species the repertoire is limited, the calls falling into a number of clearly recognizable categories with relatively few intermediates. With freely moving animals, they can be readily recorded on tape; and subsequent interpretation of the record, unlike that of film, usually requires no allowance for the orientation of the animals' body with respect to the recording apparatus. Thus in using intracranial stimulation to study the organization of behaviour in the domestic fowl, von Holst and von St Paul (1963) found that crowing, cackling and clucking provided valuable measures of the effect of the stimulation applied.

Studies of the functioning of the avian brain are as yet in their infancy. Nevertheless, as the following article by Brown (chapter 5) shows, knowledge about the neural bases of vocalizations is not inconsiderable: because of their suitability for study, rapid progress can be expected in this field. His essay contains a critical discussion of the techniques used, and also illustrates the manner in which studies of brain functioning can throw light on the organization of behaviour itself.

Vocalizations are also peculiarly useful as behavioural indices in the study of the endocrine control of behaviour. Andrew's paper (chapter 6) is concerned with a particular aspect of this—the precise manner in which gonadal hormones, which are normally secreted in increasing quantities as the bird comes into reproductive condition, influence the appearance of new vocalizations. Although his conclusions are based primarily on studies of young birds given exogenous androgen, the principles are likely to be applicable to natural seasonal changes in vocalizations. Two points of general importance for studies of physiological aspects of behaviour arise from his paper. First, his analysis illustrates the manner in which an understanding of the physiological bases of behaviour may depend on the recognition that what appear to be discrete units of behaviour may not be discrete units in terms of underlying mechanisms. By proceeding in steps, analysing first from behaviour to effector organization, coupling this with a situational analysis, and only then speculating about the nature of hormonal action, he has shown that apparently diverse call-notes are in fact interrelated, and thereby achieved considerable economy of hypothesis. The second and related point is that his study shows that diverse changes in behaviour may be accounted for in terms of only one type of change in the underlying mechanisms—though the precise nature of his postulated 'testosterone mechanism' remains a matter for speculation.

Brockway's article (chapter 7) is also concerned with the endocrine control of vocalizations, but she extends her discussion to the reciprocal problem—the influence of vocalizations on the endocrine states of other individuals. In doing so she emphasizes the complexity of the control and integration of the avian breeding cycle. In order to breed successfully, a pair of birds must perform a number of activities (e.g. pair formation, nest-site selection, nest-building, egg-laying, incubation, feeding the young) in the right order. Each activity must be synchronized appropriately with environmental events, and the behaviour of each individual must be synchronized with that of its partner. This is achieved by complex interactions between environmental stimuli, endocrine states, and behaviour: for instance gonad growth in the early breeding season is partly due to environmental influences. The resulting endocrine changes lead to changes in behaviour, which bring the bird into contact with fresh stimuli, and so on. Apart from Brockway's studies on the budgerigar, detailed analyses of such interactions are available for only two species–the ring dove (*Streptopelia risoria*, e.g. Lehrman, 1965) and the domestic canary (Hinde, 1965; Hinde and Steel, 1966).

REFERENCES

HINDE, R. A. (1965) The integration of the reproductive behavior of female canaries. In *Sex and Behavior* (F. A. Beach, ed.). Wiley, New York.

HINDE, R. A. and STEEL, E. (1966) Integration of the reproductive behaviour of female canaries. *Symp. Soc. exp. Biol.* **20,** 401–26.

HOLST, E. VON and SAINT PAUL, U. VON (1963) On the functional organisation of drives. *Anim. Behav.* **11,** 1–20, translated from *Naturwissenschaften* **18,** 409–22.

LEHRMAN, D. S. (1965) Interaction between internal and external environments in the regulation of the reproductive cycle of the ring dove. In *Sex and Behavior* (F. A. Beach, ed.). Wiley, New York.

5. THE CONTROL OF AVIAN VOCALIZATION BY THE CENTRAL NERVOUS SYSTEM

by **JERRAM L. BROWN**
Department of Biology, University of Rochester, New York

INTRODUCTION

The vocal behaviour of birds has been much studied from the viewpoints of communication, evolution, motivation and development. In the resulting motivational and evolutionary hypotheses innate neural control mechanisms have often been implicit as the structural and physiological bases for vocal behaviour. These neural mechanisms are the necessary links upon which ontogenetic and phylogenetic influences converge to produce vocalization. Yet between the early studies of Kalischer (1905) on speech in parrots and the more recent research activity initiated by von Holst and von St Paul (1957, 1958, 1960, 1963) they were essentially ignored as the subjects of research. In the last decade, however, significant progress has been made. This paper will attempt to summarize what has been learned to date.

LOCALIZATION OF VOCAL FUNCTION AS SHOWN BY STIMULATION STUDIES

Of greatest developmental and evolutionary significance is the finding that a high degree of localization of function occurs in neural tissue in regard to vocalization. The brain is not everywhere equipotential in its ability to influence vocal behaviour. Rather, particular regions, which have been delimited by experiment, have profound effects, while other regions have none or only a slight and indirect influence. The evidence indicates that these regions are similar or identical from individual to individual within a species and from species to species.

Method

The method which has proven to be the most precise and efficient for the establishment of localization of vocal function in the avian brain has been that of stimulation of populations of neurons. Fine, insulated electrodes of 0·1 to 0·3 mm diameter with uninsulated tips have been used to stimulate 'points' in the brain with minute amounts of current.

Stimulation is usually performed with a series of negative pulses, each of which may, or may not, be followed by a positive pulse—biphasic and monophasic, respectively. The pulses may be rectangular, sinusoidal or some other form. To produce a behavioural response typically requires currents of from 20 microamperes to one milliampere. Pulse durations employed in mapping studies on birds have ranged from 0·5 to 10 milliseconds; frequencies, from 30 to 50 Hz.

For a given electrode position and environmental situation the behavioural response commonly depends primarily on the total energy delivered per unit of time and on the length of time during which it is applied. A given rate of energy delivery can be equally effective over a broad range of combinations of current, pulse duration, and frequency. Consequently, less current is required to reach behavioural threshold when longer pulse durations or higher frequencies are employed—within limits. These and other parametric relationships have been studied in chickens in respect to clucking by Kramer, von St Paul and Heinecke (1964). Similarly, a longer train duration may compensate for a lower rate of energy delivery within limits, as shown quantitatively for warning calls and cackling in chickens (von Holst and von St Paul, 1960).

For the purposes of this discussion perhaps the most significant conclusion to be derived is the realization that there are many variable physical parameters in a stimulation experiment and that comparisons between the findings of different workers cannot be safely made without taking all of them into consideration. Since no two authors working independently have used the same stimulus parameters, and since some authors have not even specified completely the ones they were using, this may not be easy.

One methodological criticism may be made of most of the mapping studies published for birds: in few of them have the locations of *negative points* been adequately illustrated, that is, those at which stimulation did not evoke the response in question. It is a logical necessity in the delineation of a functionally specialized brain region to indicate its borders. In a stimulation study this can only be done by establishing the location of unresponsive as well as responsive regions. With much of the published data discussed here one can only assume that the responses were not obtained from regions where positive results are not shown: this is dangerous at best.

RESPONSIVE REGIONS FOR VOCALIZATION

Considering the variety of species and methods which have been used, the degree of ageement on main points is gratifying. The brain sites from which vocalizations can be evoked by stimulation fall into a zone which runs, probably continuously, from the anterior paleostriatum back through the archistriatum in the forebrain, connecting with septal and hypothalamic areas, and extending at least to the lateral nuclear masses of the midbrain. The remaining striatal regions, the thalamic nuclei and the tectum are unresponsive with but few exceptions (n. dorsomedialis anterior, for example, Putkonen 1966*a*, 1967).

Reproductive behaviour

Within this rather extensive region at least two basic divisions may be recognized, one for reproductive behaviour and one for protective behaviour. In the pigeon (*Columba livia*) the division is most clear, although the two fields seem to overlap in part (Åkerman, 1965*a*,*b*). In the region generally concerned with reproductive activities the evoked behaviour is characterized by the presence of bow-cooing and/or nest-cooing. This zone extends from relatively far anterior in the striatum just lateral to the ventricles back into the preoptic nuclei, where thresholds are lowest, and has its posterior representation in the brain stem under and slightly behind the anterior commissure close to the third ventricle; the posterior extent of the striatal field is found in the posterior paleostriatum, near the ventral corners of the lateral ventricles. Within this zone is found a smaller one which is characterized behaviourally by nest demonstration and nest-cooing and is located in the preoptic region and close to the ventricle under the anterior commissure just anterior and posterior to it; there are also a few striatal points. The nest-coo and bow-coo points intermingle and are in some cases identical; sex appears to have an effect on the type of response.

Protective behaviour

The second main division is characterized behaviourally by various kinds of protective behaviour. It might be still further subdivided into regions for defensive threat, attack and escape, but the available data seem too few to do this reliably. The defensive threat in pigeons is typically initiated by 'freezing', followed by ruffling of the feathers and lowering of the tail, and culminates in crouching with unilateral wing raising and a growling call, or with actual blows delivered by the wings and accompanied by a 'wao' call. Escape points, which are characterized by crouching, locomotor mobility, and absence of calls and attack, are also present in this division. The anterior limit in the brain stem for protective behaviour lies under the anterior commissure and overlaps partly with the reproductive subdivision; it extends back part way through the hypothalamus, at least. In the forebrain it runs from the posterior paleostriatum, apparently over-lapping slightly again with the reproductive zone, back into the medial archistriatum. In summary, in the pigeon the reproductive system seems to be located more anteriorly both in the forebrain and the brain stem, and the protective behaviour system, more posteriorly, with some apparent overlap at and just posterior to the level of the anterior commissure.

Other species

In other species there is also some evidence for a basic subdivision of the brain along these behavioural and anatomical lines. In chickens stimulation in the above described protective behaviour zone evoked a variety of types of agonistic behaviour: these were classified by Putkonen (1967) as (1) threat display and calling, (2) directed attack, usually preceded by threat, (3) fear-like crouching, (4) panic, (5) and various combinations of these. The mapping of attack behaviour sites in chickens seems to agree reasonably well with Åkerman's defensive attack zone, with a few exceptions. The threat and threat-attack points reported by Putkonen in chickens correspond anatomically to the 'defensive posture' in pigeons rather than to the bow-coo posture. Comparisons here are difficult since the regions of the brain which were explored in the two species were somewhat different.

The reproductive behaviour zone in chickens cannot yet be outlined in detail, but present indications are that it lies in about the same position as in pigeons. Electrical stimulation in preoptic and anterior palaeostriatal sites in chickens evoked mainly 'attention and excitement', and in some cases 'fear-like crouching' (Putkonen, 1967).

Behaviourally these are heterogeneous categories, and they include various vocalizations including, most notably, crowing. Threat and threat-attack, as described by Putkonen, were not evoked from these regions except in the region of overlap at the level of the anterior commissure.

Evidence from implants of crystalline hormones also points to the recognition of a zone for the control of reproductive behaviour in the region described by Åkerman. In capons, testosterone implants in the preoptic region induced copulatory but not aggressive behaviour (Barfield, 1965). Progesterone implants in the preoptic and neighbouring hypothalamic and palaeostriatal regions were found effective in eliciting the complex of behaviours related to incubation in ring doves (*Streptopelia risoria*; Komisaruk, 1965). Although detailed comparisons are not yet possible the data on chickens and ring doves agree with those for pigeons on the importance of the preoptic and neighbouring regions for reproductive behaviour.

In red-winged blackbirds (*Agelaius phoeniceus*) various alarm calls have been evoked from the medial archistriatum and from the preoptic area back through the hypothalamus to the midbrain (Brown, 1965*a* and unpublished). Although these data are in partial agreement with the general distribution of sites from which vocalizations can be evoked in other species, they are not suitable for subdividing the responsive zone into regions primarily concerned with reproductive and protective behaviour; they are not, however, inconsistent with such a division.

Species differences in the behaviour patterns evoked by stimulation of the same brain region are evident mainly in the behavioural details, which are, of course, species-typical. The locations of the fields for reproductive and protective behaviour in the various species appear similar from the available data, which are not really sufficiently comparable to allow the recognition of small differences.

ELECTRICALLY EVOKED SEQUENCES OF BEHAVIOUR

In general it seems preferable to treat the behavioural responses to electrical (or chemical) stimulation as sequences of events, rather than as single events, even though they may be referred to for convenience by their most characteristic acts. For example, in the bow-coo pattern of pigeons as described by Åkerman (1965*a,b*) the reaction is typically initiated with the raising of the head and body and erection of the feathers on neck and crown. Next the bird begins walking and performs the typical bowing accompanied by a characteristic type of cooing. Bow-cooing and walking alternate with periods of standing still and looking around. There is considerable variability in the sequence and the elements incorporated.

In the nest-coo pattern the reaction is initiated with movements of attention and is followed by nodding, the characteristic nest-coo (which is distinctly different from the bow-coo), and at high intensities the nest-demonstration postures with wing-vibration, pecking and displacement preening, in various alternating sequences.

The complexity of many of these sequences is impressive. In the simpler ones there may be little more than the repetition of an act or the assumption of a posture, but in the more complex ones there is an unfolding of behavioural events which corresponds to what one would observe in a normal animal with the persistence of a social stimulus, such as an introduced male or female pigeon or a predator. Consequently, the impression is often given of a build-up in the intensity of the behavioural response to a plateau, even though the intensity of electrical stimulation is kept constant from the start. The extent of the build-up is typically graded according to the level of central stimulation. Having reached a plateau, the intensity of the behavioural response may decline gradually or suddenly, or persist at the same level indefinitely, depending on the kind of response and on the stimulus parameters.

What is evoked in these cases seems to be a motivational state rather than simply a rigidly coordinated sequence of motor patterns. This interpretation, which has long been accepted for mammals, is suggested by the naturalness and plasticity of the evoked responses, the variability in details, the behaviour toward objects and animals in the environment, and the parallel way in which the behavioural responses are graded according to intensity of both natural and electrical stimuli. It is further strengthened by the finding that brain stimulation in birds at some of these sites can be used to reinforce the learning and maintenance of approach or avoidance behaviour (Goodman and Brown, 1966).

In some sequences the acts all belong to the same functionally related group of behaviour patterns, such as courtship or protective threat. In others, acts from different functional groups may be intermixed: bow-cooing in some cases was interrupted by drinking or pecking the floor in pigeons (Åkerman, 1965*a*). The defining of a functionally related group must be done on the basis of naturally occurring behaviour and is arbitrary to a degree, even with the latest statistical techniques for analysis of sequences. Care must be taken with ritualized displacement activities, such as displacement preening in pigeons during courtship (Fabricius and Jansson, 1963).

Both pure and mixed sequences typically vary with the intensity and other parameters of stimulation. At a near-threshold intensity of electrical stimulation only a few initial elements characteristic of a behavioural sequence may be present—for example, slight postural changes or incomplete locomotory movements. As the stimulation intensity for a pure sequence is raised, only elements characteristic of that sequence would be added; some might also be omitted. In a mixed sequence elements from a different pure sequence would be added. At some brain loci the threshold for behavioural response may be so high that only the initial elements of a sequence can be evoked. Such sites are common in the striatum, where thresholds seem generally higher than in the brain-stem.

The spatial representation in the brain of one behaviour sequence to another may sometimes correspond to a functional relationship between them; anatomical proximity may sometimes help to explain temporal proximity. In pigeons escape points were found to be just lateral to and bordering the defensive-threat and attack points. In chickens the representation of these responses in the brain showed much intermixing—a convenient arrangement for shifts from defence to escape. In some cases in pigeons and chickens continued stimulation at a defence point led to a transition to escape—a not uncommon result in brain-stimulated cats also. The close relationship between the fields for defence and escape might eventually prove to have functional significance in the mediation of normal sequences of behaviour. Similarly the overlap of anatomical fields for bow-cooing and nest-cooing in the preoptic and nearby parts of the pigeon brain may be related to the close functional relationship between these two behaviour patterns in the normal courtship of this species.

At the neuronal level the events which lead to well-coordinated behavioural sequences are virtually unknown in birds and even in mammals. Certainly the point-like activation supplied by an electrode must differ greatly from that arriving on natural afferent pathways. Nevertheless, the end results are quite similar. We cannot describe the pattern of activation from a normal afferent stimulation; but because of the physical nature of an electrical stimulus, certain conditions may be anticipated. As intensity of electrical stimulation is increased, three types of phenomena may be expected: (i) the region at the electrode tip in which some neurons are directly excited by physical effects of the stimulus should increase in size ('stimulus spread'); (ii) the number and density of neurons firing because of physical effects of the stimulus should increase; and (iii) the rate of firing of some of these neurons might increase. In addition, indirect biological effects should occur, such as the synaptic activation or inhibition of other neurons in or outside of the zone of direct physical influence.

Such changes on a neuron-population basis are probably responsible for the behavioural phenomena of long latencies and differing thresholds for different behaviour at the same brain locus. Some sequences observed under electrical stimulation may be due simply to the time required for differing neuronal populations to reach threshold. Such differences might be due to differing spatial location of the neuronal populations with respect to the site of activation, or they might be due more to differing thresholds of the neurons involved, or to combinations of these and other factors. Other sequences may depend on environmental cues or reafference.

Central and peripheral influences on centrally evoked behaviour

As in mammals the vocal behaviour evoked by brain stimulation is not rigidly stereotyped and may vary under the influence of many factors in addition to the physical parameters of the stimulus. For convenience, these factors can be grouped into three categories: (i) peripheral stimulation under the direct control of the experimenter; (ii) artificial central excitation or inhibition elsewhere than at the first stimulating electrode, e.g. from a second stimulating electrode or cannula; and (iii) internal events in the brain not directly attributable to (i) or (ii), such as adaptation and motivational changes.

That peripheral stimuli, such as the presence of a stuffed predator or a species companion, can affect the centrally evoked behavioural response is well known in mammals and has also been described in birds. A stuffed polecat (*P. putorius*) may strengthen the defensive threat response of a chicken (von Holst and von St Paul, 1960). A second pigeon or chicken may serve as the object of an attack by the stimulated bird (Åkerman, 1965*a*,*b*; Goodman and Brown, 1966;

Putkonen, 1966*b*, 1967). Detailed studies of the effects of peripheral stimuli on centrally evoked responses have not been reported.

Simultaneous stimulation with two electrodes at different sites was tried by von Holst and von St Paul (1958, 1960). They described seven types of interaction effects in terms of the responses elicitable from each electrode alone. These effects were found in a variety of different behaviour patterns at different levels of behavioural complexity. Few details of such interaction experiments are available yet for any species.

Motivational changes, which are not due to the central and peripheral stimuli applied during stimulation, can be an important source of variability in brain stimulation experiments. The temporary suppression of a weak centrally evoked response by fear and the variations in threshold caused by different levels of wakefulness would be examples. The importance of mood shifts was stressed by von Holst and von St Paul (1960), who demonstrated their effects by means of simultaneous stimulation with two electrodes.

TRANSECTIONS, LESIONS AND ABLATIONS

Decerebration effects

Experiments in which a part of the brain is removed, anaesthetized, or otherwise prevented from functioning are useful for revealing which parts of the brain are indispensable for particular behaviour patterns—a type of information which cannot be gained from stimulation or electrical recording. Numerous transection experiments have been done to determine the importance of the forebrain in avian behaviour. In general these studies, which have been summarized by ten Cate (1936, 1965), show that a decerebrate bird stands quietly and does not eat or drink or perform any agonistic or sexual actions unless he is grasped, in which case he struggles to escape. In a recent study of detelencephalated pigeons, calling and the behaviour patterns of fighting or courting were abolished by the operation, which was done with minimal surgical trauma by means of a proton beam (Åkerman *et al.*, 1962). We may conclude that the forebrain is indeed necessary for the natural production of most vocalizations in the species investigated. However, 'crowing' evoked by electrical stimulation of the midbrain may persist in pigeons after decerebration (Popa and Popa, 1933).

Perceptual and specific sensory deficits

The reasons why the forebrain is necessary for normal vocalization are apparently numerous, since a variety of more general behavioural functions is affected by its removal. A deficit in the visual system, for example, may cause a loss of responsiveness to visual stimuli which would normally evoke vocalization. Unilaterally decerebrate chaffinches (*Fringilla coelebs*) performed mobbing behaviour with the characteristic calling at a stuffed owl if only the contralateral eye was removed but not if only the ipsilateral one was taken (Strata, 1964). (The optic tracts are virtually completely crossed in birds.) Since the damage to the brain should have been about the same in terms of volume and particular regions destroyed, the difference between the two groups cannot be attributed to a loss of general activation of brain-stem mechanisms from the forebrain. Further, when the owl was absent, the birds 'occasionally and spontaneously presented the characteristic calls'. The results suggest a deficit in visual perception of the stuffed owl even though the birds of both groups could still see sufficiently well to feed themselves, learn a visual discrimination based on the appearance of seeds, and perform 'the menace response'. The same operation in mallards resulted in birds which were alert, moved quickly, and responded to unfamiliar sounds or touch with sleeking, crouching and vocalizing; however, they failed to respond to objects thrust near either eye (Phillips, 1964).

The evidence for participation of the avian forebrain in the analysis of visual information, especially in pattern and brightness discrimination, is considerable and is based on lesion, evoked-potential, and single-unit studies (Zeigler, 1963*a*; Phillips, 1966; Vasilevskii, 1966; and others). Zeigler has shown deficits in pattern and brightness discrimination after hyperstriate damage in pigeons. Hodos and Karten (1966) have shown deficits in brightness discrimination in pigeons after lesions in n. rotundus, which is known to receive direct connections from the optic tectum (Karten and Revzin, 1966) and to degenerate after destruction of the palaeostriatum augmentatum or ectostriatum (Powell and Cowan, 1961; Revzin and Karten, 1967).

Similar deficits in the perception of key stimuli for evoking vocalization in other sensory modalities should also be expected as a result of forebrain damage. Although the exact role of the forebrain in perception is unclear, evidence is accumulating which indicates its involvement in audition, olfaction, and other senses

(Erulkar, 1955; Vasilevskii, 1966; Adamo and King, 1965; Wenzel and Sieck, 1966; and others).

Arousal

In the forebrain of pigeons the hyperstriatum accessorium, part of the 'neopallium' of Portmann and Stingelin (1961), can be removed without seriously disrupting normal courtship and other behaviour, including vocalizations (Rogers, 1922). When in addition the hyperstriatum dorsale and ventrale were extensively damaged bilaterally in pigeons, or when a unilateral decerebration was performed, the birds became less active sexually and failed to copulate (Beach, 1951). Deficiencies in courtship behaviour, including cooing frequency, in these pigeons could in some cases be compensated for by androgen therapy if the extent of cerebral damage was not too extensive. Although gonadal insufficiency could not be ruled out by the data, Beach favoured an interpretation of these results primarily in terms of a 'general reduction of responsiveness to external stimuli' (p. 53), which could, if not too great, be made up by activation from a different source, namely the effects of a supranormal level of androgen. Evidence from single unit recordings during temporary inactivation of the forebrain above the palaeostriatum by spreading depression suggests that the dorsal striatal layers exert a tonically excitatory influence on the thalamus and mixed excitatory and inhibitory effects on other brain-stem structures (Shima and Fifkova, 1963). Zeigler (1963*b*) has demonstrated quantitatively a transient reduction in locomotor activity after partial ablation of the hyperstriatum in pigeons. Thus, a variety of rather weak lines of evidence suggests a possible role of the avian forebrain in activation of certain mechanisms of behavioural arousal.

Speech

In trained parrots surgical ablations in a restricted area on the lateral surface of the forebrain caused the loss of the ability to talk, while large lesions elsewhere had no such effect (Kalischer, 1905). This was not the same area as that where alarm calls could be evoked by electrical stimulation in the same species. Whether the deficit in speech is specific to learned vocalization or is due to interference with motor skills or perceptual mechanisms is not known. Kalischer's findings deserve modern re-examination in parrots.

Motivational deficits

In view of the experiments in which vocalizations were evoked by stimulation of archistriatum and other striatal structures, the effects of lesions in these areas are of interest. Phillips (1964) found in mallards that small lesions in the ventral medial archistriatum and the tract which connects it to the brain stem (tractus occipito-mesencephalicus) reduced escape and associated behaviour without interfering with maintenance activities or general health, while larger lesions in palaeostriatum and neostriatum were ineffective. Some brain-stem lesions in red-winged blackbirds have also had taming effects (Brown, 1965*b*).

Brain-stem lesions

Hypothalamic lesions in the stratum cellulare externum of mallards reduced escape and associated behaviour in mallards (Phillips, 1964). Lesions in the midbrain of red-winged blackbirds in the region where stimulation evoked calling have consistently abolished or reduced alarm calling to peripheral stimuli (Brown, 1965*b*). In the more successful cases no vocalizations of any kind have been heard from these birds for as long as they lived. They were in good health, did not lose weight, and behaved normally in other respects so far as could be observed. These results suggest that the region of the lesion, the n. mesencephalicus lateralis pars dorsalis and pars ventralis, is necessary for vocalizations of all sorts, although further behavioural studies will be necessary to confirm this for other behavioural contexts and in other species.

COMPARISONS WITH MAMMALS

A sufficient amount of similarity has been found in the brain regions controlling vocalization in birds and mammals to recognize the outlines of what may eventually prove to be part of a basic vertebrate plan for the neural control of vocalization and associated components of emotional behaviour patterns. Both groups have regions in the forebrain, diencephalon, and midbrain from which vocalizations can be evoked by artificial stimulation. The vocalizations evoked from these regions are in both groups components of integrated patterns of emotional expression and action.

The location of the vocalization regions within each of these major subdivisions of the brain also reveals similarities. A major cerebral region for vocalization in birds is the medial part of the archistriatum. This structure has been interpreted as being homologous to the mammalian amygdala (Ariens Kappers, 1923; van Tienhoven and Juhasz, 1962), from the medial part of which vocalizations have also been obtained as parts of alerting and defence responses (Gastaut *et al.*, 1951; and later authors).

In the diencephalon, similarity exists in that the responsive region lies in the hypothalamus in both birds and mammals. The thalamus appears to play a lesser role in vocalization, but it is involved in mammals (Hunsperger, 1963) and birds. The important role of the preoptic region and anterior hypothalamus in reproductive behaviour in both groups constitutes an additional resemblance.

In the midbrain, regions from which vocalization can be evoked are found in both mammals, the central grey matter (Magoun *et al.*, 1937), and birds, the torus and neighbouring regions (Brown, 1965*a*). These structures have certain anatomical features in common. Both have a sharply demarcated nuclear configuration; both lie internal to the optic tectum; both lie deep in the brain and border the ventricles; and both lie in close relationship to the large mesencephalic trigeminal cells (Weinberg, 1928), which receive proprioceptive afferents from jaw muscle spindles (Manni *et al.*, 1965*a,b*). But since Wallenberg's (1898) finding that the torus receives auditory fibres some authors have concluded that the torus is part of the auditory system, which would imply that it is not homologous to the midbrain vocalization region in mammals.

Considerations of homology aside, it seems likely that the midbrain vocalization zones in both birds and mammals have polysensory afferent connections. In view of the close behavioural relationship between hearing and vocalization in birds a close neuroanatomical and neurophysiological relationship between them would be adaptive. The finding of auditory evoked potentials in the avian torus (Harman and Phillips, 1967) is in agreement with this view, but raises many questions.

An apparent discrepancy between birds and mammals may be found in the presence of many responsive electrode sites in the palaeo- and neostriatum of pigeons (Åkerman, 1965*a,b*). These structures have been interpreted as being homologous to the globus pallidus and caudate nucleus respectively in mammals (Ariens

Kappers, 1923; Fox *et al.*, 1967), but they have not been implicated importantly in the activation of vocalization or other forms of emotional expression in mammals (Hunsperger, 1963). On the contrary, caudate stimulation may inhibit aggressive vocal responses in monkeys (Delgado, 1964).

NEUROCHEMICAL STUDIES

Recent studies in mammals have demonstrated the possibility of chemical coding in the brain (Miller, 1965). According to this concept the transmission properties of a synapse are determined not by the anatomical connection alone, but also by the type of transmitter substance released presynaptically and the differential sensitivity of the post-synaptic membrane to various possible transmitter substances.

Although neurochemical interpretations of avian behaviour have not yet been attempted, enough evidence is available to reveal considerable promise for this line of research. Reserpine, dartal, luminal and nembutal had differential effects on the thresholds for centrally evoked clucking and for locomotor responses (von Holst and Schleidt, 1964; von St Paul, 1965). Shy birds become tamer under reserpine and dartal; the thresholds for clucking and 'grumbling' were raised by reserpine and dartal; luminal and nembutal at certain dose levels raised the thresholds of evoked locomotor responses without affecting vocalization or wildness.

The behavioural and EEG responses of young chicks to a variety of centrally active drugs have been studied by Spooner and Winters (1966). They found 'whining noises' associated with higher doses of morphine sulphate and a 'low frequency chirping' from LSD. The striking and long known vocal response of chicks to amphetamine, a 'continuous twitter', was interpreted as being dependent on activation via the reticular formation.

SYSTEM PROPERTIES OF VOCAL CONTROL MECHANISMS

The central problem in the study of neural vocalization control mechanisms is not to locate brain regions which 'control' vocalization; it is to determine how the brain as a whole functions during the various vocalizations. This is not an isolated problem but must be viewed within the larger context of brain-behaviour relations in general. Vocalization should not be viewed as a behaviour to be studied by itself; the various vocalizations must be studied in respect to the other components of behaviour with which they are associated in time and space. The controls of vocalization are the controls of sexual, agonistic and other types of behaviour in which vocalization plays a role. Similarly, the brain regions which have been implicated in vocalization by means of brain stimulation and other experiments must not be studied alone but should be viewed in relation to other parts of the brain, especially those which have significant sensory, motor, inhibitory or excitatory functional relationships.

At the present stage of analysis one can only speculate about how a vocalization control system might look in a flow diagram. There is too little evidence now to justify even a tentative sketch. The principal cell populations participating in the vocalization process are not yet well known. The pathways between them are still largely matters of speculation. The neurochemical and single-unit characteristics of the neurons even in the already known vocalization zones are essentially unknown. And the functional interrelationships between the known vocalization zones and other neuron populations with afferent and efferent connections are unknown.

Nevertheless, the longitudinal organization of the vertebrate brain with the motor outflow and sensory input confined mainly (in birds) to the posterior end suggests a basic plan. The most direct pathways with the fewest synapses connecting vocalization zones to sensory inputs and motor outputs are likely to be found posteriorly—perhaps in the midbrain. Alternative pathways which include more anterior parts of the brain are likely to involve many more synapses and to depend ultimately on the same final efferent paths to the vocalization effector organs. How far anteriorly a common efferent pathway for vocalization can be traced is unknown. The organization of the reflex components of vocalization at the medullary and spinal levels has apparently not been investigated in birds except through comparative neuroanatomical studies of the nuclei innervating syrinx, tongue, and bill (Black, 1922; Stingelin, 1965) and is poorly known even in mammals.

Although much remains to be learned in birds about brain-behaviour relationships, knowledge gained since 1957 makes it clear that an approach based on the older concepts of drives and centres is futile. The phenomena which led to those concepts remain as challenging as ever. In terms of explanation at the neural level the surface has barely been scratched. The older theories can now be seen as too static, vague, and simplistic. What is needed are theories in which complexities of the dynamics of neuro-behavioural systems are incorporated.

ACKNOWLEDGEMENTS

This review was supported in part by Research Grant MH 07700–05 from the US National Institute of Mental Health. I would like to thank Mr John Newman for comments on the manuscript.

REFERENCES

ADAMO, N. J. and KING, R. L. (1965) Electrical responses to auditory stimulation in the chicken telencephalon. *Anat. Rec.* **151**, 317.

ÅKERMAN, B. (1965*a*) Behavioural effects of electrical stimulation in the forebrain of the pigeon. I. reproductive behaviour. *Behaviour* **26**, 323–38.

(1965*b*) Behavioural effects of electrical stimulation in the forebrain of the pigeon II. protective behaviour. *Behaviour* **26**, 339–50.

ÅKERMAN, B., FABRICIUS, E., LARSSON, B. AND STEEN, L. (1962) Observations on pigeons (*Columba livia*) with prethalamic radio-lesions in the nervous pathways from the telencephalon. *Acta physiol. scand.* **56**, 286–98.

ARIENS KAPPERS, C. U. (1923) The ontogenetic development of the corpus striatum in birds and a comparison with mammals and man. *Proc. Acad. Sci. Amsterdam* **26**, 135–58.

BARFIELD, R. J. (1965) Induction of aggressive and courtship behavior by intracerebral implants of androgen in capons. *Am. Zool.* **5**, 203.

BEACH, F. A. (1951) Effects of forebrain injury upon mating behaviour in male pigeons. *Behaviour* **4**, 36–59.

BLACK, D. (1922) The motor nuclei of the cerebral nerves in phylogeny. A study of the phenomena of neurobiotaxis. IV. Aves. *J. comp. Neurol.* **34**, 233–75.

BROWN, J. L. (1965*a*) Vocalization evoked from the optic lobe of a songbird. *Science, N.Y.* **149**, 1002–3.

(1965*b*) Loss of vocalization caused by lesions in the Nucleus mesencephalicus lateralis of the Redwinged Blackbird. *Am. Zool.* **5**, 693.

CATE, J. TEN (1936) Physiologie des Zentralnervensystems der Vögel. *Ergeb. Biol.* **13**, 93–173.

(1965) The nervous system of birds (Chap. 22 in Sturkie, P. D., ed.). *Avian Physiology* (2nd ed.), pp. 697–751.

DELGADO, J. M. R. (1964) Free behavior and brain stimulation. *Int. Rev. Neurobiol.* **6**, 349–449.

ERULKAR, S. D. (1955) Tactile and auditory areas in the brain of the pigeon. An experimental study by means of evoked potentials. *J. Comp. Neurol.* **103**, 421–58.

FABRICIUS, E. and JANSSON, A. (1963) Laboratory observations on the reproductive behaviour of the pigeon (*Columba livia*) during the preincubation phase of the breeding cycle. *Anim. Behav.* **11**, 534–47.

FOX, C. A., HILLMAN, D. E., SIEGESMUND, K. A. and SETHER, L. A. (1967) The primate globus pallidus and its feline and avian homologues: a golgi and electron microscopic study. In *Evolution of the Forebrain* (R. Hassler and H. Stephen, eds), pp. 237–48. Georg Thieme Verlag, Stuttgart.

GASTAUT, H., VIGOUROUX, R., CARRIOL, J. and BADIER, M. (1951) Effets de la stimulation électrique (par electrodes à demeure) du complexe amygdalien chez le chat non narcosé. *J. Physiol., Paris,* **43,** 740–6.

GOODMAN, I. J. and BROWN, J. L. (1966) Stimulation of positively and negatively reinforcing sites in the avian brain. *Life Sci.* **5,** 693–704.

HARMAN, A. L. and PHILLIPS, R. E. (1967) Responses in the avian midbrain, thalamus and forebrain evoked by click stimuli. *Expl. Neurol.* **18,** 276–86.

HODOS, W. and KARTEN, H. J. (1966) Brightness and pattern discrimination deficits in the pigeon after lesions of nucleus rotundus. *Exp. Br. Res.* **2,** 151–67.

HOLST, E. VON (1957) Die Auslösung von Stimmungen bei Wirbeltieren durch 'punktförmige' elektrische Erregung des Stammhirns. *Naturwissenschaften* **44,** 549–51.

HOLST, E. VON and SAINT PAUL, U. VON (1958) Das Mischen von Trieben (Instinktbewegungen) durch mehrfache Stammhirnreizung beim Huhn. *Naturwissenschaften* **45,** 579.

(1960) Vom Wirkungsgefüge der Triebe. *Naturwissenschaften* **47,** 409–22.

(1963) On the functional organization of drives. *Anim. Behav.* **11,** 1–20.

HOLST, E. VON and SCHLEIDT, W. M. (1964) Wirkungen von Psychopharmaka auf instinktives Verhalten. *Neuropsychopharm.* **3,** 22–29.

HUNSPERGER, R. W. (1963) Comportements affectifs provoqués par la stimulation électrique du tronc cérébral et du cerveau antérieur. *J. Physiol, Paris,* **55,** 45–98.

KALISCHER, O. (1905) Das Grosshirn der Papageien in anatomischer und physiologischer Beziehung. *Abh. k. preuss. Akad. Wiss.* **4,** 1–105.

KARTEN, H. J. and REVZIN, A. M. (1966) The afferent connections of the nucleus rotundus in the pigeon. *Brain Res.* **2,** 368–77.

KOMISARUK, B. R. (1965) Localization in brain of reproductive behavior responses to progesterone in ring doves. *Bul. Ecol. Soc. Am.* **46,** 186.

KRAMER, E., SAINT PAUL, U. VON, and HEINECKE, P. (1964) Die Bedeutung der Reizparameter bei elektrischer Reizung des Stammhirnes. *Biologischen Jahresheft* **4,** 119–34.

MAGOUN, H. W., ATLAS, D., INGERSOLL, E. H. and RANSON, S. W. (1937) Associated facial, vocal and respiratory components of emotional expression: an experimental study. *J. Neurol. Neurosurg. Psychiat.* **17,** 241–55.

MANNI, E., BARTOLAMI, R. and AZZENA, G. B. (1965*a*) Relationship of the mesencephalic trigeminal cells to jaw muscle proprioception in birds. *Experientia* **21,** 536.

(1965*b*) Jaw muscle proprioception and mesencephalic trigeminal cells in birds. *Expl. Neurol.* **12,** 320–8.

MILLER, N. W. (1965) Chemical coding of behaviour in the brain. *Science, N.Y.* **148,** 328–38.

PHILLIPS, R. E. (1964) 'Wildness' in the mallard duck: effects of brain lesions and stimulation on 'escape behavior' and reproduction. *J. comp. Neurol.* **122,** 139–56.

(1966) Evoked potential study of connections of avian archistriatum and caudal neostriatum. *J. comp. Neurol.* **127,** 89–100.

POPA, G. T. and POPA, F. G. (1933) Certain functions of the midbrain in pigeons. *Proc. R. Soc. Lond., B,* **113,** 191–5.

PORTMANN, A. and STINGELIN, W. (1961) The central nervous system. *Biology and Comparative Physiology of Birds* (A. J. Marshall, ed.), pp. 1–36. vol. II, Academic Press, N.Y.

POWELL, T. P. S. and COWAN, W. M. (1961) The thalamic projection upon the telencephalon in the pigeon (*Columba livia*). *J. Anat.* **95,** 78–109.

PUTKONEN, P. T. S. (1966*a*) Fear-like behaviour elicited from dorsomedial thalamus and ventromedial forebrain of the chicken. *Acta physiol. scand.* **68,** Suppl. 277, 168.

(1966*b*) Attack elicited by forebrain and hypothalamic stimulation in the chicken. *Experientia* **22,** 405.

(1967) Electrical stimulation of the avian brain. *Ann. Acad. Sci. Fenn., Series A, V. Medica* **130,** 1–95.

REVZIN, A. M. and KARTEN, H. J. (1967) Rostral projections of the optic tectum and the nucleus rotundus in the pigeon. *Brain Res.* **3,** 264–76.

ROGERS, F. T. (1922) Studies of the brain stem. VI. An experimental study of the corpus striatum of the pigeon as related to various instinctive types of behaviour. *J. comp. Neurol.* **35,** 21–60.

SAINT PAUL, U. VON (1965) Einfluss von Pharmaka auf die Auslösbarkeit von Verhaltensweisen durch elektrische Reizung. *Z. vergl. Physiol.* **50,** 415–46.

SHIMA, I. and FIFKOVA, E. (1963) Remote effects of striatal spreading depression in pigeon brain. *Jap. J. Physiol.* **13,** 630–40.

SPOONER, C. E. and WINTERS, W. D. (1966) Neuropharmacological profile of the young chick. *Int. J. Neuropharmacol.* **5,** 217–36.

STINGELIN, W. (1965) Qualitative und quantitative Untersuchungen an Kerngebieten der Medulla oblongata bei Vögeln. *Bibliotheca Anatomica, Fasc.* **6,** 1–116.

STRATA, P. (1964) Neurophysiological analysis of the 'mobbing response' in the chaffinch (*Fringilla coelebs*). *Arch. ital. Biol.* **102,** 22–8.

TIENHOVEN, A. VAN and JUHASZ, L. P. (1962) The chicken telencephalon, diencephalon and mesencephalon in stereotaxic coordinates. *J. comp. Neurol.* **118,** 185–98.

VASILEVSKII, N. N. (1966) Functional characteristics of single neurons in pigeon cortex. *Bull. exp. Biol. Med. U.S.S.R.* **61,** 7–11 (translated from Russian).

WALLENBERG, A. (1898) Die secundäre Acusticusbahn der Taube. *Anat. Anz.* **14,** 353–69.

WEINBERG, E. (1928) The mesencephalic root of the fifth nerve. A comparative anatomical study. *J. comp. Neurol.* **46,** 249–405.

WENZEL, B. M. and SIECK, M. H. (1966) Olfaction. *Ann. Rev. Physiol.* **28,** 381–434.

ZEIGLER, H. P. (1963*a*) Effects of endbrain lesions upon visual discrimination learning in pigeons. *J. comp. Neurol.* **120,** 161–82.

(1963*b*) Effects of forebrain lesions upon activity in pigeons. *J. comp. Neurol.* **120,** 183–94.

6. THE EFFECTS OF TESTOSTERONE ON AVIAN VOCALIZATIONS

by **R. J. ANDREW**
School of Biological Sciences, University of Sussex

INTRODUCTION

The first direct demonstration of an effect of a hormone on vocalization in birds was made when Hamilton (1938) injected domestic chicks with testosterone, and caused them to crow. Work on the induction of song in passerines by testosterone followed almost at once (Baldwin *et al.*, 1940, canary; Collard and Grevendal, 1946, *Fringilla montifringilla*; Poulsen, 1951, and Thorpe, 1958, *F. coelebs*; Thielcke-Poltz and Thielcke, 1960, *Turdus merula*). Indirect evidence is available for other species. Thus in the rufous-sided towhee song begins in the field at the time at which Leydig cells begin to increase in number (Davis, 1958).

It is likely that an effect of androgens on vocalization is ancient in birds. There is direct evidence for it in at least four orders other than the Galliformes and Passeriformes. *Larus argentatus* (Charadriiformes) is induced by testosterone to give the male long call (which is preceded by bill lowering and serves as an advertisement display), and two calls associated with nest site displays (Boss, 1943). *Nycticorax nycticorax* (Ciconiiformes) develops a more guttural voice after testosterone (Noble and Wurm, 1940), which may depend on structural changes. The snap-hiss courtship display given by both sexes also appears. The neck is extended forward and down, the bill is snapped and a hiss vocalization is given. Castrate male *Streptopelia risoria* (Columbiformes) give the bowing-coo and the nest call only after testosterone replacement therapy (Erickson *et al.*, 1957). Ducklings of *Anas* and *Aythya* (Anseriformes), after the administration of testosterone, develop male courtship displays in which a characteristic movement accompanies vocalization (Phillips and McKinney, 1960). In the Galliformes, besides *Gallus* itself, *Meleagris*

(Turkey) develops strutting (which includes vocalization) after testosterone (Schein and Hale, 1959).

Secondly, nearly all orders of birds have a high proportion of species with special male territorial calls. This is true even of Struiformes ('booms' of male ostrich, Sauer and Sauer, 1966) and Tinamiformes (male call of tinamou, Lancaster, 1964). This does not in itself demonstrate a direct effect of testosterone on vocalization. Intense contact calls can become frequent in the breeding seasons because males are establishing territories and so must leave social groups. However, where new *types* of calls are involved, this, together with the evidence cited above, does suggest a direct effect; and it seems likely that this was primitively present in birds, in view of its present very wide distribution. A comparable effect (on vocalization which is produced by the glottis instead of the syrinx) has evolved in the Anura (e.g. Schmidt, 1966), but there is no good evidence for anything similar in mammals, perhaps because olfactory signals were primitively more important than acoustic.

It will be seen that the new patterns of vocalization produced by testosterone are not confined to male advertisement displays, but also appear in courtship. In the domestic fowl, as will be seen below, new patterns appear in all the contexts in which vocalization is given. There is evidence for a somewhat similar state of affairs in certain passerines. In *Fringilla montifringilla*, testosterone caused the appearance of an alarm call otherwise absent in castrates, and this preceded the development of song (Collard and Grevendal, 1946). Male *F. coelebs* gives the calls 'see' and 'huit', which are otherwise confined to the spring, after testosterone (Poulsen, 1951). Marler (1956) describes both as alarm calls. They resemble in causation two alarm calls which appear in chicks after testosterone (Andrew, 1963). 'See' is given in intense alarm with freezing, and may be compared to the chick alarm trill. 'Huit' is given when mobbing, and resembles the chick juvenile cackles in the restless locomotion and continuous attention which accompany it. *Sylvia communis* (Sauer, 1954) gives a high intensity alarm call ('wad') during the breeding season. Since it reappears in the autumn with a recrudescence of song, and occurs in hand-reared birds in captivity at this season, it is probably due to physiological state (e.g. testosterone level) rather than to the restriction of special stimulus situations (like threat to the young) to the breeding season.

This article will be concerned with the causation of the changes in chick vocalization which are induced by testosterone. In particular it will be argued that, although these changes are complex, they are all generated by a simple change in a single functional system. The new patterns whose appearance have to be explained can be recognized as patterns present in adult calling (next section), so that the processes involved in their generation are probably also involved in normal development.

NORMAL VOCALIZATIONS OF THE CHICK

Any system of vocalizations can be regarded either as a number of separate (although perhaps overlapping) motor patterns, each with different causation, or as a single system which varies continuously

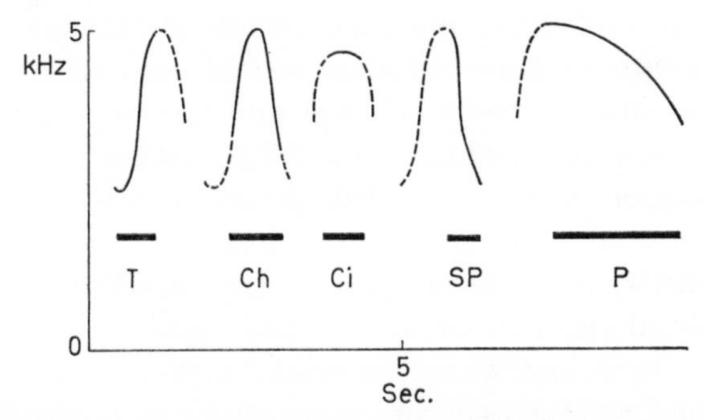

FIGURE 1. The main sequence of chick vocalizations. The dotted line indicates the unsounded position(s) of the cycle of increase and then decrease in syringeal tension, which appear to be basic to a single vocalization. The heavy line indicates the portion of the cycle during which expiration would result in the call shown. T, Twitter; Ch, chevron; Ci, circumflex; SP, short peep; P, peep.

in form with changes in causal factors common to the whole system. It is important in discussing the effects of testosterone on chick vocalizations to decide which model is nearer to the truth for the chick. In the main series of chick vocalizations, calls vary continuously in form from the lowest intensity pattern, the steeply ascending twitter, to the slowly descending peep.

These changes in form depend on two relatively independent effects (Allen, personal communication). One involves a basic cycle of pitch change, which varies sinusoidally between an upper and a lower limit, and, typically, within a call runs from a trough through a peak to a trough: this cycle is generated by cyclic changes in syringeal tension. The second effect, which depends on expiration in a way still under investigation, determines which part of the cycle is actually sounded (Figure 1). The continuum of calls from twitters to peeps, therefore, could reasonably be regarded as a single system of responses if their mode of production alone were considered.

It has been argued earlier (Andrew, 1964) that this system of responses also shares a common causation, which is very like that of alert responses: namely that it is evoked by external stimuli which are either inherently conspicuous, or have significance because of learning. Calls, ranging from twitters through intermediates to (sometimes) short peeps, are evoked by the sight of food or conditioned stimuli (CS) announcing food when the chick is hungry, by conspicuous tactile, auditory or visual stimuli (e.g. movement or grain-like objects), and by a moving object which is capable of eliciting social responses (Andrew, 1964, and this chapter). Higher intensity calls (up to peeps) are given in discomfort (e.g. due to cold), in strange environments, in high frustration, and with fear responses.

It is not possible to explain any one of the main patterns, into which chick vocalizations could be divided, as caused by a particular drive or combination of particular drives such as fear, aggression, sex or hunger (Andrew, 1964). On the other hand, if all the situations in which vocalizations occur are considered, there is a marked resemblance to the situations which induce theta waves in the cat hippocampus. Thus theta waves occur during alert responses to changes in stimulation, and also during diffuse orientation, or orientation to the CS, when an approach or avoidance task is being learnt or extinguished (Grastyán *et al.*, 1959, 1966). The alternative state of desynchronization occurs when performing a well-established conditioned response (CR), whether this involves approach to food reward or avoidance (or freezing). Theta waves thus accompany diffuse or shifting attention. This is true also of chick vocalization, which occurs when attention to any of a range of stimuli is likely, and stimuli are being scanned, with shifts of attention. More loosely, chicks call during diffuse attention, expectation, frustration and diffusely directed escape responses. Like theta waves, vocalization does not occur in intense but unshifting attention: in at least two conditions of intense attention, during response directed to a single stimulus object (full aggression and full copulation), chicks are commonly silent.

It should be noted, in passing, that such an explanation of the causation of vocalization does not conflict with the fact that a particular stimulus situation may evoke only a restricted portion of the vocalization series. Clearly on any theory of causation, the more limited the stimulus situation, the more restricted should be the range of responses evoked. Guyomarc'h (1962, 1965), for example, who has described a large number of (overlapping) patterns of calling in the chick, using both their form and the situations in which they occur, treats all calls as caused by various states of balance between 'favourable factors' and 'stressing factors', which is essentially a theory of unitary causation.

If chick vocalizations are indeed causally a single group of responses, testosterone may produce its very varied range of effects on vocalization by a single change which affects all calls. Such a conclusion, which is justified at length below, does not agree well with a first consideration of these effects. Instead they can most easily be described as the appearance of a number of new patterns, each by a separate maturational process. The new patterns may be listed as follows:

(i) *Food trill.* Given to CS announcing food, and usually consisting of very rapidly repeated short peeps separated by insufficient time for ascending limbs to develop. Adult equivalent?

(ii) *Waning bout and juvenile cackle.* Given to sudden changes in stimulation. The first involves the rapid repetition of a particular short pattern (e.g. short peep or trill of 1 to 2 cycles), with the progressive loss of its higher pitched portions (= adult alert call). A juvenile cackle is a somewhat longer, often irregularly trilled pattern, which is then repeated on each successive expiration but with the omission of nearly all the higher portions, until suddenly (after 10 to 15 seconds) it is once more fully performed (= adult cackle). The two types of call are connected by intermediates.

(iii) *Alarm trill.* A long regular trill, given to a sudden alarming stimulus (particularly a deep sound), often after a brief freeze. It can also be evoked by electric shock (adult aerial alarm call).

(iv) *Juvenile crow.* Often a descending trill, sometimes a group of modified peeps, but always recognizable by the characteristic neck elongation and thickening, into a 'ewe-neck', and erect stance, often with slight wing-raising. Repeated at intervals usually greater than 10 seconds and interrupting ongoing behaviour (= adult crow).

(v) *Elongated peep*. A group of very variable calls, all developed from peeps or bouts of intermediate calls by fusion or elongation, the most typical being long whines (= laying call, and others of adult).

It may be helpful to summarize here the main steps in the discussion which follows. The first considers three experiments which indicated that there was no general facilitation of vocalization by testosterone, as shown by rate of calling in a number of standard situations in which new patterns appeared. However, secondly, despite this set-back for a theory which postulates a single effect of testosterone on vocalization, it is equally difficult to substantiate the separate maturation of a number of new patterns. Each new pattern is a development from some part of the normal vocalization series, but the part developed may be different for the same pattern in different individuals (and even the same one at different times). Thus some juvenile crows are clearly developments from bouts of short peeps, whilst others closely resemble alarm trills. It is, in fact, not possible to classify calls into the categories listed above on form alone. The. evoking situation and associated behaviour must also be considered

Thirdly, experiments using hypothalamic stimulation of testo-sterone treated birds showed that it was possible to generate elongated peeps by the superimposition of crowing on a series of peeps. This, and other evidence, leads to a hypothesis which seems best to explain all the data. Testosterone makes much easier the activation of a single functional mechanism (the 'testosterone-mechanism') which, when active, causes the performance of an intense call, whose form is specified only to the extent that it should be prolonged, and usually trilled. The actual form of call performed is determined by the type of normal vocalization being performed at the time. The hypothetical mechanism can be activated by the same sort of stimulus situations which produce normal vocalization, but it has a higher threshold than the lower intensity patterns in the normal vocalization series, and, if strongly activated, it may continue to cause intense calls at intervals of 10 to 15 seconds for some time, independently of any continuing peripheral input.

Finally, during any period of activation of the 'testosterone mechan-ism', each call copies many details from the preceding call. Detailed examination of this copying strongly suggests that a short term memory of preceding calls influences the form of succeeding calls during such periods. It is not yet certain whether such memory is integral to the functioning of the mechanism or a by-product of it. It may well be effective over longer periods and lead to development by learning of a more fixed set of patterns. Such development has, however, not yet been studied.

A GENERAL FACILITATION OF CALLING BY TESTOSTERONE

Disturbing effects on vocalization could follow from simple changes in the functional system, which causes vocalization to vary in intensity with peripheral input. Andrew (1963) suggested that a simple facilitation by testosterone, causing the system to be activated more rapidly and to remain active longer, might be enough to disturb existing patterns. Such a facilitation should also increase general rates of calling during the development of new patterns. The effect of testosterone on calling rates was therefore studied in three standard situations. As will be seen below there is in fact no general facilitation of calling.

Expt. 1. Calls to conditioned stimuli for reward

Calls varying from twitters to short peeps are evoked by conditioned stimuli for food or drink (Andrew, 1964). In the present study, the chick had to observe the illumination of pilot lights to determine at which of three hoppers food would be available.[1] Each light remained on for a standard number of seconds until the hopper delivered food. Five normal uninjected birds (A to E) were trained in the chamber in pilot studies. Calling began or accelerated as soon as the chick noticed the illumination of a light, thus causing a peak in the second second of illumination. The more continuous the chick's attention was likely to be, the more frequent the calls evoked by the lamp. Thus the rate was higher after 18 h than 3 h deprivation (Figure 2B; also Andrew, 1964), and the peak was followed by a marked decline in bird B, which had to wait 10 sec. before feeding (Figure 2), and in the birds shown in Figure 3, which waited 5 sec. (Figure 3). In bird C, for which the lamp was on for 15 out of the 18 sec. from the end of one presentation to the beginning of the next, so that illumination had little predictive value, there was little rise during lamp-on, and no clear peak.

Early in each session in the chamber there was diffuse attention

[1] Further details of method in Andrew, 1964.

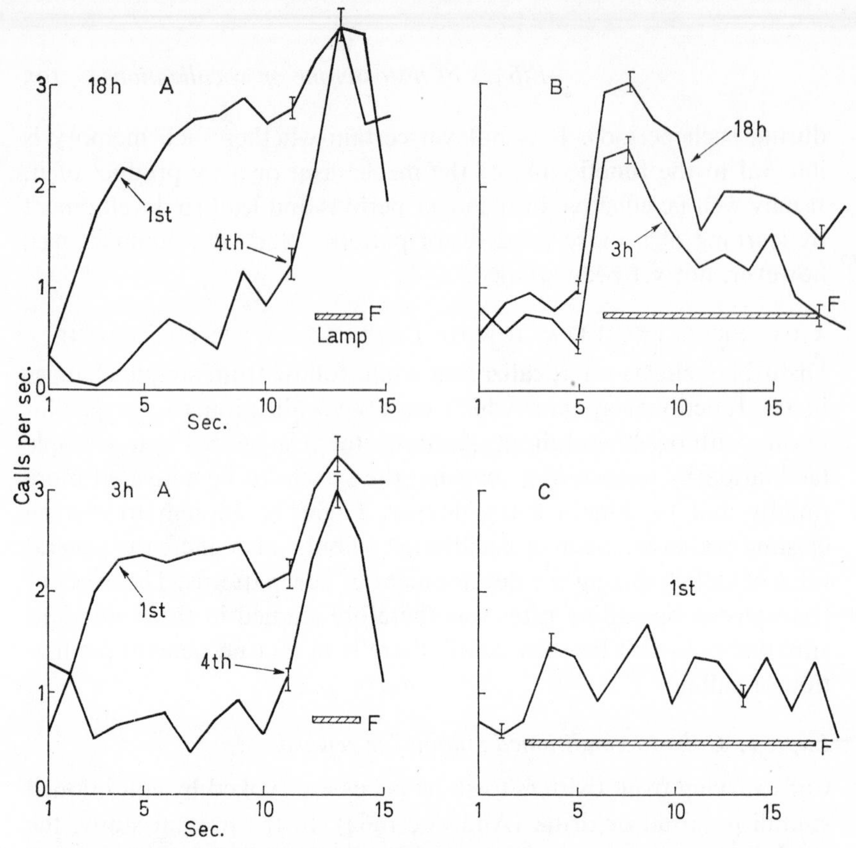

FIGURE 2. Mean calls during feeding cycles in which lamp illumination (indicated by hatched bar) indicated that food would be available for 1 sec. (F). Birds were deprived of food for 18 h, and for 3 h, on alternate days. Values are shown for the first (1st) and fourth (4th) ten feeds for bird A (18 h and 3 h deprivation), for the fourth ten feeds for bird B (18 h and 3 h deprivation), and for the first ten for bird C (18 h deprivation). The vertical lines indicate one standard deviation.

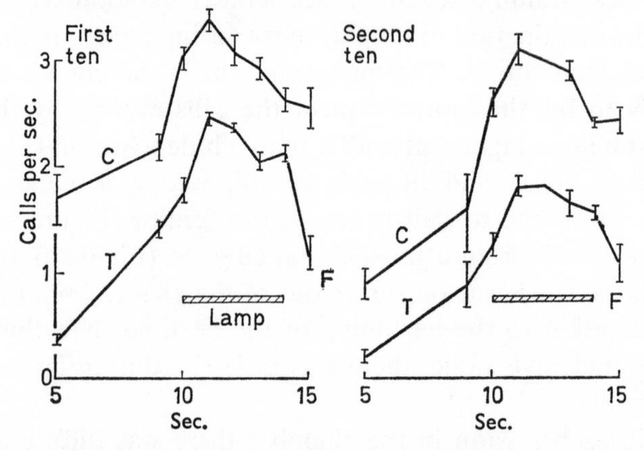

FIGURE 3. Mean calls during feeding cycles of six testosterone treated chicks (T) and seven controls (C). Lamp illumination (CS) is shown by the hatched bar and the feeding second by F. The vertical lines indicate one standard deviation.

and activity in the period between presentations. This decreased as the session progressed, and so did the calling in this period. This is shown by a comparison of mean rates of calling in the first ten, and the fourth ten trials for bird A (Figure 2). Birds D and E gave increased calling to a lamp announcing a water reward[2] (D : $P<0{\cdot}05$; E : $P<0{\cdot}01$), showing that the phenomenon was not restricted to stimuli associated with food.

In the main experimental series (shown in Figure 3) six testosterone-treated chicks (T's), and seven control (C's) were used. Despite the appearance of new higher intensity patterns (food trills) at the peak of calling (see later), T's showed an overall decrease in the rate of calling both during and between trials (Figure 3), due very largely to a decrease in the length of the bouts of twitters and short peeps. If the length of bouts beginning in the first second of the lamp is considered, (using data for the second ten trials of each bird), then the mean number of calls/bout is 2·89 for T's and 4·09 for C's (not significant). If T's and C's are compared, using number of seconds in which no bouts, bouts of two, of three and of four or more calls began, then the two groups differ very significantly ($\chi^2=42{\cdot}5$, $P<0{\cdot}001$), chiefly because of a deficiency of no bouts in the C's, and of bouts of three in the T's.

The decreased rate of calling in T's may represent decreased interest in conditioned stimuli associated with food, although T's in fact showed at least qualitatively similar interest in food. (It will also be remembered that, in bird A, 3 and 18 h deprivation produced very similar rates of calling to the lamp.) More probably, it represents a direct depressive effect on vocalization. However, in either case, the appearance of higher intensity patterns in this situation after testosterone cannot be due to a general facilitation or increase in intensity of all vocalization.

Expt. 2. *Cold chamber*

Testosterone produced marked changes in the calling (chiefly peeps) which was given by chicks placed in a test chamber for 5 min. at 0°C (Andrew, 1963). The very varied new patterns are considered later. Normal calls changed quantitatively. On the last test day full peeps were far rarer in T's than C's and bouts of short peeps, particularly of three or more, were much commoner (Table 1; Andrew, 1963).

[2] Further details of method in Andrew (1964).

This change was originally considered to be due to the replacement of peeps by new patterns (which clearly is involved to some extent) and to the prolongation of bouts by testosterone.

However, short peep bouts are now known to be typical of extinction sessions, in which pecks on a key cease to be rewarded (Andrew, in press) and also of long periods spent in a strange environment.

TABLE 1. *Numbers of chicks giving different types of calls in cold chamber on day 14*

Type of call	Call present T	Call present C	Call absent T	Call absent C	Significance P	Test
Peep	4	13	13	5	0·02–0·01	χ_1^2
Short peep bouts of 2	14	3	3	15	0·01–0·001	χ_1^2
Short peep bouts of 3	11	2	6	17	0·01–0·001	χ_1^2
Prolonged peep variants	6	0	11	18	0·02–0·01	Fisher exact test

T, Testosterone group; C, control group.

In both instances they replace full peeps when the chick ceases active locomotion or leaping and, for increasing periods, reduces the disturbing input by closing its eyes. They also precede full peeps as frustration builds up. The replacement of peeps by short peep bouts in T's thus represents a decrease in the overall intensity of calling, despite which, new and (in form) more intense patterns develop.

Expt. 3. Conditioned stimuli for punishment

Chicks (six T's and six C's) were trained daily until day 14 to avoid shock in a shuttle box by moving to the other end when a light, at the end where the chick was standing, came on.[1] All learnt by day 14 to attend to the light and to avoid shock sometimes; more usually they

[1] The shuttle-box was 3 ft × 10 in., painted black in one half and white in the other, and with a floor of pipe cleaners moistened with saturated NaCl, through one or other half of which 2·0 mA shock (Grason-Stadler shock generator) was administered at random intervals for 3·0 sec. A 60 W bulb at the end to be avoided was illuminated for 1·0 sec. before shock. In the first series six T's received 0·1 mg/g testosterone oenanthate, and six C's sesame oil, on day 2; a second series added two more T's.

ran to the opposite end on receiving shock. On day 14 calling commonly began with the illumination of the lamp, usually as peeps, although trills also occurred, and then increased in mean rate at the onset of shock, together with the appearance of alarm trills. There was no significant difference in mean rate between T's and C's (Figure 4), but again testosterone produced new patterns in T's which

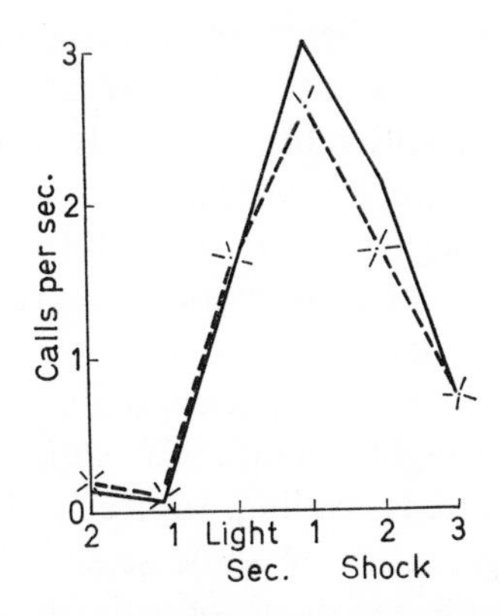

FIGURE 4. Mean rate of calling in shuttle box where a light was illuminated for one second before three seconds electric shock. Values for eight testosterone treated chicks are shown by the continuous line, and for six controls by the broken line.

were accompanied by a reduction in length and regularity of alarm trills (below), a change which suggests reduced intensity of performance of this pattern.

MAIN FEATURES OF NEW PATTERNS

The hypothesis, which will be discussed in this and the next section as an alternative to that of a general facilitation of vocalization by testosterone, is that all the new testosterone-dependent patterns represent distortions of the normal series of vocalizations by the same testosterone-sensitive mechanism (TM). It is proposed that, when active, the TM causes the performance of intense vocalization, whose form is not specified in detail beyond a greater than normal prolongation and likelihood of trilling. The call produced is, as a result, a disturbance in these particular directions of the pattern which would have been given in the absence of TM activity. Typically, a full activation of the TM is followed by a period in which its activity affects vocalization less or not at all, before a further full activation occurs. The new patterns will now be considered in turn to see to what extent this hypothesis is a reasonable one.

Food trills (expt. 1)

Some C's and T's gave only twitters to the food CS, and some, twitters and peeps. In either case the most intense form, as well as rate, of calling was reached immediately after perceiving the lamp. If twitters only were given, then bouts became unusually long, bouts of up to nine calls occurring at this time. If peeps occurred as well as twitters, then short peeps often began directly from silence at lamp-on (significance of coincidence, $P<0.001$). All such birds (2 T's and 4 C's) also gave bouts, which began with a twitter and passed into short peeps. Later in the 5 seconds of illumination, short peeps were usually replaced by twitters ($P<0.001$).

The 'food trills', which were occasionally given on the last test day by four out of the eight T's (4/8 T's vs 0/12 C's, $P<0.02$), came at this point of most intense calling. In nearly all cases their relation to the normal calls given in the same session was obvious. In the bird (1) which gave the widest range of calls, the following types were given whilst attending to the lamp: single or pairs of twitters were replaced at higher intensity by bouts which began with a twitter and passed into short peeps, or by longer bouts of short peeps (Plate 1, (7)). The trills, which commonly could be seen to be given when the bird caught sight of the lamp, consisted of short peeps repeated so rapidly that no pause in expiration could have occurred between them (Plate 1, 8). Intermediates occurred which consisted of two short peeps which were much closer than usual, suggesting the trills to be an accelerated bout. A second bird (2) gave trills, which were similarly composed of short peeps, but differed in showing a sudden increase in pitch after the first two components (Plate 1, (1)–(6)). A third bird (3) gave quite different regular trills, often with no obvious relation in form to the twitters, which were the only other call given. However, occasionally the first leg of the trill resembled a preceding twitter. The fourth bird (4) also gave trills quite specific in form. Here the cyclic change in pitch appeared to be superimposed on a rising twitter-like form (Plate 1, (10)).

Food trills are thus often modifications of accompanying normal vocalization. Their duration would agree with a single activation of the TM. Bird 3 showed a suggestion of subsequently persisting activity, like that associated with juvenile cackles (below), in that short segments of the trill were often repeated after the first full trill (Plate 1, (9)).

The input provided by the sudden perception of the food CS was, from the very sporadic delivery of food trills in any test, just on the threshold for activation of the TM. A marginal activation would explain the close relationship of the trill to the form of calling which is being given. Instead of developing a new and more elaborate pattern like juvenile cackles, which result from rather similar but more intense peripheral inputs (below), food trills simply continue and exaggerate the tendency towards rapid short peep bouts, which is already apparent in the trials of normal birds.

Waning bouts (WB) and juvenile cackles (JC)

Andrew (1963) argued that waning bouts develop from twitters. In fact a WB typically begins with one or more twitters, and then passes to higher intensity patterns, which may be calls consisting of an upper inversion point with short ascending and descending limbs, or true short peeps, or in some instances merely twitters with a more completely developed upper inversion. The waning which follows may begin with a return towards the twitter form, or the higher intensity form may persist; in either instance, the pattern thereafter changes only by the progressive omission of the later (usually higher pitched) portions, until only the very beginning of the pattern is given. WB thus can be a disturbance of any part of the whole lower intensity end of the normal vocalization series. The JC (Plate 4, (4)) corresponds to the established pattern in the second phase of a WB. Typically there are no introductory calls of increasing intensity, but all intermediates between WB and JC are found. JC are very varied in form but they usually involve several cycles of ascent and descent. After the first performance a small portion of the pattern is given on each succeeding expiration. This usually resembles a twitter or the very beginning of a twitter, and may be a limb taken from the beginning, middle or end of the pattern (Andrew, 1963). The full pattern may be repeated almost at once, but more usually it recurs only after intervals in which the (often very) incomplete portion of the pattern is given repeatedly for ten or more seconds. The behaviour, both in the phasic nature of the vocalization, the sudden extension of neck for the full pattern and the darting head movements as the chick glances around, is almost identical with adult cackling. If a WB is caused by a single activation of the testosterone mechanism, then JC represent repeated full activations separated by periods of partial activation. The level of activation in these intervening periods is presumably not constant, since the portion of the full pattern which is performed may become progressively more complete or fade away.

Elongated peeps

The patterns grouped together as elongated peeps were the main new call given by T's in the cold chamber. Such birds appear to develop a much elongated call from whatever pattern of calls were being given in this situation. If these are short calls then they are rapidly repeated until they fuse and, if they are long (peeps), then an elongated variant is interpolated. The distinction between the two types of change is somewhat arbitrary: thus short calls, consisting largely of an upper inversion point, may either fuse in the course of what is probably a single expiration or be replaced, after an inspiratory pause, by an elongated variant. Clearly this is directly comparable with the appearance of similar new patterns, either in the course of the first calls of a waning bout, or suddenly in a juvenile cackle.

Plate 2 shows changes in short calls. Rapidly delivered short peeps merged in some into long trills (e.g. bird 1). Short trills in which merging was not complete, and which resembled food trills as a result, developed in others (birds 1, 2 and 3). Circumflex calls merged into elongated whines, which retained some of their details of form (birds 4 and 5, Plate 2, (7)–(9)). A related development, which also involved repetition of the original short call, caused calls to lengthen by extension into further trilling cycles, the resulting patterns being juvenile cackles. Trill patterns associated with twitters commonly began with an ascending limb of twitter type (e.g. birds 4, 6: Plate 3, (2)), and those associated with short peeps with a short peep segment (bird 7, Plate 3, (3)).

Longer normal calls elongated by extension within their course, sometimes with superimposed trilling. In many instances the elongated peep occurred within a series of more normal peeps (e.g. birds 5, 7, 9, 10, 11, 12, 13: Plate 3, (4)–(9), (11), (12)). Circumflex calls showed similar interpolations (bird 1 : Plate 3, (10)). Two birds consistently gave a single elongated peep at the beginning of short bouts of peeps (6, 14: Plate 3, (11), (12)). This was a development of the normal beginning of peep bouts with a more gradually descending call. In other birds a related development resulted in stereotyped patterns of two (or more) calls which were repeated over and over again (15, 8: Plate 4, (1)–(2)).

All of these developments, but particularly the last, suggest (but do not prove) that new patterns arise as disturbances of previously existing ones, which are then stabilized in some way. Prolonged patterns formed by the fusion of short calls may be compared with a waning bout or a single juvenile cackle and may therefore represent a single activation of the TM. Call series reminiscent of a juvenile cackle series are also common. These begin with an elongated call, which is followed by short calls, which are at first a repetition of some part of the first pattern, usually its end (Plate 4, (3), (5), (6)). Comparison with juvenile cackles suggests the possibility that such series represent a full activation of the TM followed by a period of partial activation.

Juvenile crow

Crows may appear as disturbances of peeps. Andrew (1963) describes their appearance amongst series of peeps in the home cage, two to three days after testosterone. They were usually separated by about a second from preceding peeps, and themselves resembled extended peeps, particularly in their final descending limb. This last part was often repeated as one to three short peep-like calls, forming a 'crow bout'. Conversely, Marler *et al.* (1962) found the first crows recorded in groups of chicks after testosterone to be single squeaks. Guyomarc'h (1965) found both types: first crows were trilled and usually single, but sometimes were preceded or followed by short calls (which might be peep-like).

Two experiments in which the crows of isolated male chicks were observed provide more data. The calls which immediately preceded the first observed crow were noted in twelve of the injected chicks. In six the crow either ended a short bout of peeps, or occurred immediately after a bout had ended; the other six gave a crow which was neither immediately preceded nor followed by peeps, although peeps might occur soon after. In addition, two normal chicks which crowed in these experiments (below), gave crows of this second type. Such crows may show no resemblance to peeps: some are simple trills, for example.

The common association between peeps and crows is partly to be explained by the fact that conditions (like strange surroundings, Konishi, 1963) which appear to promote crowing also tend to produce peeps. An experiment with two groups of six injected and isolated male chicks supported this. One group spent an hour on the first and third test day in a strange cage, and the other group of six spent an hour of the second day in this way; on the other days chicks were in home cages for the hour of observation. The results supported previous observations that strange surroundings promote crowing, but were not significant (probably because even the home cage was a disturbing situation in which chicks often showed escape leaps, peeps and crows). Four males crowed only in the strange cage, whereas only one crowed only in the home cage (whilst five crowed in both).

Not only do a large number of juvenile crows show a clear relationship to peeps, but crows as a whole show such a great range of variation that any explanation in terms of independent maturation of a new motor pattern would be hard put to it to describe what the new pattern specified, beyond what has already been attributed to the testosterone-mechanism (TM). Thus crows (as identified by the accompanying postural changes and characteristic periodicity), may consist of one or several calls, be trilled markedly or not at all, and be very much longer than usual calls (1 sec.) or be of the same length as a peep (0·2 sec.). They may resemble elongated or trilled peeps, and intergrade with peeps, or be like alarm trills or even shrieks. Marler *et al.* (1962) also found great variation between individuals and an absence of strain differences in chick crows.

The only pattern of crowing that is difficult to explain as a disturbance of normal vocalization by the TM is that of a simple trill. However, such trills do resemble (sometimes very closely) the alarm trills of normal chicks, a point which is returned to later.

Juvenile crows may be given singly, but usually recur at intervals of 10 to 15 seconds, each time interrupting ongoing behaviour (which may involve a variety of normal calls). It is not uncommon for a chick to rise from a full resting attitude, in which its eyes are closed, in order to crow, and then subside once more into rest. It is clear that the mechanism which causes crowing (here assumed to be the TM) is not only independent of normal vocalization, but also can continue cyclic activations with no obvious sustaining peripheral stimulation. The resemblance to the repeated performance of full juvenile cackles is obvious.

STABILITY OF NEW PATTERNS AND THE PHENOMENON OF DRIFT

Perhaps the most striking feature of the new patterns is the strong tendency towards stability of form which may show within any one period of performance. Where calls of the normal series would change form as the intensity of calling decreases, juvenile cackles pass to more and more incomplete versions of the same pattern. This difference is especially obvious in waning bouts, in which the initial phase shows normal change of form, and the second phase repetition of the established pattern.

Such stability must depend on a specification of form by the CNS (sometimes, as will be seen, in great detail), which persists from call to call. This can be reconciled in two ways with the evidence that the new patterns are distortions of the normal series of vocalizations. First, this distortion could be a developmental process, which is already complete before experimental recording begins. The forms of the new patterns would then be fixed, and the only adjustment possible would be to select the new pattern closest in form to the normal call which would otherwise be performed. Second, the detailed form of the pattern could be established at the beginning of a series (as is suggested by waning bouts), and its repetition could depend on short-term storage, enabling the form of the preceding call to be copied. Such a process could well be a stage in the establishment of completely fixed patterns (and so of the first alternative).

It is therefore of some importance that a pattern does show variation over long series of performances, and that such variation is progressive, rather than showing random deviations from a constant standard. The 'drift' which results from such progressive variation strongly suggests that each call is most influenced by the form of the immediately preceding call, and so retains any changes from the original pattern which the latter might have introduced. The possibility that the progressive character of the changes depends on motivational changes, is made exceedingly unlikely by the great variety of features which are copied. Drift is clearest in series of elongated peeps. Plate 4, (3), (5), (6) shows one common way in which such series begin. An elongated call may be followed by short calls which are at first a repetition of the final part of the long call. In other series the features of the previous call which are repeated may lie at any part of the call from beginning to end. The pitch variations in a particular limb may be repeated in detail, sometimes with changes in overall shape. The band-width of the fundamental on the spectrogram, which probably reflects the amount of energy involved, may be copied. Finally, the form of the main ascents and descents of the call may be copied, with or without repetition of details. Four such sequences are illustrated in Plate 4, (5), (6) and Plate 5, (1), (2) (birds 6 and 8). Three are from the same bird (6), and show strong resemblances between series. Each begins with elongated peeps in which the initial descending limb is followed by a horizontal segment. Short ascending calls follow which, in two of the series (Plate 4, (5), (6)), can be seen to derive from the final segment of the last elongated peep (see also Plate 4, (3)). A short descending limb tends to appear, and then the whole call begins to flatten out, until final horizontal calls develop. In addition to these progressive trends, calls at each stage in the series show some resemblances of detail between series (as well as very obvious variation). The resemblance of such series to the crystallization of song phrases out of subsong is striking. The development of stereotyped patterns of two or more calls (above) is comparable, and suggests that the copying involved in drift may pass into long term storage.

Marler *et al.* (1962), in a study of testosterone-induced crowing in chicks, found great constancy in overall pattern in any one individual, with variation in details between consecutive crows. This variation can in fact be seen to involve drift in both single crows and crow bouts. In the series of eight consecutive crows shown in Plate 5, (3)–(12), the pattern can be divided into: (i) A short initial segment (ab), which is sometimes absent. (ii) A long segment (cde) with slight trilling, during the course of which the pitch of the fundamental clearly has two (or even three) stable values which correspond to the overtones, lying between 4 and 6 Kc, of a second mode of vibration of the sound source, which has a fundamental at about 1 Kc. As a result, different parts of the segment may show an abrupt change in pitch. (iii) A segment (f) with trilling of large amplitude. Here the position and form of the largest amplitude trills varies. (iv) A terminal segment (g or h).

The initial segment disappears after several changes and does not redevelop. The cde segment shows repetition of trilling once developed, and of a characteristic abruptly dropped segment in c. The f segment changes form markedly in Plate 5, (8). Instead of showing either one or two large amplitude trill cycles, it develops trilling which increases progressively in amplitude up to the end of the crow. This new form then persists, with progressive drift. In Plate 5, (9) the last two largest trills are still identifiable, despite a general reduction in the amplitude of trilling. In (10) this portion of the crow has become a depressed segment which persists with variation in (11) and (12).

The terminal segment is a steeply descending limb, which persists through (3)–(8). In (9) it is suddenly changed by the addition of a final ascending limb, h, and in subsequent calls this develops further with the addition of a short descending ending. In all, eight or more features may be repeated in two or more successive crows, independently of each other.

In crow bouts, in which the final segment of the full crow is repeated as a series of shorter calls, incipient drift occurs within the bout, usually as omission of some portion of the final segment, and a reduction in the steepness of descent (Plate 6, (1)). More elaborate changes such as exaggeration of a trill cycle also occur. Drift between bouts may produce striking changes, such as the incorporation of the second call into the initial crow (Plate 6, (2)).

Drift is more difficult to show in the less complex new patterns. Changes in detail do occur in successive juvenile cackles (Plate 4, (4)). It is particularly interesting that the trills given during shock and ensuing escape, or avoidance in the shuttle box experiment, show signs of copying and drift in testosterone birds. Four out of the six controls gave long trills (0·3 sec. or more) which were trilled regularly, rapidly and through large amplitude (Plate 2, (2), (3)). They were then, both in form and in that they were produced by a very startling stimulus, very like the normal alarm trills of a chick.[1] Such trills then passed into loud peeps. The other two C's gave only single trilled peeps, which passed at once into peeps. No T's were recorded (during analysis of tape recordings by ear) to give trills which were fully of this form. All gave trills during shock at some time during testing, but these were commonly irregular, of smaller amplitude, or were not trilled at all during part of their course. This probably represents a decrease in intensity of calling, comparable to that shown in the cold chamber experiment.

[1] It should be noted that the shock was set by preliminary experiment at a level just high enough to produce consistent escape and avoidance: it was not intense enough to evoke the very loud long and irregularly trilled shrieks produced by painful stimulation.

Recordings of three T's and three C's were analysed spectrographically. All three T's showed drift which involved progressive elongation of patterns within a series without trilling. Plate 1, (11), (12) and Plate 2, (1) shows a series for each T, in which a number of consecutive calls each show some further change whilst retaining some features of the immediately previous call. A fourth testosterone-treated bird (otherwise not discussed here) gave some forty trilled calls after one electric shock. Repetition was here so marked that it continued until single cycles of the trill were given as single calls.

No such changes occurred in the C's. C1 and C2 both showed sequences in which, instead of a single transition from trills to peeps, trilling might reassert and become more dominant. The intermediate calls were peeps for part of their course, and trills for the rest (usually not the beginning), and both the peep and the trill retained both general characteristics and sometimes details of performance, despite intervening calls in which that part of the pattern was not performed (Plate 2, (2), (3)). C2 sometimes repeated the general form of intermediate patterns, suggesting thereby that copying of patterns is possible in normal chicks under some circumstances.

In summary, there is strong evidence that, in a series of one of the new testosterone-dependent patterns, calls resemble the immediately preceding calls because of centrally persisting effects in a system controlling pattern. There are two possible explanations. The first is that the form of preceding calls is monitored and then copied. Konishi (1963) deafened chicks, and showed that the full repertoire of adult calls developed, which makes auditory monitoring unlikely. However, there were changes, such as a reduction in rapid frequency modulation of the crow in a cock deafened as an adult, which may have been due to interference with monitoring. Proprioceptive monitoring, moreover, might be adequate for the relatively simple operations of the chick syrinx.

The second explanation is that a period of activity of the testosterone mechanism is actually the period of central persistence of a motor command pattern, and that the changes seen in drift represent changes in this somewhat labile pattern, which may be random errors, or be the result of more specific disturbance (e.g. towards a short call). In either case, since the change is in the command pattern itself, it would be repeated.

Whichever explanation is nearer the truth, the phenomena of

repetition and drift strongly support the possibility that the new testosterone-dependent patterns, after originating as disturbances of the normal series of vocalizations, can be stabilized by a learning process consequent on repetition.

A SINGLE CAUSE FOR ALL TESTOSTERONE-DEPENDENT PATTERNS

In order to test the present hypothesis directly, it is necessary to find a way of activating the testosterone mechanism (TM), and then progressively changing the test situation so as to change the background of normal vocalizations on which the TM acts. The cold chamber situation was a first attempt. It produced a wide range of normal vocalizations and also of new patterns. (Two birds, 4 and 17, actually crowed whilst in it.) However, the sequence of behaviour and calls was too variable to allow any systematic testing of predictions.

Electrical stimulation of the hypothalamus[1] at a number of sites reliably causes a chick to pass from intermediate calls, accompanying the initial orienting responses, through calls of continuously increasing intensity, to rapid and intense peeps with escape responses and hiding. The whole sequence is usually completed within 20 seconds. The vocalizations are both normal in form, and accompany behaviour similar to that which might accompany them in their normal evocation. This is quite different from the direct activation of vocalization by electrical stimulation in the lateral mesencephalic nucleus, when calls are obtained at once, whose form and duration depend directly on the stimulation applied, and which are not accompanied (if stimulation is brief) by any persisting disturbance of ongoing behaviour (Allen and Andrew, in prep.).

If new testosterone-dependent patterns could be obtained during the period of stimulation, the TM hypothesis would predict that they would be appropriate to the type of normal vocalization, which would otherwise have been given. Further, since the full series of normal vocalizations is given, all the main new patterns should be obtainable, from waning bouts and juvenile cackles to crows and elongated peeps.

In testosterone-treated birds, but not controls, waning bouts were

[1] Pairs of 1 msec. rectangular pulses, 0·1 to 0·4 mA peak to peak, as monitored across 100 ohm in series with the chick, at 100 pairs/sec. were used. Six T's were injected on day 3, and daily thereafter, with 0·0125 mg/g testosterone propionate.

sometimes given during the initial orientation produced by electrical stimulation. One bird gave repeated juvenile cackles whilst it walked and looked around, just as when giving juvenile cackles after being suddenly startled in its home cage. In all cases testosterone-dependent calls gave way to normal peeps as stimulation continued. The only bird to give elongated peeps during stimulation did so before continuous rapid peeps developed: in other T's elongated peeps developed from rapid peeps as soon as stimulation ceased. Very intense peeps apparently do not allow the performance of elongated peeps. It seemed possible therefore that the transition to intense peeps during electrical stimulation was too rapid to allow the establishment of crows or elongated peeps in the brief period of moderate peeps when they would be appropriate. An opportunity to test this was provided by carbachol stimulation[1] of the posterior hypothalamus, which evoked the same sequence of behaviour as electrical stimulation but extending over a much longer period of time. Typically, a normal chick shows orienting responses (elongation of neck, looking about) towards the end of the first minute. These may at first be silent, but are nearly always soon accompanied by intermediate calls. The bird begins to walk or run, and at the same time or a little earlier the calls pass into peeps. Running becomes wild and may pass into leaping at the side or, in very intense responding, to directed attempts to hide under food dishes or in dark cracks far too small to enter. Calls reach full intensity in somewhat under two minutes, whilst locomotion does so in four to five minutes.

Following testosterone administration the increase in locomotion almost or quite disappears (Figures 5–7). This effect appears to be a genuine consequence of testosterone. Four out of the nine chicks which reached leaping intensity during their first trial with carbachol on the second or third day of life, then received testosterone.[2] All ceased to leap in carbachol trials two or three days afterwards, whilst all of the other five were still showing escape leaps. This

[1] Approximately 0·1 μg carbachol was given in the end of a 30 gauge tube which could be slipped inside a permanently implanted cannula. A retaining collar on the insert ensured that, when it was home, its tip was flush with that of the cannula, whose placement in the hypothalamus was subsequently verified histologically. Testing, in which the bird was placed for 10 min. in a 2 ft × 2 ft test chamber (differing markedly from the home cage), began on day 3 of life and continued on and after day 5. In the first test of each day the insert was empty; the second, with carbachol, followed 5 h later. 0·1 mg/g testosterone oenanthate was given after testing on day 3, to six T's. Six C's were also run.

[2] *Ibid.*

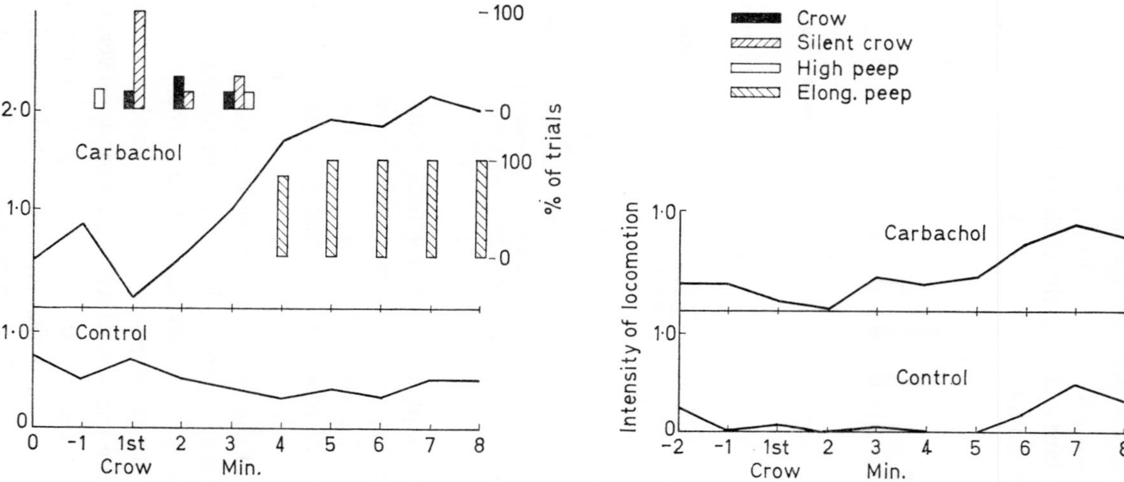

FIGURE 5. Mean values of intensity of calls and locomotion averaged from six carbachol trials and six corresponding control trials of bird 1. Intensities were numerically rated for each minute, according to the type of call or movement which dominated: 0, silence; 1, intermediate calls; 2, soft peeps; 3, intense peeps; 0, no movement; 1, walk; 2, run; 3, leap at sides. The minute in which the first crow occurred (mean 3·67 min.) is taken as the fixed point, in order to show more clearly the sequential relation between crowing and elongated peeps. The testosterone-dependent patterns of vocalization are discussed in the text.

result (which is under further investigation) does not affect discussion of the vocalizations which continued to be evoked in T's, although it prevents direct comparison between T's and C's.

Crowing or elongated peeps (or both) developed in five T's. Both appeared together in bird 1 during the carbachol trial on the fourth day after testosterone administration. Very much the same sequence

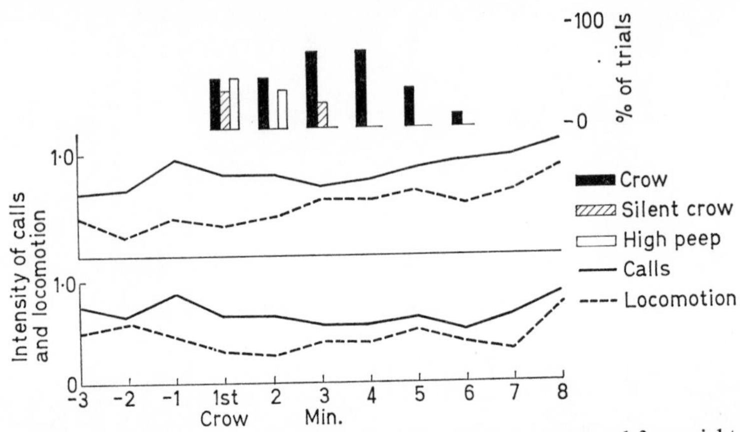

FIGURE 6. Mean values of intensity of calls and locomotion averaged from eight carbachol trials and eight corresponding control trials of bird 2 (see legend to Figure 5). The minute in which the first crow occurred (mean 3·5 min.) is taken as the fixed point in each trial.

occurred in every subsequent carbachol trial, but not in control trials (Figure 5). Crowing began at some point between the end of the second and the beginning of the fifth minute (mean, 3·37 min.). At this time, apart from crows, the animal gave calls intermediate between peeps and twitters or was silent. The drop in intensity of calling in the minute of the first crow was very marked ($P<0{\cdot}06$)[1] by comparison with the control trials. Bird 2 began to crow at almost exactly the same time in the trial as did bird 1 (mean, 3·50 min.; Figure 6). Definite elongated peeps did not develop in this bird during those ten minute trials in which it crowed; however, they did appear in two such trials, which were prolonged for another ten minutes, at about minute 18. It also gave elongated peeps in minute 9 and 10 of one carbachol trial in which it did not crow (which was therefore excluded from the data for Figure 6).

[1] Scores were ranked, and the distribution between the two test situations in the upper half and lower half of the ranking series was compared, using a Fisher exact test (Mather, 1943, p. 195).

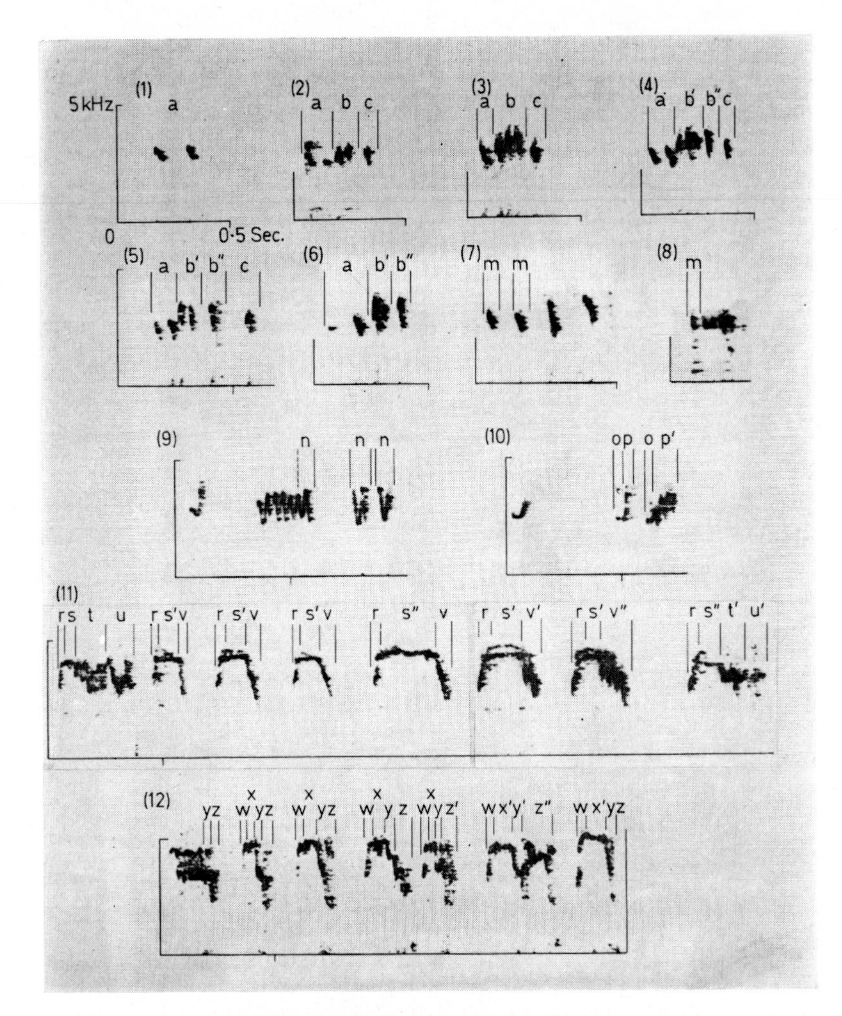

PLATE 1. (1) to (6). 'Food trills' of bird 2, arranged to show the development of a tripartite structure (a, b, c) within a bout instead of a long call. b Becomes a trill in 6. (7) and (8). The 'food trill' of bird 1 (8) develops from bouts of short peeps (7). (9). Food trill of bird 3, followed by repeated trill cycles (n). 10. Food trill of bird 4. Note relation to twitter. (11) and (12). Calls immediately following shock in T1 (11) and T2 (12). Both show drift. Similar sections are identified by the same letters.

(*facing p. 120*)

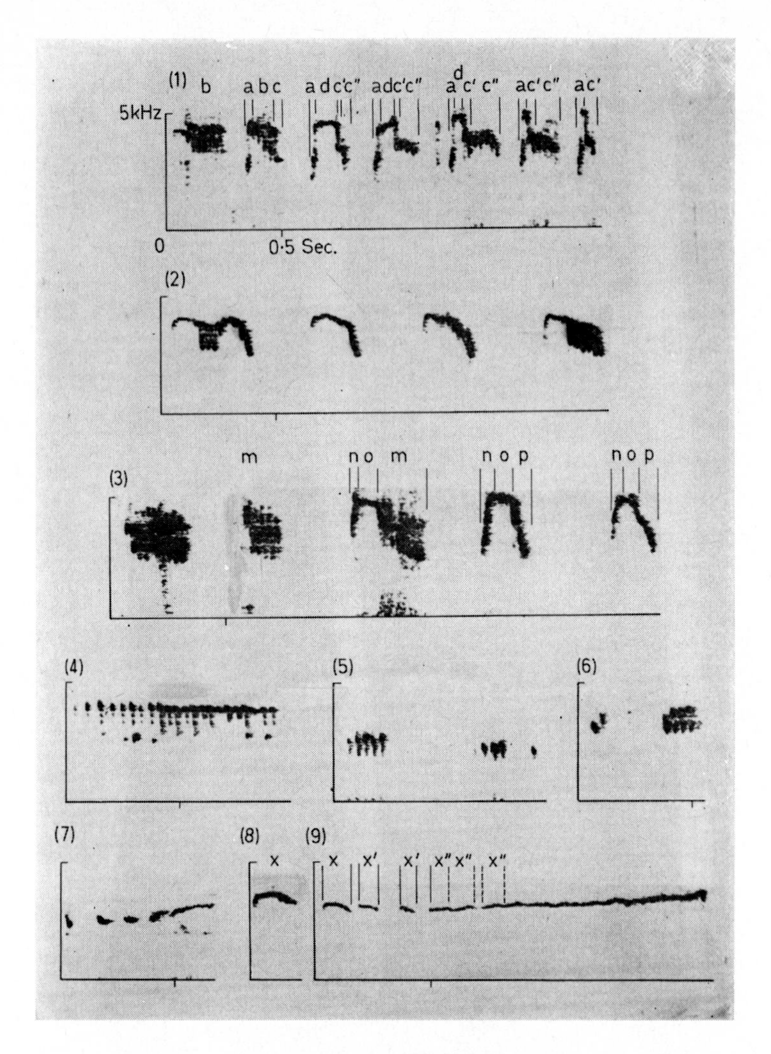

PLATE 2. (1), (2) and (3). Calls immediately following shock in T3 (1), C1 (2) and C2 (3). Drift is obvious only in (1). (4) to (9). Calls given in cold chamber by birds 1 (4), 2 (5), 3 (6), 4 (7) and 5 (8), (9). (4) to (6) Resemble most food trills in that they consist of very rapidly repeated short peeps. (8) Is a circumflex call given a few seconds before the series of calls shown in (9), which merge into a long whine. Similar sections are identified by the same letters.

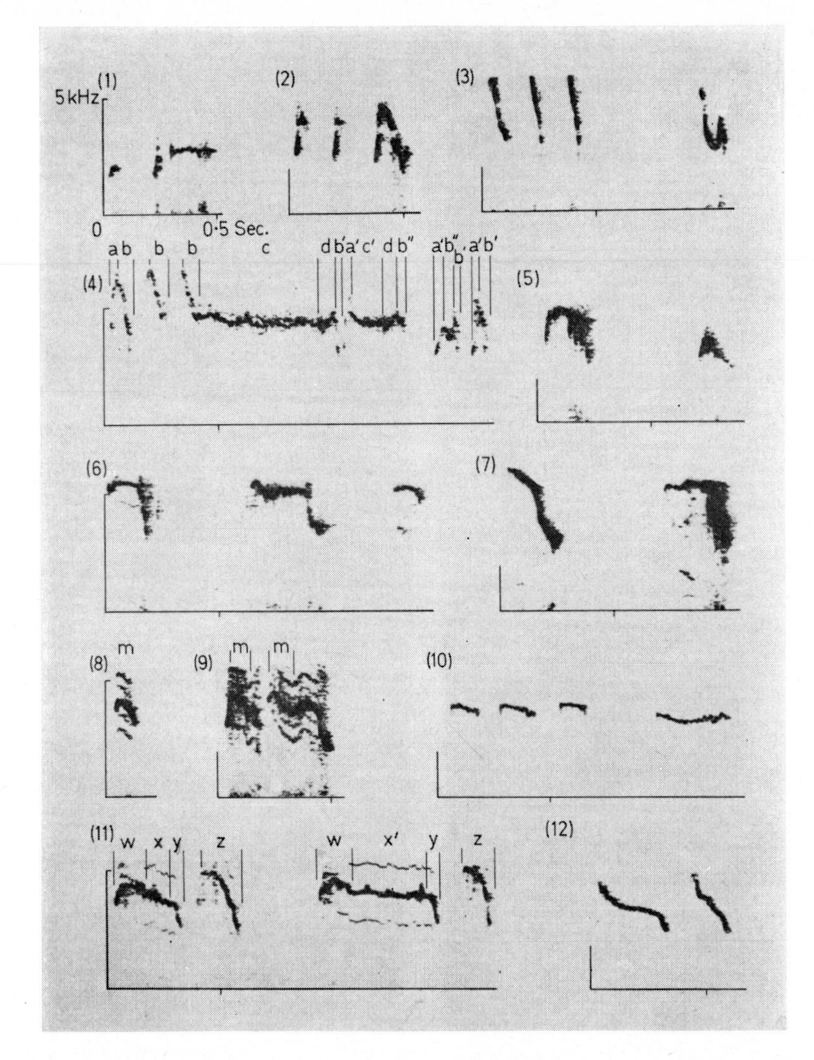

PLATE 3. Calls given in cold chamber. (1), (2) and (3) (birds 4, 6 and 7 respectively) show the interpolation of a call, which is more complex by the addition of a further segment or trill cycle. (4), (6) and (10) (elongated peeps in birds 6, 10 and 1 respectively) show the elongation of a segment of a call. This is most obvious in (4) which also shows drift. (5) and (7) (birds 9 and 11) show interpolated calls, in which the descending limb becomes trilled. (8) and (9) (bird 13) shows the interpolation of a complex double call (9) in a series of circumflex calls (8), which some parts of the double call resemble. Both (11) and (12) (elongated peeps; birds 6 and 14) show an exaggeration of the common commencement of a bout of peeps with a more gradually descending call. In (11) the pattern is fixed into that of a pair of calls, the second of which is a normal peep, whilst the first shows varying degrees of elongation in its central segment. Similar sections are identified by the same letters.

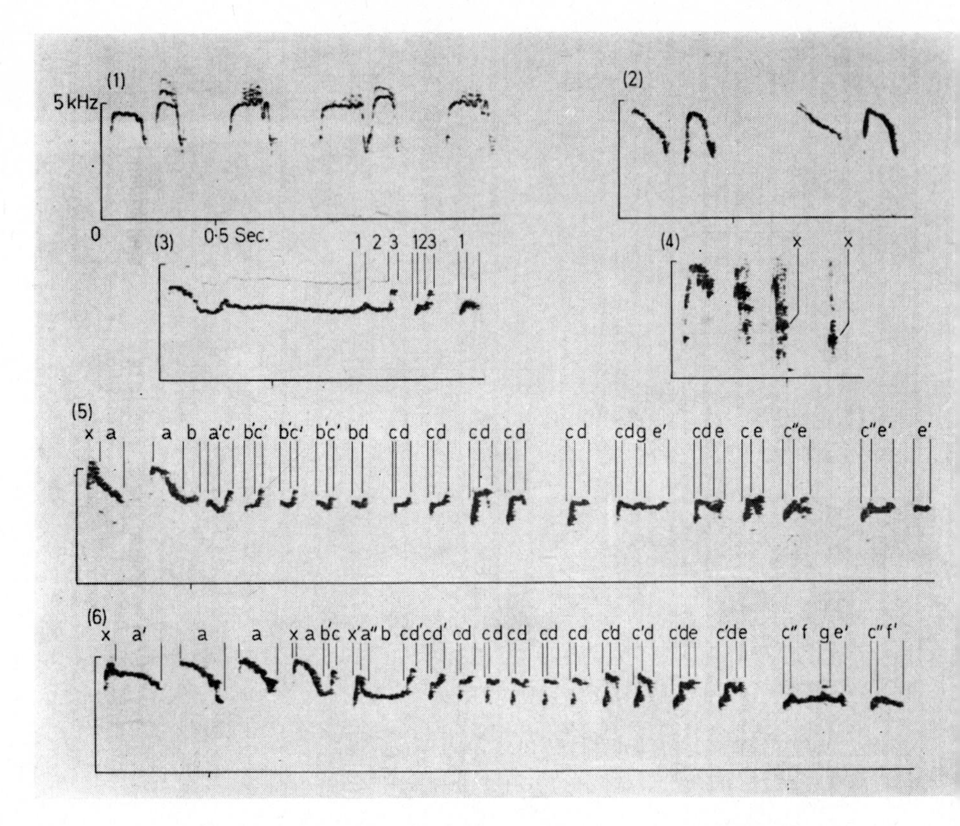

PLATE 4. Calls in cold chamber. (1) and (2) (birds 8 and 15) show stereotyped patterns of two calls repeated over and over. (4) (bird 16) shows a waning bout, which will also serve to illustrate juvenil cackles. If one of the first three calls had occurred alone, it would have been termed a juvenile cackle whilst the last call is an incomplete version (x) of the third call, like the incomplete patterns whic follow cackles. (3) (elongated peep; bird 6) shows the repetition of the final segment of a greatl elongated peep. (5) and (6) (bird 6) are two series of calls showing marked drift from the initial pee from (a), involving progressive changes in some 7 sections of the pattern. Both series (and a third Plate 5) show similar progressive changes. Similar sections are identified by the same letters.

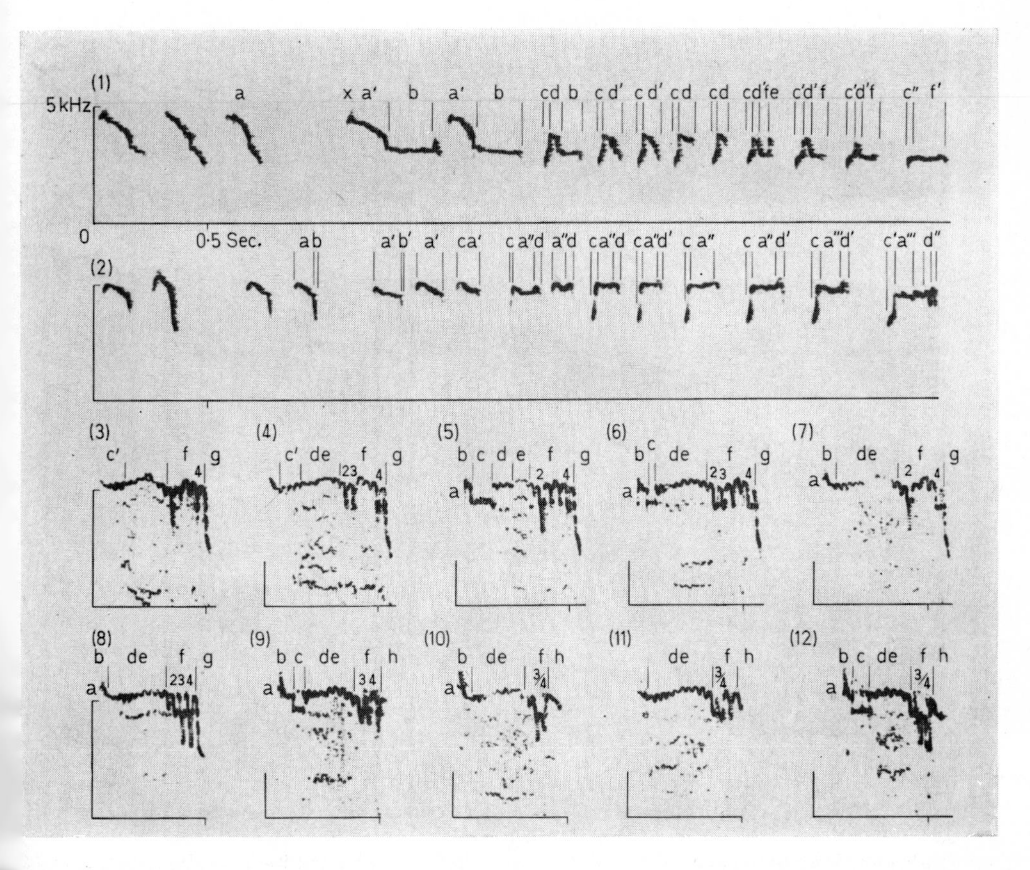

PLATE 5. (1). Third series of calls in cold chamber from bird 6. Letters as in Plate 4 (5) and (6). (2) Bird 8. A similar series, with drift. Lettering corresponds only broadly with that for bird 6. (3) to (12). Ten consecutive crows of bird. Drift is most obvious in the shift of the large amplitude trills to the end of section f, with the development of a depressed two cycle section, and in the disappearance of g and the development of a new terminal segment, h.

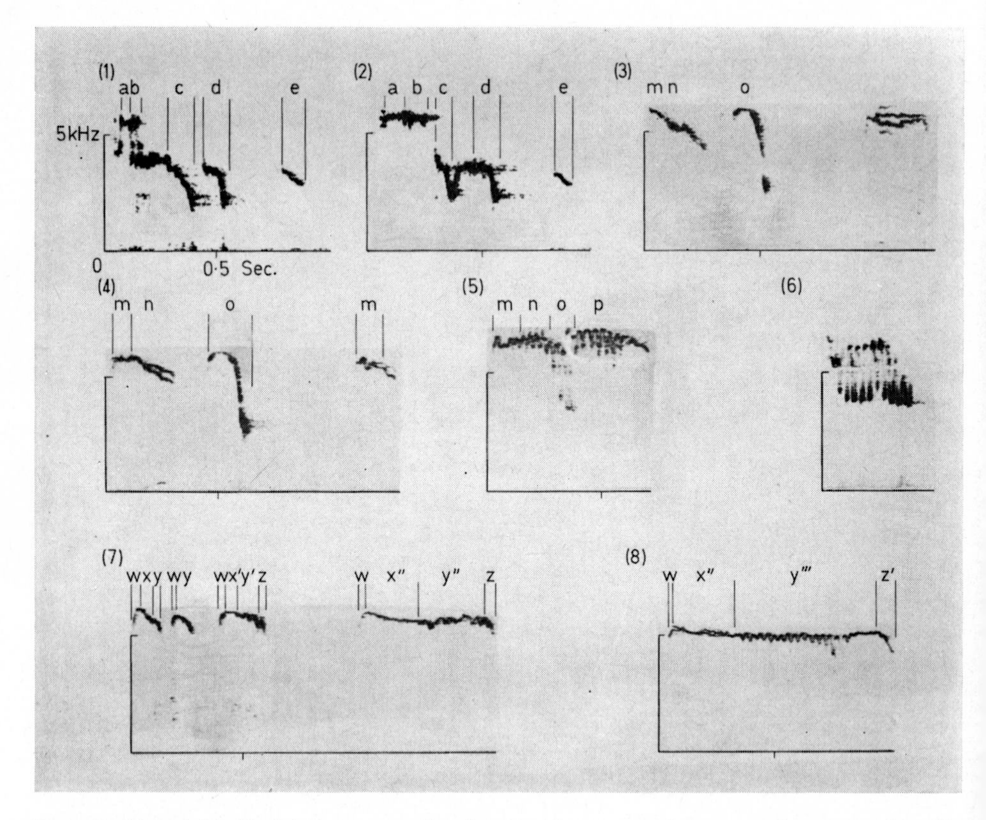

PLATE 6. (1) and (2). Fusion between calls of a crow bout to produce a complex crow. (3) to (8). Calls of bird 2 under carbachol. (3) and (4). High peep is indicated by mn, followed by a normal peep, o. The segment m closely resembles the beginning of the crow (5), (4) and (5) being separated by only a few seconds. The marked drop in the middle of the crow appears to correspond to the peep o in the stereotyped sequence m n o. Such sequences of two calls, sometimes fusing together, are also shown in Plate 4. The transition from a pair of calls to a complex pattern incorporating both, is perhaps the most extreme example of drift. (6) An earlier crow of the same bird. (7) and (8). Transition to elongated peeps during a carbachol session. Drift is present.

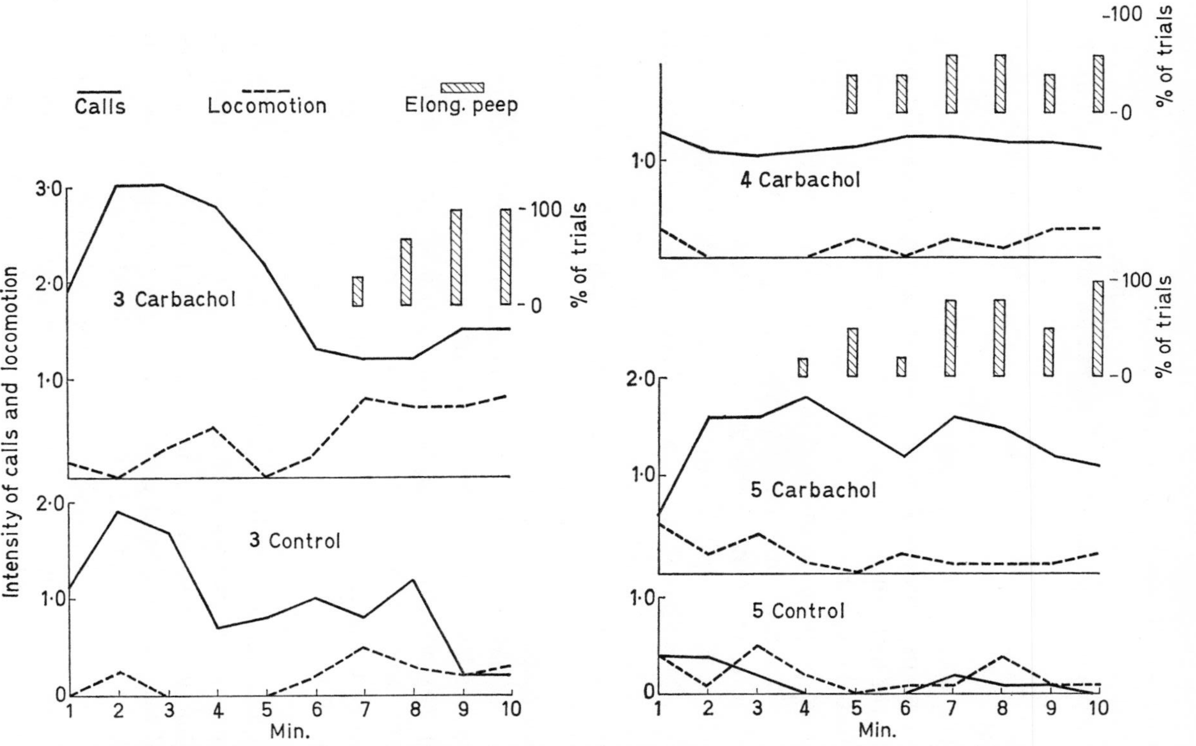

FIGURE 7. Mean values of intensity of calls and locomotion averaged for bird 3 from three carbachol and three control trials, for bird 4 from five carbachol and five control trials, and for bird 5 from four carbachol and four control trials. Bird 4, in which control figures are not shown, gave no elongated peeps in control sessions. See legend to Figure 5.

Elongated peeps developed in bird 1 as general calling rose to the level of frequent peeps (Figure 5); the increase is significant by comparison with the earlier part of the trial ($P<0.008$),[1] and with the control trial ($P<0.01$).[2] Elongated peeps appeared in the other birds at much the same point in the trial, except for bird 2 (means: bird 1, 5·9 min., 2, 18·0 min., 3, 8·0 min., 4, 5·3 min., 5, 6·5 min.), and in each case were accompanied by general calling which included peeps (Figure 7).

This last correlation was almost inevitable, since in every case elongated peeps developed only within peep bouts, usually progressively, the first calls being the most likely to remain normal peeps. The elongated peeps showed, it should be emphasized, all the features of elongated peeps given in the cold chamber test (and other situations which normally produce peeps), including drift in birds 1, 2, 3 and 5 (Plate 6, (7), (8)). Carbachol-induced crows were also fully normal (Plate 6, (5), (6)), showing the same pattern throughout a series of performances. The sequence of behaviour in carbachol trials is thus consistent with the activation of a single mechanism (the TM) which at first results in crowing but which later, when sustained peeping develops (consequent on other effects of stimulation), disturbs such peeps towards a new form.

As in peripherally evoked crowing, crows were sometimes interpolated as the last call of a series of peeps, or were followed by one or two peeps, which would agree with such a superimposition of crowing on peeps. There is also direct evidence for this. Both birds 1 and 2 often began series of crows with 'silent crows' such as precede full crowing induced in T's by strange surroundings. These involve the full neck extension, with the characteristic 'ewe-neck' arching (which is associated with air-sac inflation), the wing-raising (without extension) and leg extension of crowing, but without vocalization. The motor movements of crowing, which can thus occur independently of the vocalization of crowing, may also be superimposed on peeps. Bird 1 in two, and bird 3 in three trials held the neck extended as for a crow, either for a single peep or for several peeps in a short series; in the latter instance the neck remained extended between peeps. The resulting 'high peep' was almost level, instead of descending like a normal peep (Plate 6, (3), (4)). The neck extension was thus

[1] *See footnote p. 120* [2] *Ibid*

associated with (and very likely caused) sustained syringeal tension. Such high peeps occurred in association with silent crows, but also during periods when crows were given. In one trial the first voiced crow began as the last of a series of high peeps, which then continued into a trilled portion.

In bird 2 transitions occurred between high peeps, and both crows and elongated peeps. The first type of transition sometimes involved the development of stereotyped pairs of calls (like those already described in the cold chamber test) into crows. Plate 6, (4), (5) shows a typical example in which a pair of calls, the first being a high peep and the second a normal steeply descending peep, was followed at once by a corresponding crow. The first high peep resembled, sometimes in detail, the beginning of the crow, and the second peep corresponded to a steep descent in the middle of the crow. Bird 2 also gave crows which were far more strongly trilled, and were not associated with high peeps (Plate 6, (6)). In the second type of transition high peeps developed directly into elongated peeps, which retained details from the last high peep (Plate 6, (7)).

A final resemblance between the causation of crowing and elongated peeps is that both can continue through and interrupt maintenance activities. This is true of carbachol-induced, as well as normal crowing. In the case of elongated peeps, birds watched for 30 min. after carbachol administration continued their series of very elongated whines whilst walking normally and pecking at food.

ACTIVATION OF THE TESTOSTERONE MECHANISM IN THE NORMAL CHICK

One aspect of testosterone-dependent patterns which has not been discussed is their occasional performance by normal chicks. For some time this was only known to be true of waning bouts. However, it was found, during close observation of chicks for the first crows following injection, that one male out of six male controls receiving only sesame oil, crowed on day 4 of life, 30 h after injection. Three crows were given at intervals of perhaps a minute. In the second and third, which were seen, the characteristic neck extension and upright stance of crowing were assumed for the trilled crow. Immediately afterwards the bird subsided into a resting attitude. A careful watch was kept in all subsequent experiments for such spontaneous crowing. None of ten control males in a second experiment crowed, but one of a final group of six crowed twice in succession on day 3. Four other males were subsequently heard to crow, in the first week only and in the home cage, during casual observation as a part of various experiments not discussed here. If the other males involved which were not heard to crow are included, then at least 6/140 untreated males crowed, so that this, though rare, is not aberrant. None of the six control females in the first experiment crowed, but a crow has since been heard from one female in the first week of life. Crowing must thus be added to the list of hormone-influenced adult patterns, like copulation (Andrew, 1966), which are performed by the very young chick. Again, like copulation, crowing appears to be given, in the absence of exogenous testosterone, only infrequently even by those birds which do crow. Both of the two males mentioned above were heard to crow on only one occasion despite much subsequent observation. On the other hand, such crowing resembles testosterone-dependent crowing in that, once started, crowing recurred after intervals during which the bird assumed a resting attitude or carried out maintenance activities.

These observations, on any theory of the mode of action of testosterone, show that a threshold change, rather than the maturation of a new pattern, is involved in its effect on crowing. In the case of the theory here discussed, this raises an interesting possibility. The alarm trill, which is evoked in normal chicks by very startling stimuli, becomes more frequent after testosterone (Andrew, 1963). On the present hypothesis this change would imply an effect of the testosterone-mechanism on (TM) alarm trills. It has already been noted that one type of juvenile crow is very similar in form to alarm trills. Considered together, these two facts suggest that a long trill is the form of call produced, when the TM is fully activated in the absence of any strong tendency to perform part of the normal series of vocalizations, and that alarm trills represent single activations of the TM in normal chicks. Alarm trills are in fact commonly given after a few seconds freezing, in which normal vocalization is completely inhibited, and are then followed by silence.

CONCLUSIONS

(1) Some features of the hypothetical model advanced here to explain the effects of testosterone on chick vocalizations seem reasonably well established. This is true of the distinction between the normal series

of vocalizations and the new patterns, which are distortions of different parts of the normal series by an effect which specifies only a longer more intense pattern. The evidence is also good that food trills, waning bouts and juvenile cackles, on the one hand, and crows and elongated peeps on the other, are each a group of responses within which such distortions have a common causation. The conclusion that in both groups of new patterns the effect involves the same testosterone-mechanism (TM) rests on more indirect arguments. However, it is clear that testosterone does not act directly on normal vocalizations but on a mechanism which affects vocalization only when it is activated.

It has been necessary to postulate that the TM is activated by the same causes as produce normal vocalization. It differs in having a somewhat higher threshold (as shown by the fact that food trills appear only at times during cycles of low intensity calls when very intense twitters or intermediate calls would be expected), and by being suppressed by very intense peeps. Evidence from crowing and elongated peeps shows that the TM can also be activated by an initially sub-threshold input, if it is sustained long enough.

(2) The aspect of the data which is most likely to need reinterpretation is that concerning repetition and drift. The existence of these phenomena raises a number of separate points. First, it suggests processes which might allow the permanent storage, following repetition, of new patterns, once they have been generated. Second, it is not clear whether repetition and drift occur in normal vocalizations. If they do not, then one new effect of activation of the TM must be to maintain centrally a particular motor command pattern whilst the TM is active. Third, there are at least two ways in which this could be achieved. If repetition depends on monitoring, whether acoustic or proprioceptive, then the TM, when active, might initiate or promote monitoring. If repetition and drift depend on solely central persistence of a motor command pattern, which does not repeat itself exactly but drifts as each change is perpetuated, then TM might in some way promote such persistence.

(3) The fact that calls as well as song are affected by testosterone in passerines suggests that a causal system like that of the chick TM may be involved. This would explain the range of situations in which song may occur. Typically song, like crowing, is induced in male passerines by the disturbing absence of social fellows (Andrew,

1957*a*). It is at first unexpected to find a pattern, so caused, also performed in agonistic or sexual situations, sometimes in an abbreviated or compressed form. Thus robins may give 'a preliminary loud song-phrase' as they fly at an intruder (Lack, 1943). Male corn buntings will sing when suddenly swooped on by a rival (Andrew, 1957*b*). However, inputs of just this sudden type have been shown to activate the TM in the chick; abbreviated song may have the same relation to juvenile cackles as full song clearly does to crowing. (It should be noted that adult cocks often crow after fighting, Konishi, 1963.)

The development of subsong out of repeated call notes also suggests a condition like that of the chick, in which new patterns develop as a result of a disturbing effect of a single mechanism on *all* existing vocalizations. In *Fringilla coelebs* the songs, which develop when song-learning is prevented, indicate a mechanism which specifies only overall duration, and probably a tripartite division (Thorpe, 1958). Again, one is reminded of the chick TM. Many passerines (e.g. *Emberiza*, Thorpe and Lade, 1961) have stereotyped songs which do not depend on learning from fellows. Here, patterns may be specified in detail. However, a considerable degree of uniformity could be achieved by specifying duration, and relying on self-tutoring such as is implied by drift. The form of components of song would be strongly limited by the form of the calls from which they developed.

(4) A quite different question which should be mentioned briefly is the extent to which the *movements* of vocalization have given rise to display components. In some instances, the movements accompanying testosterone-dependent calls are probably components of nest-building by origin (*Larus*, *Nycticorax*). An association of nest-building movements (e.g. bill-lowering) with vocalization in courtship displays is widespread and presumably ancient in birds (Andrew, 1961). In species in which both the movement and vocalization are testosterone dependent, developmental studies of the origin of the association between the two would be of great interest.

The movements of crowing, on the other hand, appear to be directly involved in the production of sound. Neck extension (shown in crowing and some juvenile cackles) probably applies tension to the syrinx through the trachea. Humerus elevation suggests participation of pectoralis minor in expiration, unlike the normal condition. Wing-raising also occurs in intense vocalization produced by electrical stimulation of the nu. mesencephalicus lateralis in anaesthetized decerebrate preparations. Neck extension movements often help to make song visually conspicuous in passerines (e.g. *Sturnus*, some Estrildid finches). The neck movements of the 'burp' (*Anas acuta*) and 'head throw' (*Aythya americana*) may also have a function in vocalization, which would explain their joint facilitation with vocalization by testosterone (Phillips and McKinney, 1960).

(5) The changes in chick vocalizations produced by testosterone represent changes in a system of responses, which is dependent on external stimulation and very responsive to changes in it, by disturbing effects from a second mechanism which is much more independent of moment by moment changes in external stimulation. Accompanying this independence, is a tendency to repeat the form of a pattern through a series of performances, which results in a certain degree of stereotypy in the pattern.

Similar effects may operate on other systems of responses: there is evidence for them in a range of responses given by testosterone-treated chicks when strongly motivated to mount and copulate. It is worth emphasizing therefore that their consequences (exaggeration and stereotypy) are precisely the changes involved in ritualization. They might be expected in any behaviour, which was both at times internally activated and therefore somewhat independent of immediate environmental control, and also monitored so as to allow storage of patterns. The exaggeration of such a process during development would provide a much simpler route for the evolution of ritualized behaviour like displays, than would be supposed possible from a consideration only of the differences between the drives normally active in the unritualized and in the ritualized state. It also would make it less necessary always to postulate selective advantage for the process of ritualization itself.

(6) Finally, Andrew (1962) has discussed the misleading consequences of the assumption, which used to be a common one, that display components are in some way each caused by a different special combination of particular intensities of major drives like aggression, fear and sex. This is particularly misleading for vocalizations (Andrew 1964). It might well be objected that the causal model outlined here is an explanation of very similar type, in which one hypothetical construct, the testosterone-mechanism, produces different patterns at different intensities of a second such construct, the normal series

of vocalizations. It is worth therefore emphasizing the differences. First, the nature of the interaction between the two variables is specified in some detail. Second, both variables are new constructs. Their success in explaining a wide range of changes following testosterone shows that it is important to avoid the assumption that the main causal systems underlying instinctive behaviour have all been identified and clearly defined.

REFERENCES

ANDREW, R. J. (1957*a*) A comparative study of the calls of *Emberiza* spp (Buntings). *Ibis* **99,** 27–42.

(1957*b*) The aggressive and courtship behaviour of certain Emberizines. *Behaviour* **10,** 255–308.

(1961) A review of the displays of the passerines. *Ibis* **103a,** 315–48.

(1962) Situations evoking vocalization in the primates. *Ann. N.Y. Acad. Sci.* **102,** 296–315.

(1963) Effect of testosterone on the behaviour of the domestic chick. *J. comp. physiol. Psychol.* **56,** 933–40.

(1964) Vocalization in chicks, and the concept of 'stimulus contrast'. *Anim. Behav.* **12,** 64–76.

(1966) Precocious adult behaviour in the young chick. *Anim. Behav.* **14,** 485–500.

(1967) Intracranial self-stimulation in the chick. *Nature* **213,** 847–8.

(in press) Vocalization during intracranial self-stimulation in the chick. *Proc. N.Y. Acad. Sci.*

BALDWIN, F. M., GOLDIN, H. S. and METFESSEL, M. (1940) Effects of testosterone propionate on female Roller canaries under complete song isolation. *Proc. soc. exper. Biol. Med.* **44,** 373–5.

BOSS, W. R. (1943) Hormonal determination of adult characters and sex behaviour in Herring Gulls (*Larus argentatus*). *J. exp. Zool.* **94,** 181–208.

BROWN, J. L. (1964) The integration of agonistic behavior in the Steller's Jay, *Cyanocitta stelleri* (Gmelin). *University of California Publ. Zool.* **60,** 223–322.

COLLARD, J. and GREVENDAL, L. (1946) Étude sur les caractères sexuels des Pinsons. *Gerfaut* **2,** 89–107.

DAVIS, J. (1958) Singing behaviour and the gonad cycle of the Rufous-sided Towhee. *Condor* **60,** 308–36.

ERICKSON, C. J., BRUDER, R. H., KOMISARUK, B. R. and LEHRMAN, D. S. (1967) Selective inhibition by progesterone of androgen-induced behaviour in male Ring Doves (*Streptopelia risoria*). *Endocrinology* **81,** 39–44.

GRASTYÁN, E., LISSAK, K., MADARÁSZ, I. and DONHOFFER, H. (1959) Hippocampal electrical activity during the development of conditioned reflexes. *EEG clin. Neurophysiol.* **11,** 409–30.

GRASTYÁN, E., KARMOS, G., VERECZKEY, L. and KELLÉNY, E. E. (1966) The hippocampal electrical correlates of the homeostatic regulation of motivation. *EEG clin. Neurophysiol.* **21,** 34–53.

GUYOMARC'H, J.-C. (1962) Contribution à l'étude du comportement vocal du poussin de *Gallus domesticus*. *J. Psychol. norm. path.* **3,** 283–306.

(1965) Contribution à l'ontogenése des émissions sonores du Poussin: les cris et comportements induits par l'hormone mâle. *Bull. Soc. scient. Bretagne* **50,** 81–93.

(1966) Les émissions sonores du Poussin domestique, leur place dans le comportement normal. *Z. Tierpsychol.* **23,** 141–60.

GWYNNER, E. (1962) Über die biologicshe Bedeutung der 'zweckdienlichen' Anwendung erlernter Laute bei Vögeln. *Z. Tierpsychol.* **19,** 692–6.

(1964) Untersuchungen über das Ausdrucks—und Sozialverhalten des Kolkraben (*Corvus corax corax* L.). *Z. Tierpsychol.* **21,** 657–8.

HALL, M. F. (1962) Evolutionary aspects of Estrildid song. *Symp. zool. Soc. Lond.* **8,** 37–55.

HAMILTON, J. B. (1938) Precocious masculine behaviour following administration of synthetic male hormone substance. *Endocrinol.* **23,** 53–7.

KONISHI, M. (1963) The role of auditory feedback in the vocal behavior of the domestic fowl. *Z. Tierpsychol.* **20,** 349–67.

LACK, D. (1943) *The life of the Robin.* London.

LANCASTER, D. A. (1964) Life history of the Boucard Tinamou in British Honduras. *Condor* **66,** 165–81, and 253–76.

MARLER, P. (1956) Behaviour of the chaffinch, *Fringilla coelebs*. *Behaviour Suppl.* 5.

MARLER, P. KREITH, M. and WILLIS, E. (1962) An analysis of testosterone-induced crowing in young domestic cockerels. *Anim. Behav.* **10,** 48–54.

MATHER, K. (1943) *Statistical Analysis in Biology.* Methuen, London.

NOBLE, G. K. and WURM, M. (1940) The effect of testosterone propionate on the Black-crowned Night Heron. *Endocrinology* **28,** 837–50.

PHILLIPS, R. E. and MCKINNEY, D. F. (1960) Effect of testosterone on the occurrence of some duck displays. *Anat. Rec.* **138,** 376.

POULSON, H. (1951) Inheritance and learning in the song of the Chaffinch (*Fringilla coelebs* L.). *Behaviour* **3,** 216–28.

SAUER, F. (1954) Die Entwicklung der Lautäusserungen vom Ei ab schalldicht gehaltener Dorngrasmucken (*Sylvia c. communis*, Latham) im Vergleich mit später isolierten und mit wildlebenden Artgenossen. *Z. Tierpsychol.* **11,** 10–93.

SAUER, E. G. F. and SAUER, E. M. (1966) The behaviour and ecology of the South African Ostrich. *The Living Bird,* **5,** 45–75.

SCHEIN, M. W. and HALE, E. B. (1959) The effect of early social experience on male sexual behaviour of androgen injected Turkeys. *Anim. Behav.* **7,** 189–200.

SCHJELDERUP-EBBE, T. (1923) Weiter Beiträge zur Sozial- und Individualpsychologie des Haushuhns. *Z. Psychol.* **92,** 60–87.

SCHMIDT, R. S. (1966) Hormonal mechanisms of frog mating calling. *Copeia,* 1966, 637–44.

THIELCKE-POLTZ, H. and THIELCKE, G. (1960) Akustisches Lernen verschieden alter schallisolierter Amseln (*Turdus merula* L.) und die Entwicklung erlernter Motive ohne und mit künstlichen Einfluss von Testosteron. *Z. Tierpsychol.* **17,** 211–44.

THORPE, W. H. (1958) The learning of song patterns by birds, with especial reference to the song of the chaffinch, *Fringilla coelebs*. *Ibis* **100**, 535–70.
(1961) *Bird Song*. Cambridge University Press, Cambridge.
THORPE, W. H. and LADE, B. I. (1961) The songs of some families of the order Passeriformes. *Ibis* **103a**, 231–59.

7. ROLES OF BUDGERIGAR VOCALIZATION IN THE INTEGRATION OF BREEDING BEHAVIOUR

by **BARBARA F. BROCKWAY**
The Ohio State University, Columbus, Ohio[1]

INTRODUCTION

This chapter will consider some ways in which budgerigar vocal behaviour and neuroendocrine states interact within and between individuals in order to facilitate species propagation. The value of this material is, perhaps, best judged in view of the following information.

Regardless of the sensory modality involved, courtship stimuli may function in *at least* one of three ways. They may serve to (i) increase the proximity of potential sexual partners, or (ii) signal the availability of a suitable sexual partner or (iii) promote and integrate the behavioural performances and the hypothalamic-hypophyseal-gonadal activity of sexual partners so as to produce viable zygotes in a minimum of time. Studies of avian vocalizations have provided extensive information concerning only the first two functions (for recent reviews, see Collias, 1960, Thorpe, 1961 or Armstrong, 1963). Little is known about the third, integrative, function. For example, Collias (1960) cites only one experimentally substantiated example of an integrative vocalization: a series of squeaks used by female house wrens as an invitation to copulation. Evidence that vocalizations of conspecifics can stimulate gonadal activity has been reported for only two avian species, ring doves (Lott and Brody, 1966) and budgerigars (Vaugien, 1951; Ficken *et al.*, 1960).

Main obstacles to the study of integratively functioning sound seem to lie in its low volume, short duration and infrequent occurence. Moreover, the motivational states underlying such behaviour and the contexts in which it occurs make a performer more prone to

[1] Author's present address: 2175 Tabor Drive, Denver, Colorado.

discontinue this behaviour if disturbed, and less apt to quickly resume it afterwards, than when he (she) is performing another type of behaviour. Furthermore, should there be any tendency to ignore the possibility that behaviour may serve more than one function, the afore-mentioned difficulties would weigh against the discovery of an integrative function.

Most evidence on auditorally-evoked neuroendocrinological responses concerns mammals and non-vocal sounds. Such responses are often similar to those resulting from stressing agents and consist of increased secretions of adrenocorticotrophins and corticosteroids and decreased gonadotrophic secretions and gonadal activity (e.g., Anthony *et al.*, 1959*a,b*; Biró *et al.*, 1959; Bond *et al.*, 1963; Henkin and Knigge, 1963; Sackler *et al.*, 1960). Only one recent study suggests that, in specific instances, sound may stimulate gonadotrophin secretions (Zondek, 1967).

To date, budgerigars (*Melopsittacus undulatus*) are the only known avian species in which specific male vocal behaviour has been experimentally shown to promote gonadotrophin secretions and full ovarian activity in females (Brockway, 1965, 1967*a*), induce sexually active females to assume a soliciting-for-copulation posture (Brockway, 1964*a*) and promote gonadotrophin secretions and testicular activity in males (Brockway, 1967*b*). Consequently, studies of budgerigar vocalizations are of tripartite value. First, they increase our knowledge concerning the integrative functioning of vocal behaviour. Second, they show that an auditory stimulus can increase gonadal activity in both sexes. Last, they illustrate considerations elucidated in earlier studies by Hinde (1965) and Lehrman (1965)—namely, (i) that the complexity of the causal and functional interrelations of the various behaviour patterns which comprise a reproductive nexus require that such interrelations be considered in the study of any component, and (ii) that any given stimulus may affect the integration of sexual behaviour by producing both long- and short-term physiological effects.

GENERAL BIOLOGY

Seasonal cycles of gonad recrudescence and regression were not observed in the domesticated strains of budgerigars used, which were studied and housed under laboratory conditions. Once a male reaches sexual maturity, his testes seem to remain at a fairly high level of development, as measured by sperm and androgen production, provided that he has the opportunity to hear the vocalizations of others. When males are visually and vocally isolated from each other for three or more weeks, testes and vasa deferentia decrease to about half of their original size and weight and only about 15 per cent of the males still yield semen when milked (Brockway, 1967*b*, 1968). In adult females which can hear other budgerigars, ovaries are typically maintained at a relatively lower level of activity: most ovaries have ovocytes of 1·5 mm or less in diameter. Yet, once such females are provided with nest-boxes, maximal ovarian development (ovocytes about 9·0 mm in diameter) and oviposition can occur in as few as ten days (Brockway, 1962, 1964*a*). Relative daylength does not seem to be very critical. Spermatogenesis and oviposition occur in budgerigars which are housed either in continual darkness (Vaugien, 1951, 1952, 1953; and Brockway, 1965 and unpubl.) or in continual light (Brockway, unpubl.).

In successful breeding, reproductive behaviour consists of intrapair precopulatory and copulatory interactions that occur concurrently with a female's nestbox-oriented behaviour. In contrast to other reports (Brereton, 1966), unless females that are housed under typical aviary photoperiods (ten to fourteen hours of daily light) repeatedly perform nestbox-oriented behaviour, they seldom show significantly increased ovarian or oviductal development, even though they often sexually interact and copulate with their mates.

In their natural Australian habitat, budgerigars nest mainly in tree cavities. They do not build nests. In lieu of nest construction, females steadily perform nestbox-oriented behaviour during the one to two weeks of greatest ovarian development which precedes oviposition. Initially, a female spends much time looking into the nestbox while perched outside. At first she enters only infrequently, and remains inside for only a few seconds. As days pass, she enters the nestbox more often and remains inside for longer periods. About a week before egg-laying, she often occupies the nestbox for 40 ± 12 min/h and remains inside for periods from ten to fifteen minutes. As the day for her initial oviposition approaches, she tends to remain within the nestbox and may even perform an agonistic bill-gape and/or 'nest defence' ehh call when she is disturbed.

An individual's genetic sex does not solely determine which sexual behaviour he performs. Under appropriate conditions, all behaviour patterns except sperm emission, incubation, feeding of young less than seven days of age and oviposition may be performed by either males or females. Sexual behaviour has been described in more detail elsewhere (Brockway, 1964*a*). In summary, a male exhibits strong tendencies to flee from and copulate with a female and a weaker tendency to behave aggressively toward her. She, in contrast, shows a strong tendency to behave aggressively toward him, together with relatively weak tendencies to flee from or mate with him. When sufficiently stimulated by his vocal and visible precopulatory actions, she ceases her typically aggressive responses and either solicits or permits copulation. Visible male precopulatory behaviour includes several types of nudging-pumping displays, head-bobbing with or without courtship feeding, sidling toward and away from the female, attempts to mount, mounting and copulation. Vocal behaviour will be considered next.

TYPES OF VOCAL BEHAVIOUR

Identity and occurrence

To date, ten types of vocal behaviour have been distinguished. They are ehh, nest-defence ehh, squacks, chedelees, whedelees, tuk-tuk, loud warble (LW), soft warble (SW), loud-plus-soft-components-warble (LPSC) and a nest entrance note. All except the last two vocalizations have been described and discussed elsewhere (Brockway, 1964*a*,*b*).

Ehh, performed alone or with a bill-gape, may function as a defensive threat. Nest-defence ehh, a prolonged variant of ehh, certainly appears to serve such a function. Squacking is typically performed during and after a disturbance. This sound is highly mimetic, so its performance by one bird readily elicits squacks from others. Chedelee typically occurs after a bout of squacking subsides or upon the introduction of new individuals into an area already containing budgerigars. It, too, is highly mimetic. The nest entrance note is an extremely high-pitched 'peep', performed by the male after arriving at the nest-entrance when he is accompanied or joined by his mate. It may terminate a bout of LW. I know nothing more about this note.

Experimental evidence implicates the remaining five vocal patterns in sexual situations, so we may call them courtship vocalizations. Tuk-tuks are usually performed by an individual that is oriented bill to bill with another. They are often performed simultaneously with nudging, a visible precopulatory display. I have repeatedly induced females to assume the soliciting-for-copulation posture by imitating tuk-tuks. Whedelees often follow a rebuffed or interrupted courtship attempt. They may be performed alone or alternately with displacement head-flicks or with head-shaking. Loud warble is the chatter so commonly heard in an aviary containing budgerigars and is not restricted to precopulatory contexts. Performers typically have ruffled crown and throat feathers and show no specific orientation to other birds. Loud warble is highly mimetic and is performed stridently and rapidly with little pause between components. Soft warble is commonly performed by a bird which is oriented bill to bill with another and is restricted to precopulatory contexts. Various visible and highly sexually motivated precopulatory actions generally precede and follow it. It is performed more slowly, emphatically and softly than LW. Unfortunately, SW is difficult to hear when birds nearby are performing LW. Loud-plus-soft-components-warble is a 'hybrid' between LW and SW. It occurs in contexts that are similar to those associated with LW and contains components typical to SW interspersed with others typical to LW. The entire pattern of LPSC is exceedingly variable, but it is performed in the loud, strident, hurried and perfunctory manner of LW. We are currently analysing sonograms of all these vocalizations. Since LPSC has only been recently identified, many earlier quantitative data on causation of LW are actually data on both LW and LPSC combined.

Quantitative differences and sexual identity

Only courtship vocalizations will be discussed here. Superficially, tuk-tuks, whedelees, LW, SW and LPSC by males do not seem to differ qualitatively from those performed by females: however, definite knowledge must await audiospectrographic analyses.

In contrast, quantitative sexual differences are so apparent that courtship vocalizations generally are described as being 'typically performed by males'. Table 1 shows sample data for warbling by adult males and females that were visually isolated from members of the opposite sex but able to hear the sounds of over 200 budgerigars in an adjacent aviary. No nestboxes were available. Both intact and castrated males perform significantly more and longer bouts of LW and SW than do intact females. In contrast, tuk-tuks and whedelees seem more influenced by social circumstances than by the performer's physiological (hormonal) condition. As such, they are associated with the 'male' role in heterosexual and homosexual intra-pair interactions. Quantifying data from homosexual interactions is complicated because the pair members frequently alternate in assuming 'typically male' or 'female' roles and the amounts of such

TABLE 1. *Sexual differences in mean quantities of warbling.*[1]

Group[2]	Min./h of warbles		Bout lengths of warbles (sec.)	
	Loud	Soft	Loud	Soft
Intact males	9·2	2·5	63·6	23·9
Castrate males	4·4	2·5	53·6	14·9
Intact females	0·6	0·3	6·6	3·8

[1] All differences between females and either group of males were significant (analysis of variance $P<0{\cdot}05$).

[2] Intact males possessed active sperm and androgen producing testes. Intact females possessed an inactive, tiny ovary and oviduct. Means represent the sum of five or six separate one-hour records per bird. No. of birds per group = 6.

alternation differ from pair to pair. Since quantitative data concerning tuk-tuks and whedelees are incomplete, only warbling will be considered in the remainder of this chapter.

CAUSAL FACTORS FOR COURTSHIP WARBLING

The performance of any behaviour pattern at any one time depends upon the interactions of multiple causal factors. These factors include stimuli both from the external environment and influences from the animal's internal (physiological) environment (for recent reviews, see Eisner, 1960; Hinde, 1966; Lehrman, 1959, 1961; van Tienhoven, 1961; Young, 1961). In budgerigars, studies of internal factors have, to date, only involved hormonal manipulations. Studies of external factors have dealt only with the general warbling of the aviary colony, e.g. social vocal stimuli.

Information about hormonal influences on avian vocalizations seems mainly to concern male song. Field studies have correlated the increased performances of male song with testicular recrudesence (e.g. Davis, 1958) and experiments on species including domestic fowl have shown that androgens stimulate the performances of male sexual vocalizations (e.g. Andrew, 1963; Leonard, 1939; Noble and Zitrin, 1942; Marler *et al.*, 1962). Because of the distinctively male character of these sounds, these studies understandably focused on androgens rather than pituitary gonadotrophins. Yet, certain sexual behaviour of some vertebrates may be directly stimulated by gonadotrophins, neurohypophyseal secretions or by the synergistic action(s) of gonadal steroids plus pituitary secretions (Egami, 1959; Hoar, 1962; Pickford, 1952, 1954; Rey, 1948; Wilhelmi *et al.*, 1955).

Although most warbling is performed by males possessing active testes, some is performed by females and castrated males (see Table 1). The highly mimetic nature of LW, however, complicates determinations concerning the influences of internal factors upon warbling. Accordingly, the effects of exogenous androgenic, estrogenic and gonadotrophic materials upon warbling were determined both for isolated and non-isolated males.

Studies on isolated males

Each male's vocal behaviour was recorded while he was deprived of social vocal stimuli both before and during hormonal treatments. During treatments, his warbling was also recorded while he could hear a taped two-hour concert of warbling (predominantly LW). So as not unduly to affect the males' isolation, this exposure to social vocal stimuli was limited to a single period following the penultimate of six hormonal injections. Intact males were injected with either pregnant mare's serum (PMS, a mammalian gonadotrophic preparation containing large quantities of FSH and lesser amounts of LH), testosterone propionate or placebo solutions. Estradiol cypionate or PMS was administered to castrated males (Brockway, 1968).

During the twenty-day period of isolation preceding treatments, both intact and castrated males performed similar small amounts of warbling (1·53 to 0·0 min./h). As expected, the reproductive organs of intact males regressed to almost minimal conditions during this twenty-day period, thereby converting intact males into what might be termed essentially 'physiologically castrated' males. However, these changes in gonadal and vocal activity do not identify any cause-effect relationship.

Warbling was compared before and after treatments, or between treatment groups on any one experimental day. Also compared, were the differences between the vocal behaviour of males while they did or did not hear others on the day after injection no. 5. Evidence (see Figure 1) indicates that (i) androgens, but not gonadotrophins alone, significantly decrease a male's threshold for performing LW in the absence or in the presence of social vocal stimulation. They also

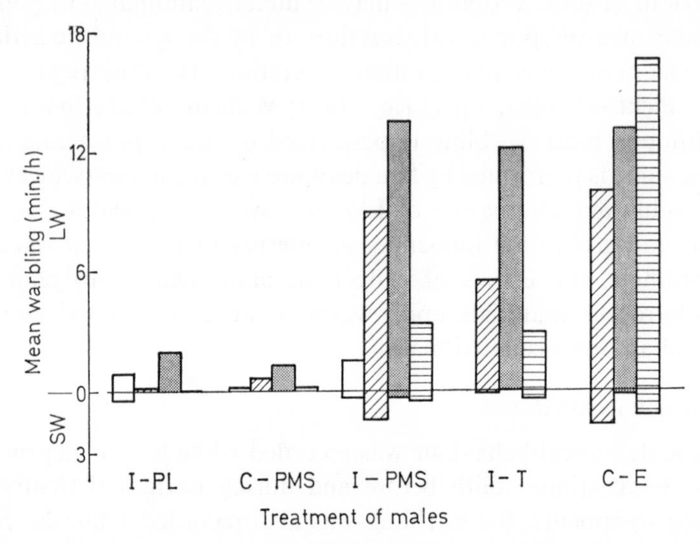

FIGURE 1. Effects of gonadal steroids and PMS on the warbling of isolated males. SW = Soft warble, LW = loud warble, I = intact males, C = castrated males, PL = placebo solutions, PMS = pregnant mare's serum, T = testosterone and E = oestradiol. Histobar code: □ = maximal group mean for any one day prior to treatments, ▨ = maximal group mean following any of the first five injections, ▩ = group mean after the fifth injection while hearing the LW of others, and ▤ = group mean after the last (sixth) injection (after Brockway, 1968).

decrease his threshold for performing SW in the absence of social vocal stimulation. These data do not eliminate the possibility of a synergistic PMS-androgen effect, however. (ii) Irrespective of treatments, SW performances were unaffected by the presence or absence of social vocal stimulation. (iii) In the absence of social vocal stimulation, estradiol stimulated males to perform LW. Yet, during the taped concert, there was no uniform response in the quantity of LW performed by males receiving estradiol. So exogenous androgenic and estrogenic materials may exert differing influences on LW, depending on the external circumstances. (iv) Androgens and social vocal stimuli may act complementarily to lower a male's threshold for performing LW, as intact males given either PMS or testosterone performed significantly more LW during the concert, compared to before or afterwards, than did the castrates receiving PMS or the intact males receiving placebo injections.

Studies on non-isolated males

Data were obtained from intact males possessing fully active testes and receiving either testosterone (Group A) or oil (Group B) and castrates receiving either oil (Group C), 0·2 mg oestradiol (Group D) or 0·4 mg oestradiol (Group E).[1] Males of Group C can be regarded as occupying a 'neutral' position between males under a progressively greater androgenic influence (Groups B and A respectively) and those under a progressively greater oestrogenic influence (Groups D and E respectively). This presumes that intact males receiving testosterone possess higher plasma levels of androgens than do intact males receiving only oil and that quantities of exogenous oestrogenic material parallel internal oestrogenic levels.[2]

Mean quantities of LW (min./h) did not differ among Groups C, D and E (4·4, 6·0 and 5·6 respectively). Yet these groups performed significantly less LW than did either Group B (9·2) or A (15·4), and Group A performed significantly more LW than did Group B. Mean quantities of SW (min./h) did not differ among Groups B, C, D and E (2·4, 2·5, 1·1 and 1·6 respectively), but Group A performed significantly less SW (0·6) than did any other group. These data suggest that whereas exogenous oestrogens do not significantly alter the quantities of either LW or SW, increased androgenic levels prompt males to perform more LW and less SW. Data, especially from Group A, also suggest that LW may differ causally from SW.

Since hearing others stimulates a male's warbling, it is pertinent to examine how hormonally influenced warbling is affected by prolonged social vocal deprivation and by the opportunity to hear others following such deprivation. To do this, the hormonally

[1] After a week's acclimation to new quarters, the six birds of each group received six injections, one every other day. Injection volumes were similar for all groups. Vocal data were obtained daily, following the third injection. No significant changes in daily vocal behaviour were noted within any group, so all daily values per bird were pooled. Data were statistically evaluated by analysis of variance tests. The null hypothesis was rejected at $P < 0{\cdot}05$.

[2] Because of differences in the possible fates of injected steroids, such presumptions may often be invalid. Exogenous steroids may be rapidly inactivated by either anabolic or catabolic processes, or they may be converted into another chemical which might have physiological actions opposite from those of the original material. Such fates depend on the nature and the amount of exogenous material and the species studied.

influenced warbling of males, continually able to hear others (non-isolated), was compared with the most warbling that isolated males performed when they could not hear others and when they could hear a taped LW-concert. Mean quantities (min./hr) of warbling were compared between isolated and non-isolated males receiving either 1·0 mg doses of testosterone (intact) or 0·4 mg doses of oestradiol (castrate). As a control, the warbling of isolated castrates receiving PMS was compared with that of non-isolated castrates receiving oil.

Mean LW performed by non-isolated controls (4·4) was not significantly different from that performed by isolated controls during the concert (1·3), but was significantly greater than that of isolated controls in the absence of social stimulation (0·4). Data for testosterone-treated males was similar. Non-isolated males and isolated males hearing the concert both performed similar quantities of LW (15·4 and 12·1, respectively). They performed significantly more LW than did isolated males in the absence of social stimulation (7·3). Thus the quantity of LW performed by males, which possess either low or high plasma levels of gonadal androgen, seems to be more influenced by ongoing social vocal stimuli than by the amount of warbling they had previously heard or performed. In contrast, isolated males receiving oestradiol performed just as much LW during the concert (13·2) as they did in the absence of social vocal stimulation (16·8). In either situation, they performed significantly more LW than did non-isolated males (5·6). This suggests that oestrogenically influenced LW is affected more by previous warbling the males had heard or performed than by ongoing social vocal stimuli: isolation tends to increase the quantity of oestrogenically influenced LW.

I cannot explain why oestrogenically influenced LW may be more sensitive to one parameter of social vocal stimulation whereas androgenically or non-oestrogenically influenced LW may be more sensitive to another parameter. Nevertheless, this emphasizes that differences in androgen and/or oestrogen levels may exert subtly different influences.

The following data suggest that one hormone can affect LW and SW differently. The non-isolated controls performed significantly more SW (2·5) than did isolated controls either during the concert (0·9) or in the absence of social stimulation (0·1). Furthermore, isolated males performed significantly more SW during the concert than in the absence of such social stimulation. Thus both the SW and LW of males with low plasma levels of gonadal androgens are affected (stimulated) by ongoing social vocal stimulation. However their SW, unlike their LW, is also influenced (increased) by the greater amount of warbling that they had previously heard or performed. The quantity of SW performed by males receiving testosterone was identical for both non-isolated males (0·6) and isolated males that did not hear others (0·6). Isolated males performed significantly less SW during the concert (0·2). There was no significant difference ($P >$ 0·20) in the quantities of oestrogenically influenced SW performed by non-isolated males (1·6), isolated males when they did not hear others (1·7) and isolated males during the concert (0·9). These data suggest that the quantity of SW performed by males with relatively high plasma levels of gonadal androgens or exogenous oestrogen is unaffected by both the amount of ongoing social vocal stimuli and the amount of warbling that they had previously heard or performed. Yet this does not explain why isolated males receiving testosterone performed less SW during the concert than in the absence of social stimulation. They may have performed less SW because the concert, the first time they had heard another budgerigar in about five weeks and a source of highly mimetic LW, plus their hormonal condition, predisposed them to perform extensive LW at the expense so to speak, of their tendency to perform SW during the concert. The decreased influence of social vocal stimuli on the SW of androgenically stimulated males does not conflict with information about social influences on the warbling of normal males.

In summary, the quantities of both LW and SW are increased by social vocal stimulation in the absence of relatively high internal androgen or oestrogen levels. Continual (daily) social stimulation is more stimulating then ongoing social vocal stimulation. The presence or absence of oestrogenic or androgenic material seems to affect which parameter of social vocal stimulation exerts a major influence on the quantity of LW performed by males. Large amounts of exogenous androgenic or oestrogenic material, in general, cause the performance of SW to be unaffected by either parameter of social vocal stimulation.

In these studies, males had little or no opportunity to interact sexually with a mate, a circumstance promoting SW. Hence, experimental circumstances (the presence or absence of social vocal stimuli) would generally affect the males' LW more than their SW. Thus data about internal causal factors always, albeit subtly, reflect concurrent external influences as well. To ignore the one for the other may produce misleading information.

Hormonally-influenced warbling by females and how it compares with that by males

Data on females are presented because they: (i) illustrate that females' thresholds for warbling are not genetically insensitive or proscribed but can, like the thresholds of males, be influenced by physiological (endocrinological) factors; (ii) may help to explain why females seem to perform little warbling; and (iii) show that some sexual differences concerning hormonally-influenced warbling do exist. Because these studies are superficial, a sexual difference may reflect genetically determined differences in the response thresholds of target organ tissues, the fate of an exogenous hormone or a certain antagonism between endogenous gonadal hormones and exogenous gonadal hormones 'from' the opposite sex.

LW and SW were recorded for females which were repeatedly injected with either progesterone or oil or with different quantities of testosterone, oestradiol or oestradiol plus progesterone. All females were able to hear other budgerigars of both sexes and were caged without nestboxes. In forming conclusions, hormonally treated birds were judged against baseline (control) groups of oil-injected individuals whose gonads were either absent (males) or regressed and relatively inactive (females). Meaningful comparisons are possible so long as one remembers that male and female control groups cannot, *a priori*, be considered as physiological equivalents. Data are presented in Figure 2.

Females, receiving large quantities of either testosterone or oestradiol, performed significantly more LW than did either the females of control groups or castrated males receiving oil or oestradiol, and performed as much LW as did intact males receiving either oil or testosterone. Seemingly, females are more stimulated by a given dosage of oestradiol to perform LW than are males.

The SW of females was prolonged only by oestradiol treatments and, unlike LW, was further lengthened by combined oestradiol-progesterone treatments. However, neither females nor castrated males receiving oestradiol performed more SW than did untreated intact or castrated males. Perhaps, once a certain level (min./h) of

SW performance is attained, exogenous oestradiol cannot increase it further, regardless of the individual's sex. Large quantities of testosterone did not significantly decrease the SW of females as it did with males.

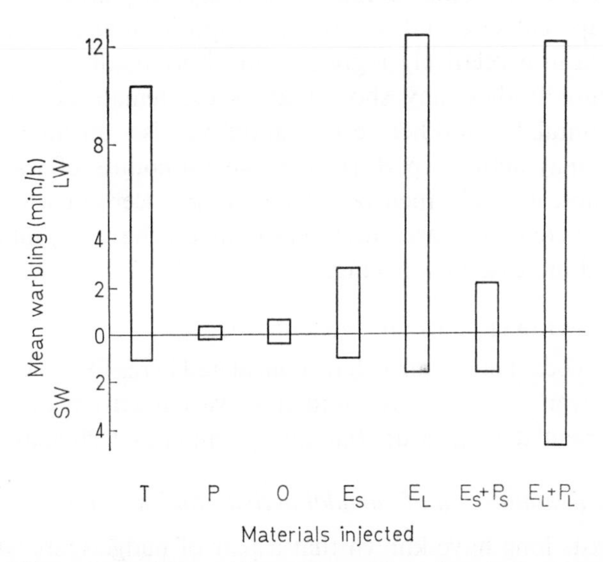

FIGURE 2. Effects of gonadal steroids on the warbling of nonisolated females. SW = soft warble, LW = loud warble, T = testosterone (1·0 mg), O = oil placebo, E_S = oestradiol (0·2 mg), E_L = oestradiol (0·4 mg), Ps = progesterone (0·05 mg), and P_L = progesterone (0·1 mg). Dosages refer to the amount per injection. Females were given six injections, one every other day. Mean amounts of warbling are based on six hours of data obtained after the third injection (Brockway, unpublished).

In summary, ovarian steroids seem to stimulate more warbling by females than by males. Indeed, both androgenic and oestrogenic materials may prompt females to perform quantities of LW and SW that are either similar to, or greater than, those performed by similarly-treated males. However, testosterone injections do stimulate the ovarian activity of some birds (van Tienhoven, 1961). Should this be true for budgerigars, then the testosterone-induced warbling of females might actually be due to the influence of increased ovarian secretions. A definite answer to this awaits tests on females receiving testosterone plus an oestrogen inhibitor, such as MER-25.

If we consider only those hormonal influences on warbling which are of primary importance in the normal life of a budgerigar, we see that the thresholds for warbling are lowered in males by androgens and in females by oestrogens. Thus budgerigar warbling, which is traditionally regarded as typical male behaviour, may be more often associated with males because they secrete relatively large quantities of androgens throughout the year. Conversely, females may be less associated with warbling because the ovary only attains rapid and full development several days prior to oviposition and remains more of the time in a relatively regressed and hormonally inactive state. If so, future studies may show that as the female becomes more sexually stimulated and her oestrogen output rises during a breeding cycle, she may actually perform extensive amounts of warbling. If she does not, it would then be interesting to ascertain what, if any, other factors (e.g. increased nestbox-oriented behaviour) inhibit such a predicted increase in warbling.

FUNCTIONAL ASPECTS OF WARBLING

Only male vocal behaviour will be considered here. Evidence suggests that male warbling not only stimulates ovarian activity and females' nestbox-oriented behaviour, but also stimulates testicular activity.

Social vocal stimulation of gonadal activity and sexual behaviour

Aviculturists long have known that a pair of budgerigars will rarely breed if it is isolated from others. Vaugien (1951) showed that vocalizations of breeding pairs stimulated the ovarian activity and egg-laying of unpaired females which were kept in continual darkness without nestboxes. He recognized that darkness may potentiate a female's physiological responses to vocal stimuli, for continual darkness might be a superoptimal substitute for the gloomy conditions to which females are exposed daily while occupying their nestboxes prior to egg-laying.

Accordingly, I tested the effects of hearing males, and of darkness within the nestbox, upon females which were provided with nestboxes, able to see and hear each other, and caged under fourteen-hour daily photoperiods (Brockway, 1962 and unpublished). Females, able to hear but not to see males, performed nestbox-oriented behaviour, attained full ovarian and oviductal development, and laid eggs within eighteen days. Similar results occurred when nestboxes with translucent lids were substituted for the usual box having a wooden lid and a dark interior. Controls, unable to hear males, showed little nestbox-oriented behaviour or ovarian and oviductal development even after five to eight weeks. Thus, male vocal behaviour appears to be more important for ovarian activity than are lighting conditions in general or within the nestbox. Since females caged under ten- to fourteen-hour daily photo-periods show no ovarian or oviductal development in response to male vocal stimulation unless they can perform nestbox-oriented behaviour, perhaps it is more correct to say that continual darkness may serve a potentiating function in lieu of the opportunity to perform nestbox-oriented behaviour.

To determine why a heterosexual pair typically fails to breed when visually and vocally isolated from others, pairs able to both see and hear others, only hear others, only see others and neither see nor hear others were compared (Brockway, 1964*c*). Results showed that the sounds and, to a lesser extent, the sight of other pairs prompted: (i) the maintenance of high levels of testicular activity (sperm and androgen production) and male courtship behaviour, and (ii) full ovarian activity, oviposition and female courtship behaviour. The four experimental situations differed vocally only in the quantity of male warbling. Males of isolated pairs performed virtually no warbling. In pairs able only to see others, distinctive quantitative individual differences in male warbling and visible courtship behaviour were closely correlated with gonadal activity. Approximately fifty per cent of these males performed moderate amounts of warbling, the remainder performed virtually none. Only the former group of males performed significant amounts of visible precopulatory activities, continued to possess active sperm- and androgen-producing testes and had mates which laid eggs. Females whose mates performed virtually no warbling and visible courtship displays, were physiologically and behaviourally indistinguishable from females of isolated pairs.

These studies show that (i) the vocal and visible behaviour of a single male is sufficient to stimulate sexually females that can see other pairs, and (ii) tendencies to warble differ markedly among individual males.

Male warbling as a stimulus for testicular activity

Casual studies suggest that both increased levels of plasma androgens and social vocal stimuli act, complementarily, to increase a male's warbling. Indeed, androgens may stimulate even an isolated male to warble extensively. This evidence, however, does not resolve the question of why a male's vocal behaviour and testicular activity decrease in the absence of social vocal stimulation.

Does the vocal behaviour of conspecifics stimulate testicular androgen secretion by (i) directly stimulating a male's hypothalamo-hypophyseal-gonadal axis or (ii) stimulating a male's warbling (through social facilitation or inductive mimesis?) which, in turn, stimulates his hypothalamo-hypophyseal-gonadal activity?

To investigate these alternatives, I compared the reproductive tracts of surgically 'devocalized' and sham-operated males able to hear others (Brockway, 1967*b*). 'Devocalized' males uttered only hoarse, rasping mono- and polysyllabic 'burp' and 'keek' sounds during the study. At best, these sounds vaguely resembled a budgerigar's threat, alarm or tuk-tuk notes: they did not resemble warbling or other budgerigar vocalizations. Cages were placed so that both groups of males could hear these abnormal utterances of the 'devocalized' males. On the day of surgery, there was no significant difference ($P > 0{\cdot}20$ or $0{\cdot}10$) between the 'devocalized' and sham-operated (control) groups regarding testicular or vasa deferentia dimensions or the number of males (all 'devocalized' males and 71 per cent of the controls) yielding semen upon milking. Three days after surgery, all controls had resumed their typical vocal performances including much LW. Birds were examined 21 and 35 days after surgery: similar results were obtained at both times. The testes and vasa deferentia of controls were of the same sizes as they had been before surgery ($P > 0{\cdot}20$), whereas those of 'devocalized' males were significantly reduced ($P < 0{\cdot}005$). Furthermore, none of the 'devocalized' males but 86 per cent of the controls yielded semen upon milking.

These results indicate that only those males which performed typical budgerigar sounds, maintained high levels of testicular activity. Hearing the vocalizations of others was not, in itself, a sufficient stimulus. However, these results are currently open to two possible interpretations. Either a male which is abruptly able to perform only abnormal vocal behaviour ('devocalized' males) undergoes a stress sufficient to result in testicular regression, or some factor stemming from a male's vocal performance(s) is more directly responsible for promoting his testicular activity than is social vocal stimulation. Although not conclusive, the following evidence favours the latter interpretation: 'devocalized' males showed no overt symptoms of stress (e.g. decreased food and water consumption, loss in body weight, or abnormal feather postures or visible behaviour patterns). Also, both 'devocalized' and sham-operated males performed some homosexually oriented visible precopulatory behaviour, an activity typical of healthy male budgerigars which are caged without females.

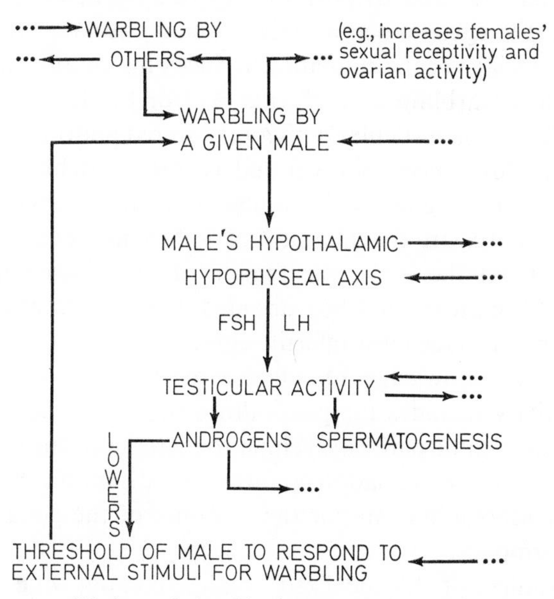

FIGURE 3. Some possible interrelationships between social vocal stimulation (warbling), a male's vocal performance and a male's testicular activity. Any ellipsis (. . .) denotes additional and/or unknown factors which may be inhibitory or stimulatory in effect.

Of the vocalizations performed typically by males that possess large active testes, only LW was unanimously and uniformly performed by the sham-operated males: SW, tuk-tuk and whedelee were not. Moreover, LW is mainly influenced by a social circumstance (warbling by other budgerigars) that stimulates full gonadal activity, whereas SW, tuk-tuks and whedelees are influenced more by a social circumstance (precopulatory interactions) that is not a requisite for full gonadal activity. Accordingly, the performance of LW may well be the source of stimuli which promote the performer's testicular activity.

Possible interrelations between social and internal factors affecting male sexual activity are shown in Figure 3. The warbling of others (perhaps LW) stimulates a male's vocal performance which, in turn, both adds to the social vocal stimulation of the group and promotes his own testicular androgen output and spermatogenesis. Androgens lower a male's threshold for performing visible and vocal courtship displays—including warbling. These displays stimulate a female to participate in copulation and stimulate full ovarian activity, nestbox-oriented behaviour and oviposition. Spermatogenesis results in the laying of fertile eggs. When a male is vocally isolated from others, the lack of social vocal stimulation reduces his output of warbling. Either the less warbling he performs, and/or the less he hears as a result, decreases his testicular activity. Lowered androgen levels then reduce his performances of vocal and visible courtship displays. As his vocal output decreases, it results in further reduction of his testicular activity. In the absence of sufficient vocal and visible stimulation from the male, the female of an isolated pair is not stimulated to perform nestbox-oriented behaviour, participate in copulation or produce even infertile eggs.

Our current knowledge about the mechanisms for increasing gonadal activity includes the possibilities that vocal behaviour may act (via hypothalamic pathways) either to increase the production and secretion of pituitary gonadotrophins or to disinhibit either the release of gonadotrophins and/or the response of the gonads to these pituitary hormones.

These studies of 'devocalized' males contain one additional implication that is worth noting. Previous reports that the performance of specific behaviour patterns prompts anterior pituitary gonadotrophin or gonadal hormone secretions in the performer concern only parental behaviour (for review, see Lehrman, 1961). Other reports of behaviourally increased gonadal activity concern interactions in which the sexual behaviour of one individual affects the endocrine activity of the other (e.g. Brockway, 1965; Burger, 1942, 1953; Erickson and Lehrman, 1964; Lehrman *et al.*, 1961; Matthews, 1939; Polikarpova, 1940; Shoemaker, 1939; Warren and Hinde, 1961). If the testicular regression of devocalized males is not a 'stressogenic artifact', then this experiment with budgerigars may be the first to illustrate the self-stimulation or disinhibition of an individual's gonadotrophin secretions or gonadal activity by his performance of a non-parental behaviour. It would be interesting to discover if any sounds of other birds might, in part, have a similar function.

Male warbling as a stimulus to female reproductive activities

Being able to hear males is required for full ovarian activity in females which are caged in continual darkness without nestboxes or in females which are caged with nestboxes under ten- to fourteen-hour daily photoperiods (Vaugien, 1951; Ficken *et al.*, 1960; Brockway, 1962). No other male-female interactions are needed. Qualitative and quantitative parameters of such vocal stimulation were investigated by playing tape recordings of different male vocalizations to individually-caged females, able to hear each other. All females initially possessed small inactive reproductive organs. Vocalizations were played for twenty-one consecutive days. On day 22, I examined the ovary and oviduct of non-laying females.

Initial studies. Females with prior breeding experience were caged in continual darkness without nestboxes. Groups were exposed to solos of specific vocalizations for six hours daily. Results, shown in Figure 4, indicate that only SW greatly stimulated ovarian activity and egg-laying. The ovarian activity of females hearing other vocalizations, including LW, was no different from that of control groups hearing either all or no vocalizations.

This suggests that LW and SW may have different functions as stimuli to gonadal activity: the former affecting males and the latter stimulating females. Also, hearing the vocalizations of only one male at a time appears to be sufficient to stimulate (a female's) ovarian activity. Thus if a male were sufficiently stimulated, he alone would be capable of stimulating the female: this is supported by data on pairs that were able only to see others (Brockway, 1964*c*).

To explore further its stimulating characteristics, SW was arbitrarily divided into a part containing mainly high-pitched components and one containing mainly low-pitched components. Each part, containing similar amounts of raspy and melodious sounds, was played for either three or six hours daily. It was discovered that either component half of SW similarly stimulated ovarian activity ($P>0{\cdot}10$) but that daily presentation for six hours was more effective than for three hours ($P<0{\cdot}025$). There was no significant effect due to any interactions between components and duration ($P>0{\cdot}20$) (Brockway, 1965).

These initial studies raised the following kinds of questions. Would the various vocalizations be more or perhaps less effective when played for longer than six hours daily? Would hearing more than one male at a time be more stimulating than hearing solos? Would the response of virgin females be equal to that of experienced females? Which, if any, male vocalizations might stimulate the ovarian activity

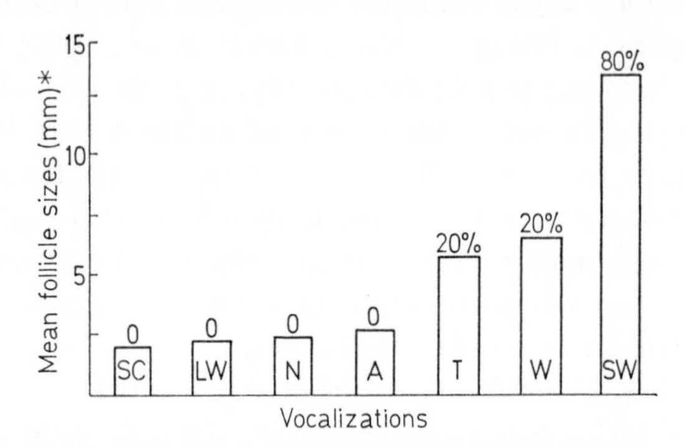

FIGURE 4. Ovarian response to taped male vocalizations. Histobar codes: SC = Squacks and chedelees, LW = loud warble, N = no vocalizations, A = equal amounts of all vocalizations, T = tuk-tuks, W = whedelees, and SW = soft warble. The percent above each histobar notes the number of females, in that group, laying at least one egg before the experiment ended but it includes one female of Group SW who had an ovum ready to ovulate on the last experimental day. Group SW differed significantly ($P<0.01$) from any other group. None of the other groups differed significantly ($P>0.20$) from each other (after Brockway, 1965).

and the nestbox-oriented behaviour of females that were housed under more natural ten- to fourteen-hour daily photo-periods? Answers, to date, are presented in the remainder of this chapter.

Experiential factors. Only females kept in continual darkness without nestboxes were studied, for two reasons. The vocal stimulation of ovarian activity is best conducted in the presence of nestboxes unless females are kept in continual darkness. Few virgins perform nestbox-oriented behaviour and lay eggs as soon as do experienced females in response to male vocal stimulation under twelve-hour photoperiods.

After fledging, all females were prevented from seeing males or nestboxes. As adults, however, experienced females interacted sexually with a mate, performed nestbox-oriented behaviour, possessed a fully active ovary, laid eggs and raised young. Virgin females lacked all such experiences.

Six groups of virgin and experienced females were exposed to either solos or quartets of SW, LPSC or LW. There was no significant difference in ovarian activity (diameters of ovarian follicles or of oviducts) between virgin and experienced females which were exposed to the same vocal stimulus, regardless of the nature of that stimulus. Thus, factors involved in prior breeding experience do not seem to affect the physiological response of females to taped male vocal stimuli, under these experimental conditions.

Interactions between qualitative, quantitative and photoperiodic factors affecting ovarian activity

These studies compared the effects of hearing taped solo or quartet versions of each warble (LW, SW or LPSC) for either six or twelve hours daily upon the ovarian activity of females which were caged either in continual darkness and without nestboxes (D-females) or under twelve-hour daily photoperiods and with nestboxes (experienced L-females). Data are summarized in Table 2. First, let us consider solo warbles.

When solos were played for 6 h/day, only SW significantly increased ovarian activity in D-females. This replication supports earlier data (see Figure 4). When solos were played for 12 h/day, however, all three warbles were equally effective in increasing the ovarian activity of D-females. Moreover, SW, which was more efficacious when played for 6 rather than 3 h/day, was no more effective in stimulating ovarian activity when it was played for 12 rather than 6 h/day. This suggests that SW may attain maximal stimulatory value near the quantity of 6 h/day. So further increases in its daily quantity may produce no greater effect in regard to the ovarian activity of D-females. The ovarian activity of L-females was unaffected by solos of any warble played for 6 h/day, and it was only significantly increased by LPSC when solos were played for 12 h/day.

Only LPSC was effective when quartets were played for 6 h/day. LPSC quartet stimulated the ovarian activity of both L- and D-females. Comparing warbles played for 6 h/day to both L- and D-females indicates that a quartet of SW was less stimulating than a solo version whereas a quartet of LPSC was more stimulating than a solo version: this applies to both L- and D-females. Furthermore, a quartet of LPSC and a solo of SW were equally effective in promoting ovarian activity.

When D-females heard quartets for 12 h/day, the results were similar to those obtained when they heard solos for 12 h/day. Namely, all three warbles were equally effective in increasing ovarian activity. But results for L-females were quite different. Whereas solo LPSC

TABLE 2. *The relative efficacy of different types and quantities of taped male warbling in stimulating the ovarian activity of females housed under two different photoperiodic conditions for 21 days.*[1]

Taped stimulus	Solo 6 h/day		Solo 12 h/day		Quartet 6/day		Quartet 12/day	
	D[2]	L[2]	D	L	D	L	D	L
Loud warble	—	—	S	—	—	—	S	S
Soft warble	S	—	S	—	—	—	S	—
Loud-plus soft components-warble	—	—	S	SS	S	S	S	—
None	—	—	—	—	—	—	—	—

[1] Ovarian activity was represented as the sum of the diameters of the two largest follicles per bird. Mean values were statistically treated by analysis of variance tests and individual degree of freedom comparisons. Data are represented by the following symbols: S, Ovarian activity was significantly greater ($P<0.05$) than that of control groups hearing no taped vocalizations; –, ovarian activity was not significantly different from that of control groups. Furthermore, all '–' values were not significantly different from each other and likewise all 'S' values were not significantly different from each other. The single mean value represented by 'SS' was significantly greater than any 'S' value. N per group = eight to twelve adults.

[2] D, females caged in continual darkness without nestboxes; L, females caged with nestboxes under 12-hour daily photoperiods.

was the only effective solo stimulus, a quartet of LPSC was ineffective. Surprisingly, LW was the only effective quartet when played for 12 h/day to L-females.

Data for all quantities of LPSC and SW for 6 h/day seems to illustrate roughly the famous adage that 'more of the same thing is not necessarily better'. A more erudite or biologically appropriate explanation must await further knowledge concerning the temporal alterations in the relative quantities of each warble during typical breeding cycles. Indeed, only for D-females can one generalize that presenting a vocalization daily for twelve hours is more effective than for half this amount.

In summary, the efficacy of these vocal stimuli seems to depend upon interactions among quality, quantity and the environment in which the tested individuals are housed.

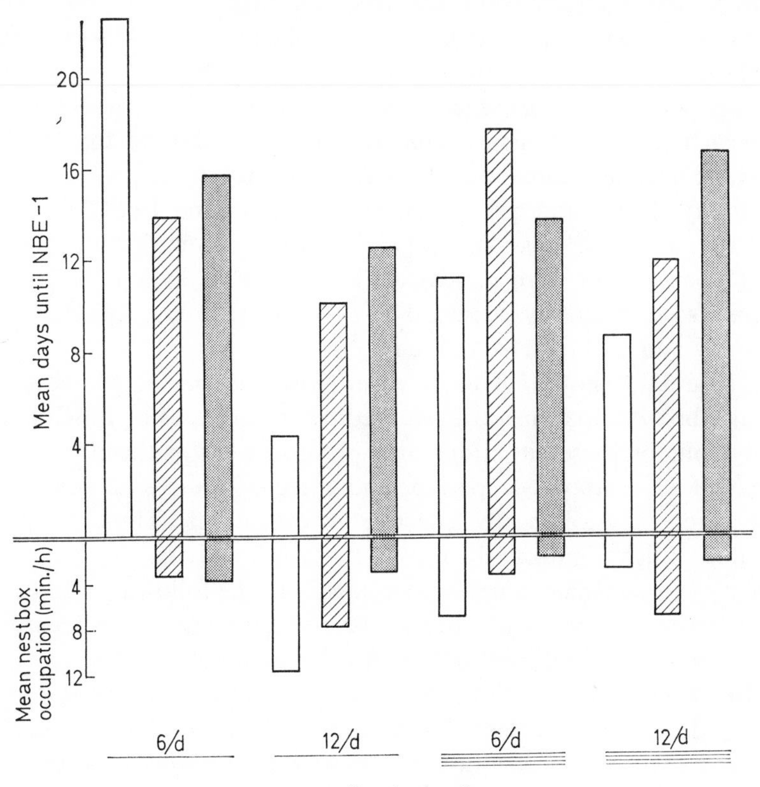

FIGURE 5. Effects of different types and quantities of taped warbling on the start and amount of nestbox occupation by females. Ordinates: above, mean number of experimental days that elapsed before a female first enters the nestbox; below, mean values for the average amount of nestbox occupation (min./hr) per female for the first 20 experimental days. Abscissa: the quantities of taped warble heard by females were solo (—) or quartet (☰) versions played for six (6/d) or twelve (12/d) hours daily. Histobar codes: □ = Loud plus soft components warble, ▨ = Loud warble, and ▒ = Soft warble (Brockway, unpublished).

Warbling and nestbox-oriented behaviour

To investigate a possible cause-effect relationship between vocally stimulated ovarian activity and nestbox-oriented behaviour, nestbox-oriented behaviour was recorded daily for females in all L situations of the study described in the previous section. The ovarian activity in females of L situations was significantly increased by only three vocal stimuli. These stimuli were LPSC solo for 12 h/day, LPSC quartet for 6 h/day and LW quartet for 12 h/day. They will henceforth be termed 'physiologically effective' vocalizations.

Main results (Figure 5) concern the mean dates of NBE-1 (the first time a female enters the nestbox) and group means of the nestbox occupation (min./h) averaged over the first twenty experimental days for each female. Statistical analyses of results showed that NBE-1 was significantly stimulated by LW solo for 6 h/day, all solos for 12 h/day, LPSC and SW quartets for 6 h/day and LPSC and LW quartets for 12 h/day. Amounts of nestbox occupation during the first twenty days were increased most by LPSC solo 12 h/day but were also significantly increased by LW solo 12 h/day, LPSC quartet 6 h/day and LW quartet 12 h/day.

These data show that eight vocal situations promoted the onset of nestbox occupation, as evidenced by mean dates of NBE-1. Yet only four of these situations also prompted a significantly greater amount of nestbox-occupation, and three of these situations were the three 'physiologically effective' vocalizations. Thus the vocal stimuli which significantly increase ovarian activity also stimulate the larger quantities of nestbox occupation. The following additional information helps to identify a cause-effect relationship regarding ovarian activity and nestbox-oriented behaviour.

Injections of oestradiol cypionate with or without progesterone will induce nestbox-oriented behaviour by females which are unable to hear any male vocalizations (Brockway, unpublished). In sufficient amount, these hormones can provoke just as much nestbox occupation as the 'physiologically effective' vocal stimuli did. Such information suggests that vocal stimuli may induce nestbox-oriented behaviour by acting to increase plasma levels of oestrogen and, perhaps, progesterone. This suggestion is further supported by recent evidence showing that injections of MER-25 (a competitive inhibitor of oestrogenic activity) to twelve females, hearing LPSC solo 12 h/day for twenty-one days, prevented all females from performing any notable amount of nestbox-oriented behaviour whatsoever (Brockway, unpubl.).

The start of nestbox occupation, NBE-1, seems to be encouraged by a variety of vocal situations involving all three warbles, including some that did not significantly increase the sizes of ovarian follicles.

oviductal diameters or ovarian and oviductal weights. Perhaps such an early phase of nestbox-oriented behaviour may be stimulated by such a small increase in ovarian activity that this activity could not be adequately revealed by gross measurements such as oviductal diameters and weights. Since the extensive nestbox occupations were, for the most part, stimulated by those vocal situations that also induced significantly increased ovarian activity, perhaps such a degree of occupation requires relatively greater plasma levels of ovarian steroids, primarily oestrogen.

GENERAL CONCLUSIONS

Budgerigars are stimulated to warble by hearing others. Under appropriate environmental conditions, the hearing and/or performance of warbling promotes full gonadal activity in both sexes. Increased titres of circulating gonadal steroids lower the thresholds for: (i) warbling by both sexes, thereby increasing the quantity of vocal stimulation; (ii) performing other, visible, courtship behaviour which culminates in successful copulation and thereby produces zygotes, and (iii) engaging in nestbox-oriented behaviour by females. Under ordinary photoperiodic conditions, full ovarian activity depends upon a female's performance of nestbox-oriented behaviour and hearing extensive quantities of warbling. Thus, such vocal-endocrinological feedback interactions in both sexes play an essential role in the breeding of this parrot species.

ACKNOWLEDGEMENTS

Research was supported by a grant from the Chapman Memorial Fund of the American Museum of Natural History, a grant-in-aid from the Sigma-Xi RESA Fund, a National Science Foundation institutional grant from Western Reserve University and a National Science Foundation grant (GB-3191). Seed was generously provided by the R. T. French Company of Rochester, N.Y. The generous supply of MER-25 by the William S. Merrell Company is also appreciated. Gratefully, I thank Drs Alan P. Brockway, William C. Segmiller and the late William E. Burkhart for their invaluable help.

REFERENCES

ANDREW, R. J. (1963) Effect of testosterone on the behaviour of the domestic chick. *J. comp. physiol. Psychol.* **56**, 933–40.

ANTHONY, A., ACKERMAN, E. and LLOYD, J. A. (1959*a*) Noise stress in laboratory rodents, I: Behavioral and endocrine response of mice, rats and guinea pigs. *J. acoust. Soc. Am.* **31**, 1430–7.

ANTHONY, A. and HARDERODE, J. E. (1959*b*) Noise stress in laboratory rodents, II: Effects of chronic noise exposures on sexual performance and reproductive function of guinea pigs. *J. acoust. Soc. Am.* **31**, 1437–40.

ARMSTRONG, E. A. (1963) *A Study of Bird Song*. Oxford University Press, New York.

BIRÓ, J., SZOKOLAI, V. and KOVÁCH, A. (1959) Some effects of sound stimuli on the pituitary-adrenocortical system. *Acta Endo.* **31**, 542–52.

BOND, J., WINCHESTER, C. F., CAMPBELL, L. E. and WEBB, J. C. (1963) Effects of loud sounds on the physiology and behavior of swine. *U.S. Dept. Agric. Res. Serv. Tech. Bull.* **1280**, 1–17.

BRERETON, J. (1966) The evolution and adaptive significance of social behaviour. *Proc. Ecol. Soc. Aust.* **1**, 14–30.

BROCKWAY, B. F. (1962) The effects of nest-entrance positions and male vocalizations on reproduction in budgerigars. *Living Bird, First Ann.*
(1964*a*) Ethological studies of the budgerigar (*Melopsittacus undulatus*): *Cornell Lab. Ornith.* **1**, 93–101.
reproductive behavior. *Behaviour* **23**, 294–324.

(1964*b*) Ethological studies of the budgerigar (*Melopsittacus undulatus*): non-reproductive behaviour. *Behaviour* **22**, 193–222.

(1964*c*) Social influences on reproductive physiology and ethology of budgerigars (*Melopsittacus undulatus*). *Anim. Behav.* **12**, 493–501.

(1965) Stimulation of ovarian development and egg laying by male courtship vocalization in budgerigars (*Melopsittacus undulatus*). *Anim. Behav.* **13**, 575–8.

(1967*a*) Interactions among male courtship warbling, photoperiodic and experiential factors in stimulating the reproductive activity of female budgerigars (*Melopsittacus undulatus*). *Am. Zool.* **7**, 215.

(1967*b*) The influence of vocal behavior on the performer's testicular activity in budgerigars (*Melopsittacus undulatus*). *Wilson Bull.* **79**, 328–34.

(1968) Influences of sex hormones on the loud and soft warbles of male budgerigars. *Anim. Behav.*, **16**, 5–12.

BURGER, J. W. (1942) The influence of some external factors on the ovarian cycle of the female starling. *Anat. Rec.* **84**, 518.

(1953) The effect of photic and psychic stimuli on the reproductive cycle of the male starling, *Sturnus vulgaris*. *J. exp. Zool.* **124**, 227–39.

COLLIAS, N. E. (1960) An ecological and functional classification of animal sounds. In *Animal Sounds and Communication* (W. E. Lanyon and W. N. Tavolga, eds), pp. 368–91, Washington D.C.

DAVIS, J. (1958) Singing behaviour and the gonad cycle of the rufous-sided towhee. *Condor* **60**, 308–36.

EGAMI, N. (1959) Preliminary note on the induction of the spawning reflex and oviposition in *Oryzias latipes* by the administration of neurohypophyseal substances. *Annot. Zool. Jap.* **32**, 13–17.

EISNER, E. (1960) The relationship of hormones to the reproductive behaviour of birds, referring especially to parental behaviour: a review. *Anim. Behav.* **8**, 155–79.

ERICKSON, C. J. and LEHRMAN, D. S. (1964) Effect of castration of male ring doves upon ovarian activity of females. *J. comp. physiol. Psychol.* **58,** 164–6

FICKEN, R. W., VAN TIENHOVEN, A., FICKEN, M. S. and SIBLEY, F. C. (1960) Effect of visual and vocal stimuli on breeding in the budgerigar (*Melopsitacus undulatus*). *Anim. Behav.* **8,** 104–6.

HENKIN, R. I. and KNIGGE, K. M. (1963) Effect of sound on the hypothalamic-pituitary-adrenal axis. *Am. J. Physiol.* **204,** 710–14.

HINDE, R. A. (1965) Interaction of internal and external factors in integration of canary reproduction. In *Sex and Behavior* (F. A. Beach, ed.), pp. 381–415. John Wiley and Sons, New York.

(1966) *Animal Behaviour: A Synthesis of Ethology and Comparative Psychology.* McGraw-Hill Book Co., New York.

HOAR, W. S. (1962) Reproductive behavior of fish. *Gen. Comp. Endo. Suppl.* **1,** 206–16.

LANYON, W. E. and TAVOLGA, W. N. (eds.). (1960). *Animal Sounds and Communication.* Washington, D.C.

LEHRMAN, D. S. (1959) Hormonal responses to external stimuli in birds. *Ibis* **101,** 478–96.

(1961) Gonadal hormones and parental behaviour in infrahuman vertebrates. In *Sex and Internal Secretions* (W. C. Young, ed.), pp. 1268–382, vol. II. Williams and Wilkins Co., Baltimore Md.

(1965) Interaction between internal and external environments in the regulation of the reproductive cycle of the ring dove. In *Sex and Behavior* (F. A. Beach, ed.), pp. 355–80. John Wiley and Sons, New York.

LEHRMAN, D. S., BRODY, P. N. and WORTIS, R. P. (1961) The presence of the mate and of nesting material as stimuli for the development of incubation behavior and for gonadotrophin secretion in the ring dove (*Streptopelia risoria*). *Endocrinology* **68,** 507–16.

LEONARD, S. L. (1939) Induction of singing in female canaries by injections of male hormone. *Proc. Soc. exp. Biol. Med.* **41,** 229–30.

LOTT, D. F. and BRODY, P. N. (1966) Support of ovulation in the ring dove by auditory and visual stimuli. *J. comp. physiol. Psychol.* **62,** 311–13.

MARLER, P., KREITH, M. and WILLIS, E. (1962) An analysis of testosterone-induced crowing in young domestic cockerels. *Anim. Behav.* **10,** 48–54.

MATTHEWS, L. (1939) Visual stimulation and ovulation in pigeons. *Proc. R. Soc. B,* **126,** 557–60.

NOBLE, G. K. and ZITRIN, A. (1942) Induction of mating behaviour in male and female chicks following injection of sex hormones. *Endocrinology* **30,** 327–34.

PICKFORD, G. E. (1952) Induction of a spawning reflex in hypophysectomized killifish. *Nature* **170,** 807.

(1954) The response of hypophysectomized male killifish to purified fish growth hormone, as compared with the response to purified beef growth hormone. *Endocrinology* **55,** 274.

POLIKARPOVA, E. (1940) Influence of external factors upon the development of the sexual gland of the sparrow. *C. R. Acad. Sci. U.R.R.S.* **26,** 91–5.

REY, P. (1948) Sur le déterminisme hormonal du réflexe d'embrassement chez les mâles de batraciens anoures. *J. Physiol. Paris* **40,** 292A–293A.

SACKLER, A. M., WELTHAM, A. S. and JURTSHUK, P. Jr. (1960) Endocrine aspects of auditory stress. *Aerospace Med.* **31**, 749–59.

SHOEMAKER, H. H. (1939) Effect of testosterone propionate on behavior of the female canary. *Proc. Soc. Exp. Biol. N.Y.* **41**, 299–302.

THORPE, W. H. (1961) *Bird Song*. Cambridge University Press, Cambridge.

VAN TIENHOVEN, A. (1961) Endocrinology of reproduction in birds. In *Sex and Internal Secretions*, pp. 1088–172 vol. II (W. C. Young, ed.), Williams and Wilkins Co., Baltimore Md.

VAUGIEN, L. (1951) Ponte induite chez la Perruche ondulée maintenue à l'obscurité et dans l'ambience des volières. *C. R. Acad. Sci.* **232**, 1706–8.

(1952) Sur le comportement sexuel singulier de la Perruche ondulée maintenue à l'obscurité. *C.R. Acad. Sci.* **234**, 1489.

(1953) Sur l'apparation de la maturité sexuelle des jeunes perruches ondulée mâles soumises à diverses condition d'eclairment: Le développement testiculaire est plus rapide dans l'obscurité complète. *Bull Biol.* **87**, 274–86.

WARREN, R. P. and HINDE, R. A. (1961) Does the male stimulate oestrogen secretion in female canaries? *Science* **133**, 1354–5.

WILHELMI, A., PICKFORD, G. E. and SAWYER, W. (1955) Initiation of the spawning reflex in *Fundulus* by the administration of fish and mammalian neurohypophyseal preparations and synthetic oxytocin. *Endocrinology* **57**, 243.

YOUNG, W. C. (1961*a*) (ed.) *Sex and Internal Secretions*, vol. II. Williams and Wilkins Co., Baltimore Md.

(1961*b*) The hormones and mating behaviour. In *Sex and Internal Secretions*, (W. C. Young, ed.), pp. 1173–239, vol. II, Williams and Wilkins Co., Baltimore Md.

ZONDEK, B. (1967) Effects of auditory stimuli on reproduction. In Ciba, Foundation Study Group No. 26, Wolstenholme and Connor, pp. 4–19, Little, Brown and Co., Boston Mass.

PART D
FUNCTIONAL ASPECTS

INTRODUCTION

Each item of behaviour forms a link in a nexus of events which precede and succeed it. If we focus on the behaviour, we think of the former as causes and the latter as consequences. If we shift our attention to, say, the neural events, they become causes and the behaviour a consequence. Reciprocally, the behaviour may be a cause of what follows. Brockway's article (chapter 7) treated behaviour as both consequence and cause: in this section we shall discuss further aspects of the consequences of avian vocalizations.

The consequences of behaviour can in principle be classified according to their adaptiveness in terms of natural selection: some are neutral, some harmful, some beneficial. We shall be concerned primarily with the latter—namely with the functions of avian vocalizations. It is worth emphasizing, perhaps that the word 'function' is used in two ways in the study of behaviour. Sometimes, and particularly by psychologists, it is used to refer backwards in time: 'Behaviour Y is a function of X' means that a (mathematical) relation exists between X and Y, and is a sophisticated way of referring to causal relations without implication about the extent to which those relations are direct or indirect. Evolutionarily-minded biologists, on the other hand, use 'function' to refer forwards in time: the function of a type of behaviour refers to a particular type of consequence—usually one through which natural selection acts to maintain the behaviour in the repertoire of the species.

It is a matter of opinion whether all beneficial consequences are to be regarded as functions, or whether the term should be limited to beneficial consequences through which selection acts. To take a hypothetical example, the body feathers both act as insulators and carry colour markings—insulation and colouration are both beneficial consequences of having feathers. But if the natural variability in the number of feathers is within the range which would affect insulation but not such as to affect colouration, natural selection presumably does not act through the latter to maintain the number of feathers around the present level. Again, the preservation of biological requirements and the prevention of epidemic diseases have both been suggested as beneficial consequences of territorial behaviour. If the variability in territory size is sufficient to affect the availability of necessities, but not the spread of disease, then it could be said that the former is a function and the latter is not (Hinde, 1956). This difficulty disappears if we use the term function only for differences, real or implied. Thus when we speak of 'The function of cliff-nesting in kittiwakes', we imply 'as compared with the ground-nesting of other gulls'; and when we speak of the function of a certain call, we imply a hypothetical population not possessing it. On this view, the prevention of disease could be a function of territorial behaviour if the differences between territorial and hypothetical non-territorial populations were considered, but not in the circumstances described above.

Be that as it may, a function is usually taken to be established if a beneficial consequence has been proven. This is sometimes no easy task. It is one matter to presume that, because a pattern of behaviour occurs, it must be adaptive; but quite another to nail down the precise consequences which make it so. Territorial behaviour is an obvious case: the precise consequences which should be labelled as functions are often obscure, and clearly differ between species. Bird song is another example, and there have been those who would regard it, or some of its attributes, as outside the sphere of natural selection.

The first indication of function often comes from contextual information. Just as the observation that birds use their wings primarily when airborne implies a function to do with flight, so does the association of a particular call with aggressive encounters raise the possibility that it functions in communicating with rivals. Mulligan and Olsen (chapter 8) use this type of argument in their study of canary calls: as they point out, it is necessary to distinguish the 'message', inferred from aspects of the calling bird, its behaviour and the context in which it is calling, from the 'meaning', which must be inferred from the behaviour of the recipient. Mulligan and Olsen also show that the calls have limited variability—a characteristic which presumably favours a simple signalling system (Morris, 1957). The field study by Hooker and Hooker (chapter 9) also uses contextual information, but in a rather different way. The peculiar nature of antiphonal singing, as well as the close habitat and the permanence of the pair-bond of the species in which it is found, imply a function in relation to that pair-bond. The implications of this work for studies of development have already been mentioned (page 22).

The descriptive method used in these studies is always necessary to provide ideas about both the causation and the function of behaviour, and sometimes it is sufficient in itself (Tinbergen, 1967). Sometimes, however, experimental verification is both desirable and possible. For instance, that selection of the proper background is adaptive for cryptic animals can be shown by varying the background and assessing differential predation (references in Cott, 1940). The value of egg-shell removal by gulls as an anti-predator device has been shown by comparing the predation on unhatched eggs with and without a nearby egg-shell (Tinbergen, 1967). Brockway's (chapter 7) study involved such an approach in that it included experimental verification of the role of certain budgerigar vocalizations in inducing reproductive development in the sex partner.

Falls (chapter 10), like Tinbergen, has carried this approach into the field, and used it not so much to identify the consequences of song on other individuals as to identify the separate roles of the different aspects of the song. Using information in its every-day (and therefore non-quantitative) sense, Marler (1956, 1961; see also Morris, 1946, and Cherry, 1957) has argued that it is possible to determine the 'information content' of a signal from a knowledge both of the response of other animals to that signal and of the other circumstances in which that same response is given. Thus since chaffinch song is given primarily by unmated males within their territories, and attracts unmated females, Marler argues that the male's song conveys to a female information that the singer is '*an unmated male chaffinch in reproductive condition, in possession of a territory* (within which nesting will take place), *who is close to a location occupied by the female at the same time as she is there*' (Marler, 1961). Since females later come to respond preferentially to the song of their own mates, they presumably learn the song's individual characteristics, and the song then carries the additional information that the singer is a *particular* individual male. The information conveyed to a male could be spelled out similarly. Although such abstractions should not be taken to imply that the bird makes an analysis in similar terms, they have proved a useful tool for understanding the systems of communication used by animals. This is borne out by the experiments of Falls and his colleagues (chapter 10), which are concerned with relating the particular aspects of the songs of white-throated sparrows to the different items of information carried. By playing artificial songs, or songs whose characteristics varied within the range normally found in the field, and assessing the responses of wild males to these songs, they identified which aspects of the song indicated the species of the singer, which its individual identity and which its motivational state. This study supports the view that the individual differences in bird sounds, which are known to occur in a variety of species (e.g. Hutchison *et al.*, 1968) are not a mere by-product of development, but do have adaptive significance.

Vince's (chapter 11) study also uses an experimental approach in the study of function, showing that the sounds made by the embryos of certain species promote synchronous hatching—a consequence presumably beneficial at least in nidifugous birds. Her study provides a link with those of experimental embryologists and those interested in the early development of behaviour (e.g. Hamburger and Oppenheim, 1967).

One aspect of functional studies of vocalizations which receives little attention in this volume must be mentioned briefly. If the consequences of a song or call are beneficial, then natural selection can operate on the form of the sound so that it can perform its function to best advantage. Detailed examination of avian vocalizations (e.g. Marler, 1957, 1959) has revealed the extent to which this is so—warning cries have characteristics which make them difficult to locate, unlike contact notes which are usually easy; and song carries over long distances while pre-copulatory notes have a relatively short range. This theme of the adaptedness of sound signals for their particular functions recurs again in the essay of Hall-Craggs (chapter 16).

REFERENCES

CHERRY, C. (1957) *On Human Communication.* Wiley, New York.

COTT, H. B. (1940) *Adaptive Coloration in Animals.* Methuen, London.

HAMBURGER, V. and OPPENHEIM, R. (1967) Prehatching motility and hatching behaviour in the chick. *J. exp. Zool.* **166**, 171–204.

HINDE, R. A. (1956) The biological significance of the territories of birds. *Ibis* **98**, 340–69.

HUTCHISON, R. E., STEVENSON, J. G. and THORPE, W. H. (1968) The basis for individual recognition by voice in the sandwich tern. *Behaviour* (in press).

MARLER, P. (1956) The voice of the chaffinch and its function as a language. *Ibis* **98**, 231–61.

(1957) Specific distinctiveness in the communication signals of birds. *Behaviour* **11**, 13–39.

(1959) Developments in the study of animal communication. In *Darwin's Biological Work* (P. R. Bell, ed.). Cambridge University Press, London.

(1961) The Logical analysis of animal communication. *J. theoret. Biol.* **1**. 295–317.

MORRIS, C. W. (1946) *Signs, Language and Behavior*. Prentice Hall, New York.

MORRIS, D. (1957) 'Typical Intensity' and its relation to the problem of ritualisation. *Behaviour* **11**, 1–12.

TINBERGEN, N. (1967) Adaptive features of the Black-headed Gull. In *Proc. 14th Int. Orn. Congr.* (D. W. Snow, ed.). Blackwell, Oxford, pp. 43–59.

8. COMMUNICATION IN CANARY COURTSHIP CALLS

by **JAMES A. MULLIGAN, S. J. and KENNETH C. OLSEN**
Department of Biology, Saint Louis University, Saint Louis, Missouri

INTRODUCTION

The study of language, of stereotypy in the communicatory signals of animals, and of the motivation associated with these signals, continue to occupy a central place in ethological research. These aspects of communication, of special significance in social behaviour, are especially amenable to quantitative and experimental study in birds. However, the field recording and quantitative description of the vocal repertoire of bird species is still at an early stage. This fact is readily understood by anyone who has attempted to collect the data required in this type of study. To obtain an adequate sample of the various calls used by a particular species is a sufficient task in itself, but to interpret the significance of these calls in addition requires a great deal of specific information on the current activities of the individuals concerned, their social status, and so forth. The scattered studies to date on this subject have shown the size and general character of the vocal repertoire in many species, together with information of various kinds on the context of the sounds. The excellent reviews of this subject in Thorpe (1961) and Armstrong (1963) lead to the formation of interesting hypotheses, many of which can be tested simply by a better description of the sounds and more detailed information on their environmental and behavioural context.

In spite of certain drawbacks the common domestic canary is an excellent subject from which to collect the above kinds of data. This is a highly vocal species which breeds well in captivity and behaves in undisturbed fashion under observation in experimental conditions. It is not known how blunted the canary responses have become under several hundred years of association with man, but it is not really tame, and there is no reason to think that its calls and their functions have undergone any drastic change. Indeed, since selection pressure for breeding ability has never relaxed in this species one would think that perhaps courtship vocalizations have not greatly changed. However even if they had, it would be of real interest to record these calls, determine the extent of variation in their physical structure and assess their communicatory value. This is particularly true since the breeding behaviour of the canary is coming to be well understood (Hinde, 1965).

The limited study of canary courtship vocalizations which is here described in summary form was designed to answer the following questions: How stereotyped are these calls? Is there intergrading between them? What information do they transmit? As an indication of the need for quantitative data in this matter, even when the answer to the first question is obtained, an appeal to the literature for comparative information shows that stereotypy is seldom expressed in quantitative terms. Therefore it is of some value to obtain an objective statement of variation based on the physical analysis of a statistically useful sample. The second question follows. It has been said (Marler, 1965) that bird sound signals, unlike those characteristic of primates, tend to vary in a limited fashion within what may be called discrete types. If this is so, as it appears to be, quantitative verification and specific detail on this question are needed. This information is useful both for comparative and evolutionary studies and for an understanding of the stimulus or communicatory value of each call type. The question can here be pursued under ideal conditions. Smith (personal communication), following Morris (1946), has proposed that the study of communication value can be pursued by dividing and characterizing independently the 'message' of the sender and the 'meaning' taken by the receiver. This approach, used by Smith (1966) in his study of communication and vocal displays in Tyrannid flycatchers, has the advantage of splitting the empirical determination of signal function into two aspects. These are first the behaviour and total context in which the signal is given, and second, the response observed in the receiver. Therefore information on the first can be tabulated without the second, which is sometimes very useful; and the meaning of a particular message can vary in this scheme according to the context. Thus message and meaning are not necessarily coextensive. The approach adopted in this study was not inspired by this thinking but the data were obtained in a manner which lends itself to this type of analysis. Conditions for obtaining data on canary courtship were arranged as follows.

OBSERVATIONS AND ANALYSIS

Thirty-four pairs of canaries, most of them from Japan, were obtained from local dealers. The canaries were kept in glass-fronted boxes of standard breeding cage dimensions (33×65×29 cm), lined with acoustical tile. These were housed in a walk-in controlled temperature chamber, observations being made from the outside through a window.

The observations were carried out by the junior author each day from 3 February to 14 July 1965. Each pair was watched for an hour daily from pairing until the day the first egg was laid, which was at about seven days. Two hundred and seventy observation hours were logged in which the sequence of behaviour patterns was recorded by handwritten notes while the vocalizations were simultaneously tape-recorded. A partial list of the eighteen code symbols used to record the behaviour is given below. These were noted by hand as they occurred on thirty-second time-scaled paper. Approximately 1,000 calls were instrumentally analysed, grouped, and characterized according to their physical dimensions. These groups were then related back to the behaviours and conditions in which they occurred.

Before the calls were classified, the following measurements were made: the dominant frequencies of each call and of one or more of its harmonics, if present; the duration of the whole; and the duration of intervals between the repeated parts of calls. The latter are called syllables, and the term 'figure' is used in this paper, instead of 'note', to refer to any sound producing a continuous tracing on an audio-spectrograph (Davis, 1964). Subsequently the calls were grouped according to overall form, component frequencies and duration.

After matching the calls to the contexts in which they took place, calculations of the percentage of occurrence of these conditions were made to clarify the message and meaning of each. In the case of a single call or brief sequence, the behaviours of each of the birds occurring immediately before, during, and immediately after the vocalization were noted. Where the vocalizing bird emitted long sequences of the same call, only the behaviours occurring in the first 90 seconds were noted. Additional information pertinent to the context which was recorded were the following: the sex of the vocalizing bird, the quarter of the courtship and nesting cycle, the stage of nest construction and the ultimate presence or absence of an egg.

The behavioural code

The code letters used in the tables, with descriptions of the behaviours they represent, are as follows:

A_1: The bird often crouches facing its mate with the head thrust forward. The wings are slightly raised and there is a sleeking of the feathers.

A_2: The bird is chasing and pecking its cage mate.

PTP: The bird is flying from perch to perch or hopping about on the floor.

R: The bird is immobile, or moves body parts in one place.

F: The bird is eating or drinking.

NJ: The bird jumps rapidly in and out of the nest cup.

NB: The actual placing of material in the nest cup, or nest-moulding.

CF: Courtship feeding; the male regurgitates food into the female's mouth.

CC: The male and female both assume copulatory position.

NS: The bird sits passively in the nest cup.

B: The female assumes a crouched, precopulatory position.

RESULTS

Classification of the calls on the basis of aural and instrumental analysis led to the identification of twenty-six preliminary types. This number was subsequently reduced by the elimination of rare forms, and in the end a total of fourteen calls was judged sufficiently numerous for statistical purposes. No. 19 below is an example of one of the rare forms which were omitted from this study. The preliminary categories based on structure fell naturally into two groups, namely, single and repeated calls. Each of these was subdivided as follows:

(i) Single call
- (a) consisting of a single figure: Nos 3, 4, 18;
- (b) having a complex figure, with several harmonics and an upslurred onset: Nos 5, 6, 16;
- (c) having a complex figure, with several harmonics, lacking an upslurred onset: Nos 14, 15, 19.

(ii) Repeated call
  (a) with few harmonics, rhythmic sequence: Nos 21, 22, 23;
  (b) with few harmonics, arhythmic sequence: Nos 24, 25;
  (c) with hiss and many harmonics, and arhythmic sequence: No. 26.

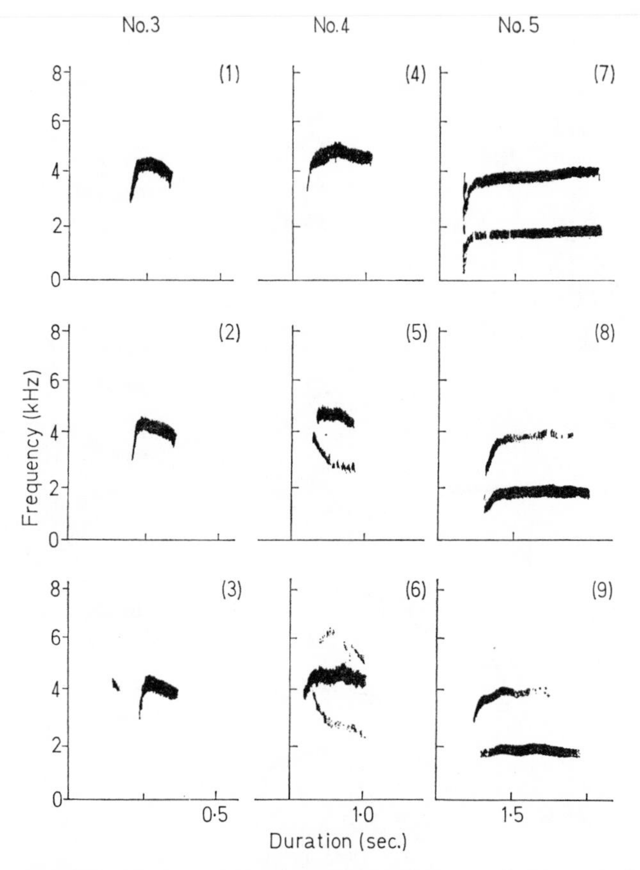

FIGURE 1. Wide band sonograph recordings of Call Nos 3, 4, 5. Three examples of each are given to illustrate typical variation.

The illustrations of these calls in Figures 1 to 5 give three examples of each call to provide an indication of the variability found in each type. The differences between certain subgroups are not great, but are consistent and very useful in pursuing the ethological or contextual analysis which follows.

The complete tabulation of the results of quantitative measurements and analysis of both physical characteristics of the calls and their ethological context is found elsewhere (Olsen and Mulligan, ms). Tables 1 and 2 provide a sample of the data on the physical characteristics of audiospectrograms of the calls. The Coefficient of Variation (V) in these tables is computed by dividing the standard

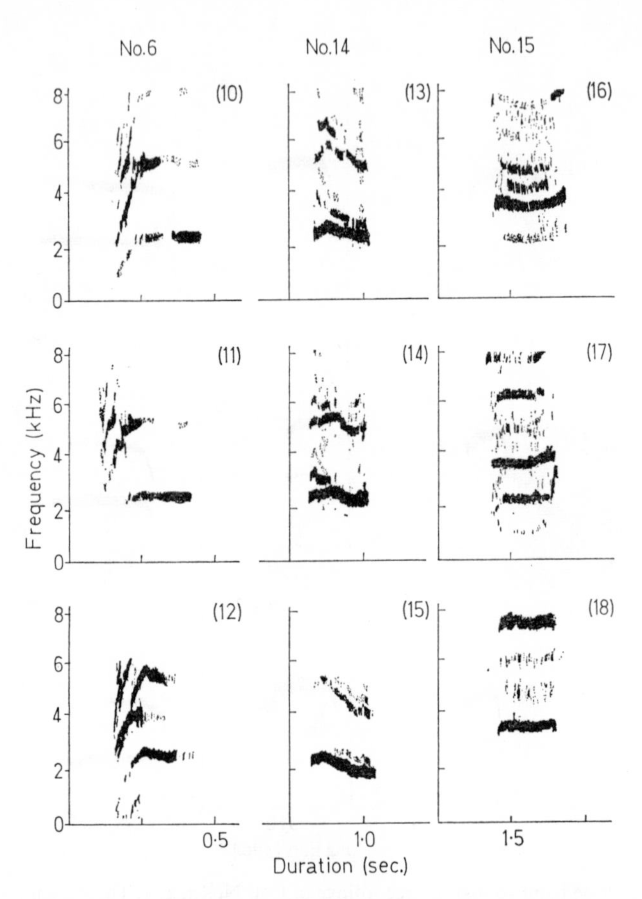

FIGURE 2. Call Nos 6, 14, 15.

deviation by the mean, and changing this ratio to a per cent. This ratio is very useful for comparison of the extent of variation in different population parameters (Simpson *et al.*, 1960). Frequency measurements show that this is clearly the most stable character. The duration of the elements or syllables of repeated calls is also very consistent. The shape of each figure, the harmonic structure, and duration of each call are also very useful features for classification purposes. Often, however, the presence of more than one harmonic on the audiospectrogram is too variable to be used for measurements.

After characterizing the calls according to their physical parameters the behavioural analysis of the occurrence of each was carried out.

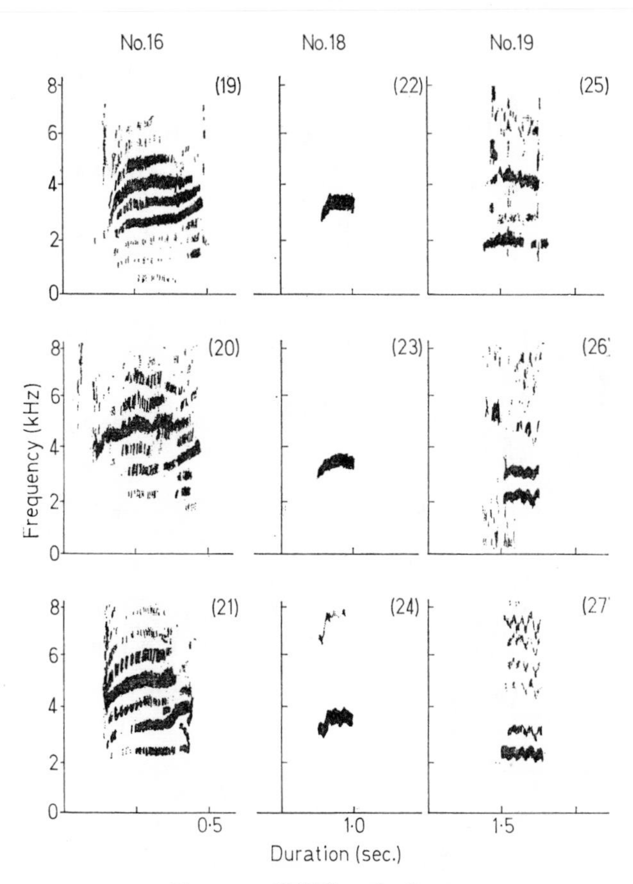

FIGURE 3. Call Nos 16, 18, 19.

The most important results of this are given in Tables 3 to 6, with the remainder being summarized briefly in the discussion of the call groups which follows.

The contextual analysis was carried out by noting the behaviour which occurred immediately before, during, and after the emission and reception of the calls. The 'message' is related to but not identified with the specific behaviour of the emitter, and the 'meaning' with that of the receiver. The frequency of occurrence of the two most prominent of these behaviours, listed earlier with their abbreviations, is given in the tables. At most about half or two-thirds of the total

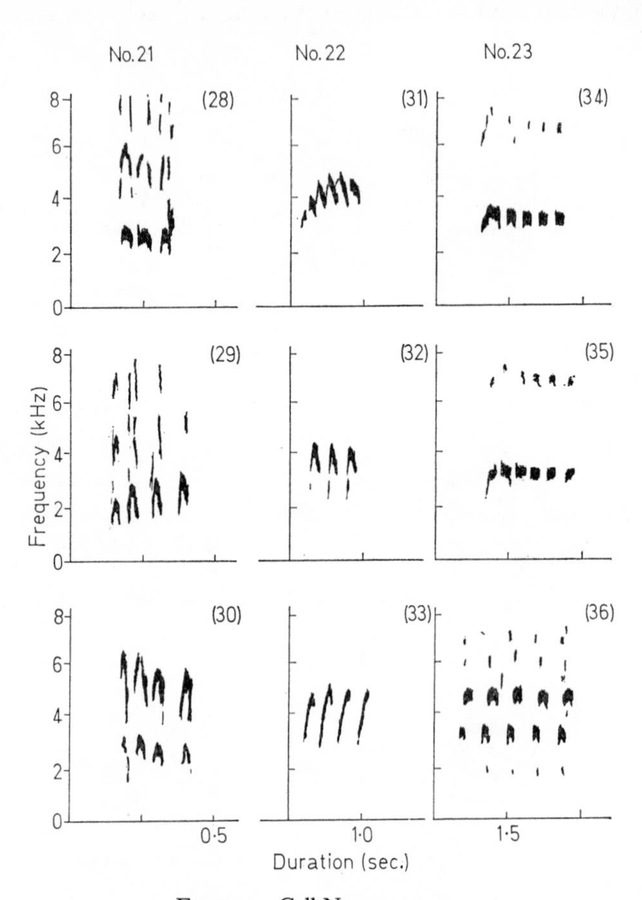

FIGURE 4. Call Nos 21, 22, 23

occurrences are usually accounted for here. The other half are typically distributed among five or six other behaviours, each with a relatively small percentage of occurrence.

In general similar calls are found in similar contexts, as expected, but there are interesting exceptions, as in the case of the strongly motivated attack calls. It should be noted here that it is rarely possible to specify clearly either the message or meaning of a call by reference to the categories of behaviour which were chosen beforehand for the study of information transfer. Further work, using the experience gained here and asking more specific and appropriate questions, will undoubtedly clarify the communication content and behavioural effect of these calls.

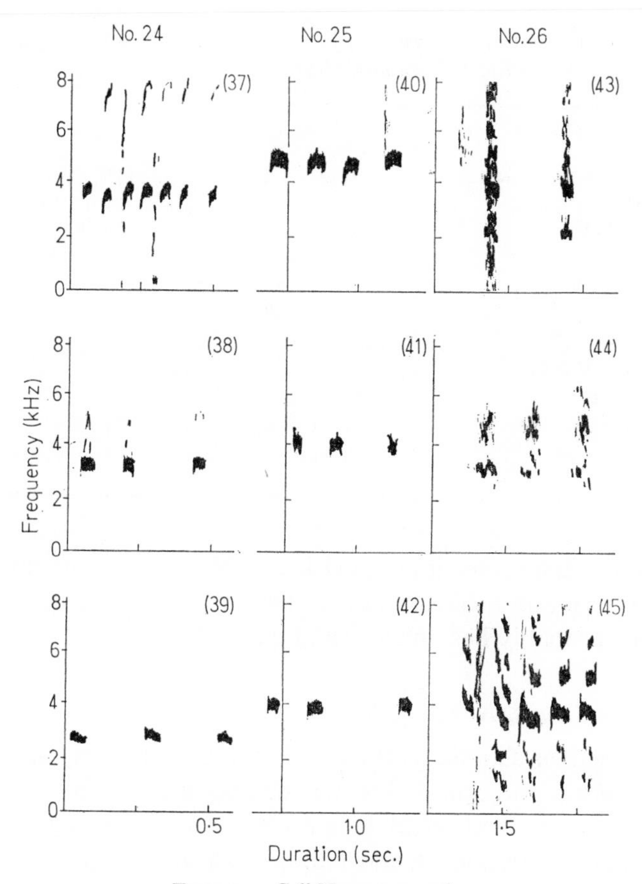

FIGURE 5. Call Nos 24, 25, 26.

The classification and naming of behaviour patterns is of critical importance because it can so easily lead to unconscious bias in subsequent study. For this reason the procedure adopted here is to adhere throughout to the original and unambiguous structural classification indicated by the numbers. The contextual groupings which follow are secondary reclassifications which have a more or less firm basis according to the consistency with which each call is associated with a particular behaviour. Thus, in order to appreciate the meaning of such blanket, summary terms as 'anxiety', 'distress', etc., one must refer to the tabular evidence and its discussion in the body of the paper. In the discussion of the individual calls which

TABLE 2. *Physical Characteristics, Repeated Calls.* (The duration of intervals between syllables of the call are also given.)

Call no.	21	22	23	24	25	26
Total no. of measurements	106	92	83	170	157	194
Durat., syll. $\overline{X}$ (msec)	27	32	35	39	42	43
Durat., syll. *V*	29	34	23	33	34	27
Durat., total $\overline{X}$	191	224	289	—	—	—
Durat., total *V*	49	31	31	—	—	—
Durat., interv. $\overline{X}$	37	26	33	—	—	—
Durat., inten. *V*	46	48	42	—	—	—
Dom. Freq. $\overline{X}$ (kHz)	3·3	4·9	4·4	—	—	4·0
Dom. freq. *V*	12	8	11	—	—	12
Number males	6	4	8	4	3	11
Number females	3	6	3	12	10	6

follows, the arrangement adopted is one based on a combination of both physical and behavioural features, rather than the latter alone, since this probably has greater validity.

Mild arousal calls, Nos 3, 4, 18

Because of their simplicity, these calls can be accurately specified in physical terms, and thus differentiated. They are contextually closely related, however, and could perhaps be grouped as one call. They differ mainly in frequency and No. 4 has a distinctive second voice (Figure 1, Sonagram Nos (5), (6)). They all have a pure, brief, plaintive 'pee' sound. These calls were not common, and were emitted mainly by the hen in the early part of the breeding cycle, less often at the end of the nest construction period. There is no close relationship with any particular type of behaviour, and a direct communicatory value cannot be specified. The incidence of perch to perch (PTP) flying and activities associated with the nest cup indicate a state of mild activity and a certain readiness to breed.

TABLE 1. *Physical Characteristics, Single Calls.*

Call no.	3	4	18	5	6	14	15	16
Total no. of measurements	29	33	20	41	158	28	46	15
Duration, $\overline{X}$ (msec)	138	210	115	344	245	201	184	282
Duration V	19	24	29	36	19	18	24	11
Dom. freq. $\overline{X}$ (kHz)	4·6	5·2	3·8	2·1	3·0	2·9	4	4·2
Dom freq. V	5·1	3·1	3·6	7·2	11·6	4·2	4·2	11·1
Number males	3	—	1	6	7	1	—	4
Number females	2	4	2	2	7	4	4	1

Mean duration is given in milliseconds, together with its coefficient of variation (V). The dominant frequency is given in kilohertz. For each call the number of birds recorded, and their sex, is given.

TABLE 3. *Behavioural Context of Call Nos* 3, 4, 18, 16.

Call no.	3						4						18						16					
	BEF		DUR		AFT		BEF		DUR		AFT		BEF		DUR		AFT		BEF		DUR		AFT	
Behav. emitter	NB	R	NB	F	F	B	R	PTP	PTP	R	R	PTP	R	PTP	PTP	G	PTP	R	A_1	A_2	A_2	A_1	A_1	A_2
No. occur.	2	2	3	3	5	4	6	2	6	4	6	3	2	1	1	1	2	2	3	3	4	1	3	1
% Occur.	17	17	25	25	42	33	54	18	54	36	54	28	50	25	25	25	50	50	43	43	55	14	43	14
Behav. receiver	R	F	R	F	R	F	F	R	G	PTP	F	NJ	G	F	PTP	A	R	PTP	A_1	PTP	PTP	A_1	PTP	A_1
No. occur.	7	4	7	3	7	4	5	3	3	3	4	2	3	1	2	1	2	1	4	2	5	2	5	2
% Occur.	59	33	58	25	59	3	45	27	27	27	36	18	75	25	50	25	50	25	57	29	71	29	71	29
Total no.	12						11						4						7					

The two behaviours most frequently occurring before, during and after the call is emitted and received are listed with the number and per cent occurrence of each. The symbols and definitions of the behaviours were given earlier.

TABLE 4. *Behavioural Context of Call Nos* 5, 6, 14, 15
(For explanations, see Table 3)

Call no.	5						6						14						15					
	BEF		DUR		AFT		BEF		DUR		AFT		BEF		DUR		AFT		BEF		DUR		AFT	
Behav. emitter	F	PTP	PTP	CC	PTP	F	PTP	F	PTP	A_2	PTP	F	PTP	NB	PTP	CF	PTP	NB	PTP	F	PTP	R	PTP	F
No. occur.	3	2	3	2	4	2	13	7	13	11	10	9	9	9	9	7	7	7	7	5	9	3	7	3
% Occur.	33	22	33	22	44	22	28	15	28	24	22	20	33	33	33	26	26	26	44	31	56	19	44	19
Behav. receiver	PTP	B	PTP	CC	R	F	A_1	PTP	PTP	A_1	PTP	A_2	R	PTP	CF	R	PTP	F	PTP	NS	R	PTP	NS	F
No. occur	2	2	3	2	3	3	7	11	14	15	15	8	8	7	7	6	9	7	4	3	5	3	5	4
% Occur.	22	22	33	22	33	33	15	24	30	33	33	17	30	26	26	22	33	26	25	19	31	19	31	25
Total no.	9						46						27						16					

TABLE 5. *Behavioural Context of Call Nos* 21, 22, 23
(For explanations, see Table 3)

Call no.	21						22						23					
	BEF		DUR		AFT		BEF		DUR		AFT		BEF		DUR		AFT	
Behav. emitter	PTP	R	PTP	R	PTP	R	PTP	F	PTP	NB	PTP	NS	PTP	R	PTP	R	PTP	R
No. occur.	9	5	17	9	12	10	10	8	20	5	15	7	9	7	12	4	8	8
% Occur.	26	14	49	26	34	28	26	21	51	13	39	18	30	23	40	13	27	27
Behav. receiver	PTP	NJ	PTP	F	F	PTP	R	F	F	PTP	PTP	A_2	PTP	NS	PTP	NS	PTP	NS
No occur.	8	5	9	8	8	7	11	9	7	6	7	7	8	8	7	5	7	7
% Occur.	23	14	26	23	23	20	28	23	18	15	18	18	27	27	23	17	23	23
Total no.	35						39						30					

TABLE 6. *Behavioural Context of Call Nos* 24, 25, 26
(For explanations, see Table 3)

Call no.	24						25						26					
	BEF		DUR		AFT		BEF		DUR		AFT		BEF		DUR		AFT	
Behav. emitter	NS	NB	CF	NS	NS	NB	NS	NB	CF	NS	NS	CC	R	F	A	CF	PTP	R
No. occur.	17	9	22	19	10	8	10	9	18	13	11	6	10	10	24	9	9	8
% Occur.	37	20	25	22	22	17	27	24	33	24	30	16	20	20	48	18	18	16
Behav. receiver.	R	PTP	R	CF	R	CF	R	F	CF	F	CC	PTP	F	R	PTP	CF	PTP	F
No. occur.	15	8	23	22	9	7	13	12	18	11	7	6	13	11	13	9	16	11
% occur.	33	17	26	25	20	15	36	32	33	20	19	16	26	22	26	18	32	22
Total no.	88						55						50					

Anxiety calls, Nos 5, 6, 14, 15

The members of this group are more distinctive in physical structure than the previous, but Nos 5 and 6 in particular are closely related to the latter in having a similar 'pee' sound, a clear tonal quality, an upslurred onset, and a context related to distress. No. 14, a not-so-clear 'pee-oh', and No. 15, a harsh 'shree', have a complex structure with second voices and even hiss, in No. 15.

A conflict between aggressive and flight tendencies, with the emphasis on the latter, and with mild arousal, is indicated by the behavioural activities associated with No. 6. This is a very common call, which occurs in a large number of situations mainly in the early period of pairing; it is readily elicited by any disturbance and it frequently itself elicits a similar call from other birds present. An incubating hen often gives this call when she comes off the nest to feed and continues to do so until again settled on the eggs. In this case the call has nothing to do with the activity of the mate and is simply an expression of discomfort during temporary absence from the nest. It may be that No. 16, discussed separately, is an extreme form of this call, since they are related in structure.

Nos 5, 14, and 15 may best be interpreted as indicating conflict between sexual and flight tendencies of varying degrees. No. 5 is a rare call, given by the cock eight out of nine times, early in the pairing period. The others are mainly female vocalizations, with the harsh form, No. 15, occurring mainly late in the breeding cycle. Because it is harsh and occurs late, this call probably signals higher level motivation than 14; this assertion cannot be argued convincingly from the contextual data, but there is more perch to perch flying and less activity at the nest cup associated with No. 15.

Agitation trills, Nos 21, 22, 23

The subdivision of this group into three calls is based on small but real differences in structure, and on differences such as sex of the emitter and the time of emission during the breeding cycle. It is unlikely that these distinctions are important, since the contextual properties of each call are so similar, and therefore nothing prevents their being treated as a single type. The whole group is characterized by the strict rhythm, or precisely timed intervals between the individual sounds or syllables. This timing varies between calls but not within them; the term trill is used because the repetition rate is rapid and the sound answers to the usual understanding of this term in bird song. This call is produced under pressure as a burst of brief sounds, as it were, and it strikes the ear when continually repeated as disturbing and indicative of stress. When one separates a mated pair, for example, the long continued repetition of this little trill can be distressing to anyone listening, at least if he is conditioned to its usual context.

The quantitative analysis of associated behaviour shows a high incidence of perch to perch flying. No. 21 was given mainly by females in the early part of the cycle, No. 22 by females both early and late, and No. 23 early and late by males. In the latter case the hen was usually sitting on the nest, and the call was indicative of sexual arousal and of conflict associated with this state. In general these calls are heard together with the previous group of anxiety calls although they are not related in structure. It is possible that the trilled calls represent a higher motivational state, and that within the latter, degrees may be reflected by a faster syllable repetition rate. This cannot be seen in the contextual data.

Attack calls, Nos 16, 26

These two calls have different structural lineage, but converge in the possession of high energy, many harmonics, and hiss. No. 16 is a harsh and concentrated blast containing an initial hissing upslur, six to ten closely spaced harmonics, and occasional second voices, which are not distinct. It is related in structure and context to the anxiety call group, but is set apart by its very high energy form, with an emitter context of strong aggression and a response of either return aggression or flight. It was found only seven times in the sample of calls analysed by audiospectrograph; six of these were given by males, all of them in the first stage of pairing, before nest construction was begun. In five of these cases the hen did not ultimately lay an egg, and therefore the occurrence of this strongly aggressive call at the outset was an indication of subsequent incompatibility.

No. 26 is a repeated arhythmic short sound sequence related in structure to the group which follows, the mating calls. It is given mainly by the male, prior to nest construction, and it is not a soft sound but a harsh, rapid 'tch, tch, tch' occurring very often with aggression in the emitter. Sexual motivation is often also present as evidenced by the incidence of courtship feeding, but verges in the male into attack instead of courtship feeding or copulation. Of all the calls studied, No. 16 is the clearest in its signal function; the sample size, however, is rather small. It is significant that in both these calls the high energy input is correlated with an obviously high motivational level.

Mating calls, Nos 24, 25

This is a class of sounds which are heard only at close range, and which cannot be recorded except by a microphone at the nest or else at a few centimetres distance from the mated pair. The sounds are a series of rapid 'pee's', very different from the trilled calls in their lack of strict rhythm and continuity, and in being barely audible. The structural variety in this class is very striking and exceptional, such that even the same individual uses several forms of this call, differing considerably in structure and frequency. The study of this variation in individuals is hampered by the fact that both members of a pair emit this call, and there is often no external movement to show the source. When the two birds are close together it is impossible to say which is making the call; sometimes both are doing so. Excessive variation in structure and tempo is the reason for the omission of certain values for these calls in the tables of physical characteristics. The illustrations in the figures are more homogeneous than the actual sample.

These are mainly (90 per cent) female calls given late in the period of nest construction, and are clearly associated with nest-building, nest-sitting, courtship feeding and copulation. They can be called incitement calls, for they often lead either to courtship feeding or copulation. This is not more clearly indicated in the quantitative contextual evidence because the total duration of calling is often rather long in this case and the responses just mentioned were not included as such in data tabulation. The cutoff point for recording a response after the beginning of a long continued call was 90 seconds. These, therefore, are mainly calls made by the hen in or near the nest inviting her mate to courtship feeding or copulation, which often enough follows.

The results above can be summarized by answering the questions posed at the outset. First, the number and kind of calls used in canary courtship are about five, if this number is based on functional groupings, and two or three times that, depending on one's inclination to lump similar types, if based on physical structure. The variation found in certain of the physical parameters of these calls is quite small; for example, the frequency of the prominent parts of the sounds is often so limited that the coefficient of variation of the mean is less than ten. The overall stability of the call structure is similarly rather narrowly defined. The duration of the calls is more variable; the standard deviation of the mean in this case approaches one third of the mean. Thus, there is a high degree of stereotypy in these calls, found in the frequency, general structure, and duration. Variation within and between individuals was not fully documented in this study, but it is clear that some calls remain the same between individuals, some vary between individuals and one or two vary within the same individual.

The five functional groups described represent an interpretation of the data, combining in partly arbitrary fashion the physical and contextual properties of the calls. Other arrangements are possible, particularly since the number of different situations in which the calls occur is quite large, and the evidence of communicatory value is compelling in few cases. Thus a clear message or meaning can usually not be assigned these calls. In keeping with this they undoubtedly have their effect in ways other than by an immediate response in the context of emission.

Degrees of arousal or of motivational level are certainly registered by these calls, but in general this is not done by a graded series which constitutes a continuum. The evidence concerning variation above shows that, for the parameters measured, continually varied sequences are lacking. However, it is possible that rate of emission of repeated syllables, or relative energy content in series of similar calls, can be a measure of motivational level. It was pointed out above that Nos 16 and 26 can be regarded as exaggerated forms of simpler calls, thus representing a graded intensity in partially discrete forms.

Significantly, there was a definite sequence in the peak occurrence of the different calls during the observation periods, providing additional evidence on their function. In the first part of the cycle the most prominent vocalizations are Calls 5, 6, 16, and 26. These are either aggressive or distress signals. Also frequent at this time are Calls 4, 14, 18, and the trills 21, 22, and 23. These are associated with some stress and movement. In the second part of the cycle, the trills, together with active flight back and forth are continued. This movement by both members of the pair may, as Tsuneki (1961) suggests, serve the purpose of separating the pair from other members of the flock. Calls 15 and 19, associated with flight, are also heard more often during this time. The greatest change is in the last quarter of the cycle, when the short, soft begging calls, Nos 24 and 25, supplant the others in the hen. During this time the cock seldom vocalizes, giving mainly the plaintive trills as he flies back and forth while the female sits on the nest. The five basic contextual situations in which the fourteen physical call types were used can be termed anxiety, distress, agitation, aggression and sexual incitement. The first three of these can alternatively be described as three levels of arousal, namely: mild, medium, and high level. This latter suggests a certain lack of specificity or of conceptual content which is in keeping with the data.

DISCUSSION

The classification of canary calls given here is open to revision, but it is safe to say that there are ten or more rather discrete sound types used in canary courtship. It is not safe to attempt to specify clearly the function or communicative role of each; this is probably due in part to our failure to appreciate adequately the birds' point of view, and partly to the experimental conditions of this study. The sounds used range from simple pure tones to cries that are harsh and coarse with most of the energy in the region from three to six kilohertz. This repertoire appears to be basically similar to those of other finches (Andrew, 1957; Gompertz, 1961; Marler, 1956), but there are no doubt significant differences whose study would be rewarding.

The strain of canary used for this study, locally called a 'chopper', is an outbred form of heterogeneous genetic background by comparison with the roller or other carefully bred 'type' canaries. There are well known strain differences in voice in these birds, the comparison of which is under way in our laboratory. The pitch of the voice in these has been notably altered by breeding selection. Hence it is interesting that in the mongrel chopper this aspect is the most stable feature of the calls, as it is also of the song. Since strong selective forces act on the latter, the calls are no doubt influenced at the same time. Continued pressure on breeding ability has also been present during domestication and the courtship vocalizations are an important factor in breeding success. Deafened canaries are retarded in breeding, as a recent experiment in this laboratory has shown. It may be then that positive selective forces have preserved certain features of canary calls in their ancestral form without much change.

Stereotypy and graded intensity

The use of the coefficient of variation to compare the variation in canary calls with that of a wild species, shows that many of these are highly uniform in their measurable features. The value of V for several hundred recordings of an extremely uniform call of the song sparrow, *Melospiza melodia*, was slightly less than ten for frequency measurements and just below twenty for those of the duration. Therefore a value of V less than twenty-five in these characters indicates low variation. The values in the tables show that most of the canary calls exhibit this low variation. Thus they are too discrete and unvarying to allow for much expression of graded intensity within them, at least in the parameters measured. Levels of intensity can still be expressed, of course, by amplitude changes, different calls, and by rates of calling or repetition in call subunits.

A remarkable exception to this general uniformity is the wide variation in the precopulatory and courtship feeding calls, which are not even consistent, apparently, within an individual. No doubt this call is used when there is no need for specific or even individual distinctiveness, since it occurs at close range between familiar mates.

In conclusion it seems that pure tone sounds are used to express low excitation and distress, harsh sounds excitement and attack tendencies, and soft, short sounds the intimacies of sexual motivation. Stereotypy is certainly present in many calls, but the possibility of expressing intensity of arousal remains open and should be studied. The typical sequence of call types used during breeding is a valuable means of following breeding readiness, and familiarity or compatibility between pairs. Sensitive measures of response to particular displays and conditions can be developed if the suggestion is verified that graded intensity is expressed by total energy input, by call sequences, and rate differences within them.

REFERENCES

ANDREW, R. J. (1957) A comparative study of the calls of *Emberiza* spp. Oxford University Press, New York, **99**, 27–42.

ARMSTRONG, E. A. (1963) *A Study of Bird Song*. Oxford University Press, London. 335 pp.

DAVIS, L. I. (1964) Biological acoustics and the use of the sound spectrograph. *S. West. Nat.* **9**, 118–45.

GOMPERTZ, T. (1961) The vocabulary of the great tit. *Brit. Birds* **54**, 369–418.

HINDE, R. A. (1965) The integration of the reproductive behaviour of female canaries. In *Sex and Behavior* (F. A. Beach, ed.). John Wiley & Sons, New York.

MARLER, P. (1956) The voice of the chaffinch and its function as a language. *Ibis* **98**, 231–61.

(1965) Communication in monkeys and apes. In *Primate Behavior* (I. DeVore, ed.). Holt, Rinehart and Winston, New York.

MORRIS, C. W. (1946) *Signs, Language, and Behavior*. Prentice-Hall, Englewood Cliffs, New Jersey.

OLSEN, K. and MULLIGAN, J. A. (ms.) Instrumental and ethological analysis of courtship calls in the common canary.

SIMPSON, G. G., ROE, A. and LEWONTIN, R. C. (1960) *Quantitative Zoology*, rev. (Harcourt Brace, ed.). New York.

SMITH, W. JOHN (1965) Message, meaning and context in ethology. *Am. Nat.* **99**, 405–9.

(1966) Communications and relationships in the Genus *Tyrannus*. *Publ. Nuttall Ornith. Club*, No. 6. 250 pp. Cambridge, Mass.

(In press) Message/meaning analysis of animal communication. In Sebeok, T. A. (ed.) *Communicative behavior in animals—some theoretical approaches* University Indiana Press, Bloomington.

THORPE, W. H. (1961) *Bird Song*. Cambridge University Press, London.

TSUNEKI, K. (1961) Breeding in a dense population of the caged canaries. *Jap. J. Ecol.* **11**, 98–104, 142–6.

9. DUETTING

by **T. HOOKER and BARBARA I. HOOKER (LADE)**
Canford School, Dorset

INTRODUCTION

In the classic cases of territorial species with which ornithologists in temperate zones are familiar, song is restricted to the male. He seasonally proclaims his territorial rights by loud singing, thereby warning off other males and attracting a female. She is usually quietly coloured and does not sing—though her responsiveness to the species song, and later to her mate's particular version of it, probably help to keep her near him and thus to maintain the pair bond.

In some tropical parts of the world a quite different system pertains. Where the seasons are not so strictly demarcated, many of the indigenous birds remain in the same general area all their lives. Many such birds pair for life and hold the same territories for long periods. Instead of the male singing alone, both members of the pair sing, either more or less together or antiphonally, each making a distinct contribution to produce the duet. This duetting continues throughout the year, though the intensity of singing is much increased when the birds are nesting. This high development of singing behaviour probably contributes to the maintenance of the pair bond throughout the year.

Antiphonal singing, a specialized form of duetting in which male and female use different notes and sing alternately, often with marked precision in timing, has recently been studied by Thorpe (1963) in a number of East African species (barbets, *Trachyphonus*; fantail warblers, *Cisticola*; and certain shrikes, *Laniarius*). Our own work has been concerned with the last of these genera.

THE TROPICAL BOU-BOU SHRIKE

One of the most interesting species is the tropical bou-bou shrike (*Laniarius aethiopicus*) in which the flute-like notes can conveniently be rendered in musical notation. The species is widely distributed throughout tropical Africa, extending from west coast to east coast. The race *major* of this species, on which most of the work has been carried out, occurs in Sierra Leone, Nigeria and the Cameroun, the Sudan, Uganda, Kenya, Tanzania, northern Angola, the Congo, Zambia and Malawi.

FIGURE 1. (*a* to *q*). *L. aethiopicus major*. A series of separate duet patterns produced by a single pair. The contributions of the two birds, designated 'X' and 'Y', are inserted where the evidence is clear. Sex is not indicated. The figure '8' above the treble clef indicates that the entire staff is raised one octave. (From Thorpe, 1966.)

The tropical bou-bou inhabits thick cover of almost any type from highland forest to scrub in dry river beds and gardens with dense shrubs. The general behaviour of the birds is cryptic in that they tend to keep to dense cover.

As a result of recordings of *L. aethiopicus major* made in Uganda and Kenya, Thorpe and North (1965) established a number of essential points. Male and female have a considerable basic vocabulary involving a wide range of sounds, all distinctly recognizable as those of the species (Figure 1). Although juveniles occasionally practise alone, the duets proper are worked out by a pair of adults in a territory. In the course of their practice the basic components become grouped into duet patterns. Each pair has a number of alternative patterns, at least some of which may be peculiar to that pair, while others may be very similar to those of neighbouring pairs, or widespread in a local population. Either partner can initiate a duet series.

Thorpe and North (1965) recorded and analysed more than 135 duet patterns of the race *major* from birds widely separated in Kenya and found not only a wide range in repertoire but also considerable variation in the duration of duets, irrespective of the number of notes. In some instances a third bird was involved, interpolating one or more notes to make a trio.

The manner in which such behaviour could contribute to the maintenance of the pair bond is apparent. In order to develop and maintain a duet pattern, each contributor must have learnt its partner's contribution in relation to its own. The mate is that bird which answers with the correct contribution and with the appropriate time interval (Thorpe and North, 1965).

Our own records were made in an area limited to the Kitale district of Kenya between September and December 1966. The observation period thus overlapped the short rains (October to December), when some breeding occurs, though the main breeding peak is in the long rains (March to June; T. J. Barnley, personal communication). While several observations and recordings were made outside the main study area, most of our records were made on one population of established pairs, many of which had been colour-ringed and kept under observation by T. J. Barnley since the previous April. Some of these had fully-fledged young, but only one nesting pair was observed.

As might be expected from the more limited extent of our study area, our records showed much less variety in duet structure than those from the wider area studied by Thorpe and North. All duets recorded fit into the summary classification illustrated in Figure 2. It is interesting to note that birds obtained from this same area, some as hand-reared juveniles and others as adults and transferred to aviaries at Madingley (Cambridge), developed very complex duets, trios and even quartets (Thorpe, personal communication). Nothing of this complex nature was observed with wild birds in the field.

Though the general behaviour of the birds was cryptic, during territorial disputes they became conspicuous both visually and vocally. They sang vigorously at dawn and more briefly at dusk, even though they were not nesting.

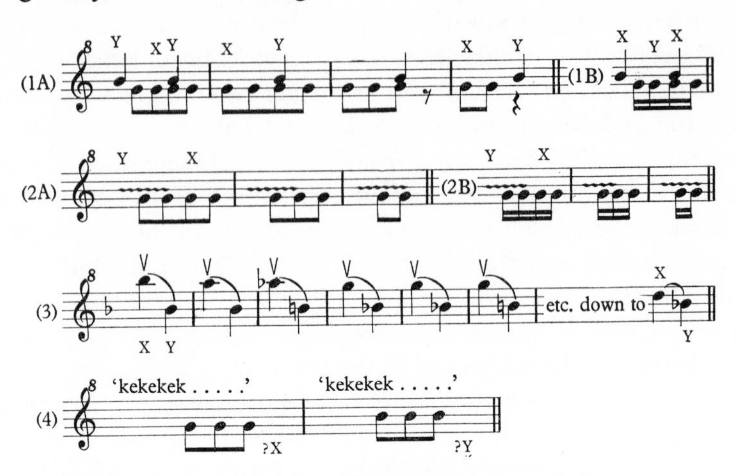

(5) 'Snarling' duets: a click and hiss followed by a snarl:
'Zizzwizzwizz' ------(Always very aggressive)

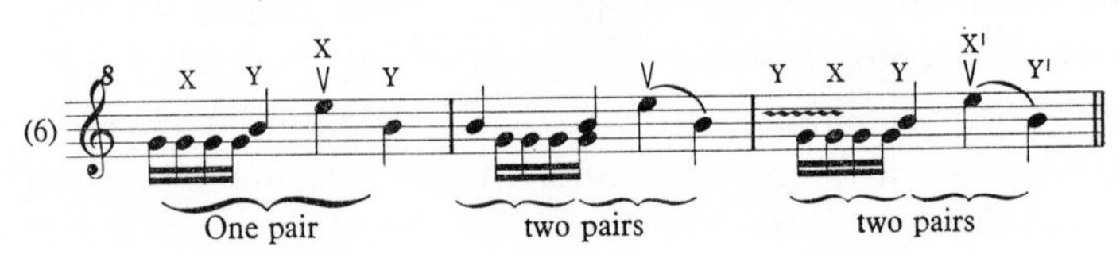

FIGURE 2. *L. aethiopicus.* Summary classification of duet patterns recorded from wild birds near Kitale.

(1A) (B) 'bell note' and 'bou-bou' duets;
(2A) (B) 'snarl' and 'bou-bou' duets;
(3) high, piercing note and 'bell note' duets;
(4) chattering call given by birds of either sex and all ages;
(5) snarling duets, both birds very aggressive;
(6) compound duets, e.g. (1B)/3.

The note values are only approximate. Time intervals are indicated by the relative positions of notes, dots and rests being omitted.

'X' and 'Y' here indicate the constant repertoires of the sexes; but as yet there is no anatomical confirmation that 'X' is female and 'Y' male. Other information is as for Figure 1.

The nesting pair was located during nest-building and was watched daily, from shortly after dawn for two or more hours, until the two eggs hatched ten days after the first was laid. When a neighbouring pair of bou-bous approached the territorial boundary, both members of the nesting pair flew to the 'disputed' tree and joined in a bout of aggressive counter-duetting.

On one visit, playback of duets of a neighbouring pair at the territorial boundary of the nesting pair elicited response from one bird only (later deduced to be the male) which sang its own contribution to a duet pattern commonly used by this pair, and did not give the complete pattern in the absence of its mate. When we reached the nest-site, the mate was found to be sitting. Throughout the period of observation the incubating bird did not engage in duetting while still sitting.

The incubating sessions were seldom longer than 65 minutes in more than thirty relief changes observed. On occasions when passing labourers disturbed the sitting bird, it left the nest without calling and also returned silently. Duetting occurred at nest-relief, either by the sitting bird leaving the nest and initiating a duet pattern—apparently calling up its mate—or by the relieving bird. In the latter case the sitting bird always left the nest before joining with the response appropriate to complete the duet. The bird being relieved almost invariably moved directly from the nest clump to a tree about twenty-five yards away and duetted briefly with the incoming bird before moving sometimes as far as 150 yards from the nest to feed. This distance was still within auditory range of the sitting bird for we often heard calls from the bird near the nest being answered immediately by the foraging partner. The observer and tape recorder were seven yards from the nest and the time lag between 'call' and 'answer' —the response time of the second bird—decreased as the partner approached. The question of auditory reaction time will be considered further in the discussion of the next species, the black-headed gonolek (*Laniarius erythrogaster*).

Our observations on these birds and on the other pairs in the area enable us to make the following generalizations:

(i) In the duet repertoire of any pair of adults, each individual has its own parts; while duetting the birds do not exchange parts.

(ii) By comparing many pairs in the same area which sing similar duet patterns, the contribution is also apparently diagnostic of sex. The conclusion that the sex of a bird can be predicted from a knowledge of its vocal repertoire is based on observations of aggressive behaviour, the behaviour of a nesting pair, and of colour-ringed birds in captivity. Although anatomical confirmation has not yet been obtained, the circumstantial evidence for the validity of this conclusion is thus strong. In a test case with colour-ringed captive birds, one of the authors correctly identified, by song alone and without previous knowledge of the birds' behaviour, the sex of a pair which had nested and laid eggs.

(iii) In any given duet pattern, the timing varies (a) between different pairs sharing the same pattern of notes; (b) the distance between the duetting birds; (c) the current activity of the responding bird: for example, if the second bird is in flight, or preening, it either does not respond, or its contribution may be 'late'. The tendency to respond appears to be strong in that the response is frequently given through a beakful of live food; or again preening may be interrupted just long enough for the response and then be immediately resumed.

(iv) Duetting only occurs regularly between adults of an established pair. Duetting between juveniles and their parents was seldom heard—though during the period of our observations most juveniles had dispersed. No duetting 'practice' was heard among juveniles.

(v) Although duetting does not occur between members of adjacent territories, in cases of aggressive interaction at inter-territorial boundaries the same or a related duet pattern may be synchronized or sung antiphonally by the pairs. During the early stages of aggressive duetting the response time of the partner is usually minimal. If the aggression becomes intense the timing may fall out of phase so that all the birds are contributing more or less at random. The most frequently used call during aggression is the hiss and snarl (Figure 2, example (5)).

(vi) There appears to be no signal, other than the production of the first note, for initiating a duetting sequence. Frequently the initiator sings unanswered for several notes, though just as often the first note is answered with more or less perfect timing irrespective of the lapse of time since the previous bout of duetting. This variation is probably the result of differences in activity of the birds at the time.

(vii) Any duet pattern may be initiated by either sex—though duet patterns (2) and (5) in Figure 2 may be exceptions. Both sexes have some degree of 'choice' in the selection of the pattern: for example there are several possible responses to a 'snarl' (Figure 2, examples (2A) and (B)). In all cases, whichever partner initiates the sequence, the call used is from its own repertoire so that the duet pattern is effectively reversed in some cases at the start of the sequence.

(viii) The pattern may change abruptly but fluently during a bout of duetting, either partner effecting the change.

(ix) Although any disturbance appeared to increase vocal activity, the birds' selection of particular patterns could not be correlated with environmental events or situations.

(x) Either sex may terminate a duet sequence simply by not singing; but again we have been unable to detect any signal for stopping. The other bird may continue for several incomplete duets or stop synchronously.

(xi) Although the parts of the individuals in each duet are usually fixed, on rare occasions one will complete the pattern in the absence of its partner. In one such case a captive bird whose partner had been removed sang the absent individual's part. On the return of the partner the birds either duplicated in perfect time or, more usually, sang antiphonally again (Thorpe and North, 1965). In another case, a wild bird, able to see its mate, sang a simple duet pattern of three notes while its mate remained silent (personal observation).

(xii) No true trios, i.e. patterns sung by three birds each with a different but regular contribution, were heard by us in the field. In one territory three birds, one of which was believed to be the offspring of the other two, sang a 'duet' where two birds duplicated the single response note in a three-note pattern.

(xiii) There were noticeable differences in repertoire between pairs recorded on a transect of fifteen miles across the foothills of the Cherangani Hills at approximately 6,500 feet above sea level. Differences were of pitch interval rather than timing or number of notes, the patterns fitting into the summary given in Figure 2. The habitat along the transect was discontinuous, being split up by open farmland.

(xiv) The simple, short duets shown in Figure 2 examples (1) and (3), can be combined to make a longer duet as in example (6). This compound duet can either be sung at the outset or by transition during a sequence from pattern (1) to (1) plus (3), sometimes with a further change to (3) alone.

(xv) The number of consecutive duets in a sequence, whether all are of one pattern or with transitions to other patterns, can vary from a single duet to more than seventy-five duets without a break. Such monotonous performances often resulted from playback of recordings of the birds' own duets.

(xvi) Calls shared by male and female are normally confined to the alarm ('tuck-tuck-tuck') and the 'kek-kek-kek'—a rapidly repeated, chattering call also given by juveniles and usually answered by 'bou' notes.

THE BLACK-HEADED GONOLEK AND THE GONOLEK

Another species of the same genus, the black-headed gonolek (*L. erythrogaster*) was investigated by Thorpe (1963). It is distributed from Lake Chad to Eritrea and south to Tanzania and the eastern Congo. Like the tropical bou-bou it inhabits dense bush and other thick cover. Thorpe considered this species to have an extremely simple form of duet pattern—the first bird, presumably the male, uttered a 'yoik'-like sound which is immediately followed by a tearing hiss-like sound made by the second bird. The timing is so perfect as to sound like one bird (Figure 3).

Thorpe supposed that the individuality of the duet was expressed, not in peculiarity of pattern (as he believed that the birds were unable to vary this), but in a very precise and exactly maintained time interval between the contributions of the two sexes. The precision of this timekeeping at once raised questions about the auditory reaction time as an expression of the powers of temporal discrimination in the avian ear. There is circumstantial evidence for supposing that the speed of response of the avian ear must be of the order of ten times that of the human ear (Pumphrey, 1961; Schwartzkopff, 1962). As yet very little is known about avian reaction times.

Thorpe (1963) observed that the black-headed gonolek frequently duetted from extremely dense foliage when the members of the pair were presumably out of visual contact. In such cases the complicating element of visual cues was thus ruled out.

Figure 3 shows eight consecutive duets of a pair of black-headed gonoleks and Table 1 shows the timing of these duets. From column 2 it can be seen that the calling of bird A is not markedly regular—the figure in brackets gives the time-lapse between the beginnings of successive call notes of A—but the response time[1] of bird B remains extraordinarily constant. Bird B must therefore be taking its time cue from the start of the call of A. The mean response time of B in this case was a little more than 144 msec., with a standard deviation of 12·6 msec. This was the most rapid response of any duetting bird

[1] The use of the term 'response time' has been preferred in the text to 'reaction time' because the latter implies a minimum, whereas in many duets the response appears slower than the minimum possible.

then investigated. The minimum shown in Table 1 is 125 msec. though faster responses may well be possible. Once a pair gets under way in a series of duets, the second bird maintains a constant response

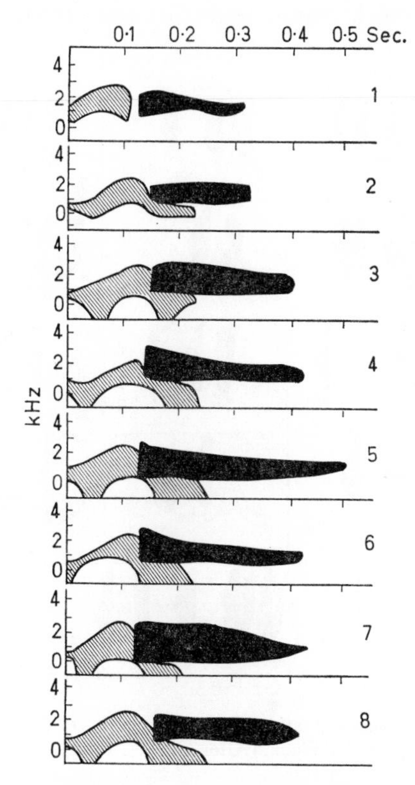

FIGURE 3. *L. erythrogaster*. Diagrammatic sound spectrograms of eight consecutive duet patterns. Bird 'A' cross-hatched; bird 'B' black. The apparent sub-zero frequencies are due to distortion and interference below 50 cycles. (After Thorpe, 1963.)

time even though this may be very much longer than the minimum possible: e.g. one pair had a mean response time of 425 msec. with a standard deviation of 4·9 msec. This extremely small deviation indicates that this species has a remarkably accurate time sense which does not vary even if the response time is extended by a factor of four. These figures for response time have not been corrected for the time taken for the sound to travel from bird A to bird B. Corrections can be made where the distance between the birds is known and the tape recorder is in line with the two birds and beyond A. No correction is required if the recorder is in line and beyond B. Duets recorded with birds separated by different distances can then be compared directly for response time.

Thorpe's preliminary studies on the auditory reaction time of the black-headed gonolek stimulated Grimes (1965, 1966) to make

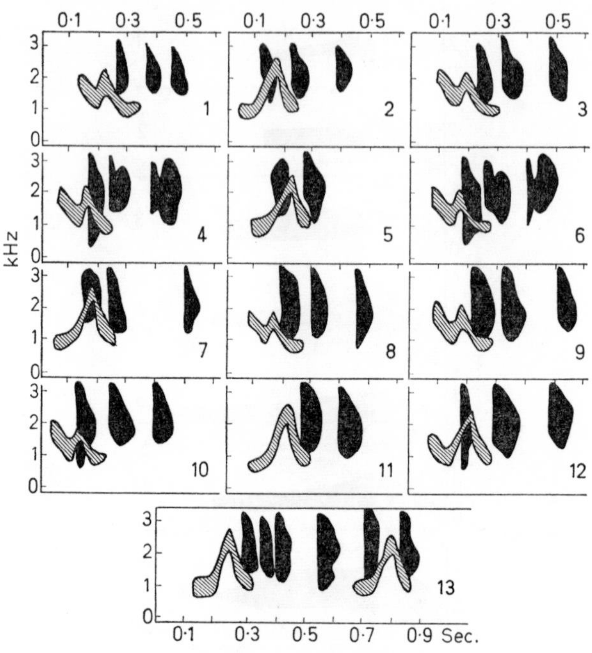

FIGURE 4. *L. barbarus*. Diagrammatic sound spectrograms of thirteen consecutive duets. Bird 'A' cross-hatched; bird 'B' black. (After Grimes, 1965.)

similar analyses of tape recordings of the gonolek (*L. barbarus*) from Ghana. This species, occurring in the eastern Congo and Uganda, has a more restricted range than the black-headed gonolek. It is usually associated with swamps and clumps of papyrus where the vegetation is extremely dense. The habitat, behaviour and type of duet strikingly resemble those of the black-headed gonolek. The initiating call—again presumably made by the male—is answered by the female with (usually) three 'clicks' (Figure 4). Analysis of the thirteen duets illustrated for the response time of bird B makes interesting comparison with Thorpe's figures for the black-headed gonolek (see Table 1). Grimes finds the mean response time for the duets in Figure 4 to be 118 msec. with a standard deviation of 30 msec. Other pairs of this species were found to show different response times which could not be accounted for on the basis of distance between the birds or position of microphone. In addition to these differences in timing, there are also different versions of the 'whee-u' call uttered by a single bird (Figure 5). But, as with different duet patterns in the tropical bou-bou, there is considerable overlap between pairs, and

TABLE 1. Timing of Eight Consecutive Duets of a pair of *Laniarius erythrogaster*.

Duet no.	Time (sec.) at which note of bird 'A' commences	Response time of bird 'B' (= time (msec.) between start of the note of 'A' and that of 'B')
1	0	125
2	2·5 (2·5)	145
3	5·7 (3·2)	150
4	10·5 (4·8)	160
5	15·0 (4·5)	150
6	19·3 (4·3)	150
7	23·7 (4·4)	125
8	29·3 (5·6)	150
		S.D. of reaction time 12·6

The figures in brackets show the time in seconds since the previous note of bird 'A'. This could be regarded as the 'fore-period' for bird 'B'. (From Thorpe, 1963.)

Figure 6 shows similar 'whee-u' calls of eight different pairs of gonolek. However, these do differ in frequency and duration to some extent.

It is possible that both auditory response time and qualitative differences in repertoire are utilized for maintaining the pair bond in all these shrike species. After all, in all strictly antiphonal duets it is only the initiating bird that can 'estimate' the response time of its partner, and not vice versa. In the gonolek species the female probably recognizes the male by his vocal individualities; the male his mate by her response time. The latter could pose problems for the male in recognizing his mate at different distances, unless her vocal contribution also has recognizable peculiarities! In the tropical bou-bou and slate-coloured bou-bou (*L. funebris*) either sex may initiate a duet. Our observations on colour-ringed black-headed gonoleks showed that in this species the same bird (A) always initiated duetting

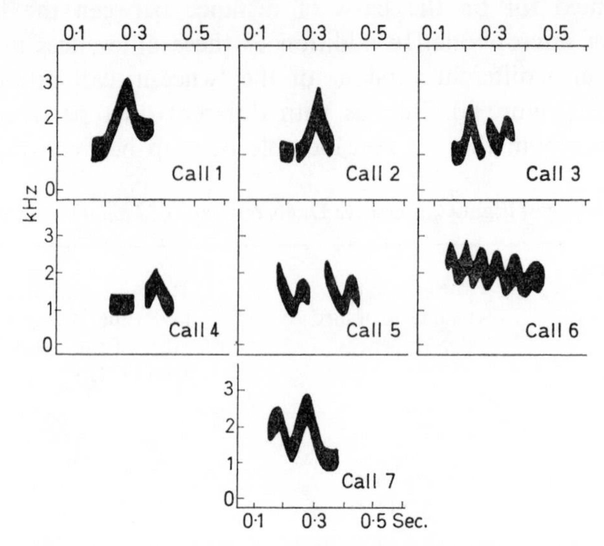

FIGURE 5. *L. barbarus.* Diagrammatic sound spectrograms of a series of different 'whee-u' notes of one pair recorded during an interval of about 1 h. (After Grimes, 1966.)

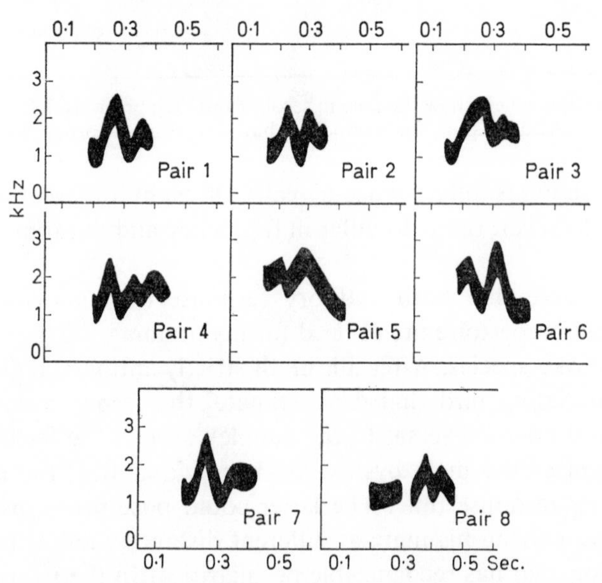

FIGURE 6. *L. barbarus.* Diagrammatic sound spectrograms of similar 'whee-u' notes of eight different pairs showing frequency and time interval differences. (After Grimes, 1966.)

and the roles were never reversed: though 'whee-u' calls frequently went unanswered, 'sneezes' were never heard alone.

We extended the field observations on the black-headed gonolek in a belt of riverine forest bordering the Suam River in north-west

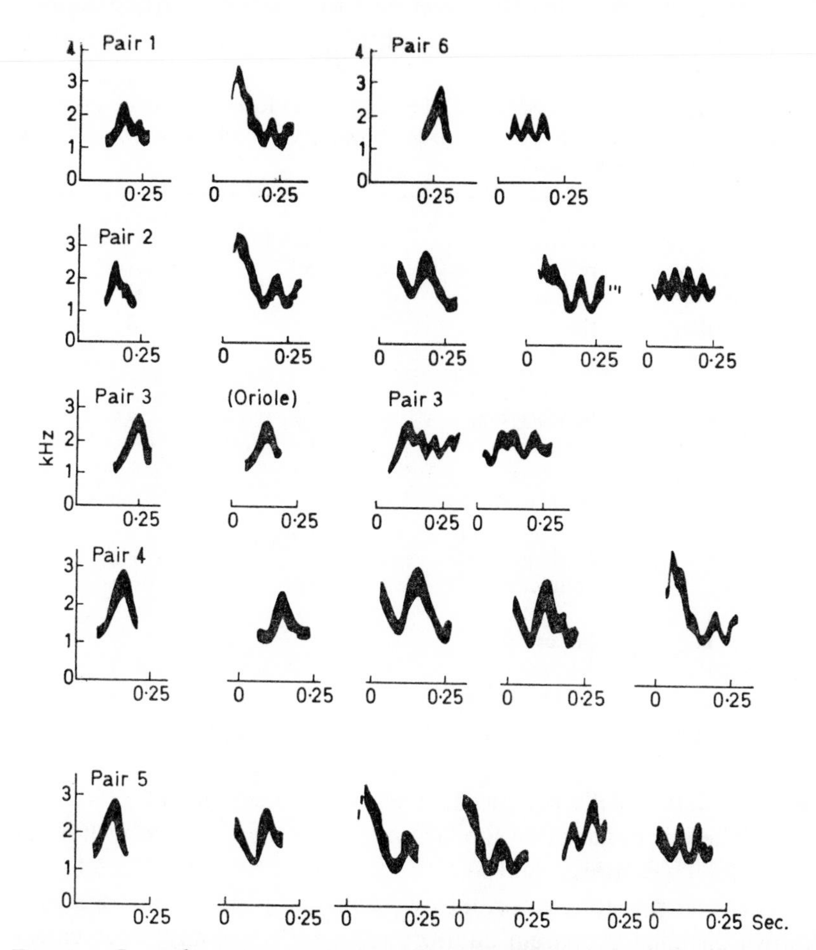

FIGURE 7. *L. erythrogaster*. A selection of diagrammatic sound spectrograms of six pairs to show variations within each pair and similarities between pairs. The oriole shown with pair 3 was apparently mimicking the first call of this pair.

Kenya. Six adjacent pairs were studied; six birds from four of these pairs being colour-ringed. The vegetation in this particular area, besides being dense, was predominantly of species covered by thorns or prickles. The birds were very difficult to see unless disturbed.

During aggressive duetting, resulting from playback of duets or in territorial disputes, the birds became most conspicuous and colour-rings showed well against the brilliant scarlet underparts.

Study of the ringed birds for a period of ten consecutive days—dawn to dusk—revealed that, contrary to Thorpe's earlier impression, there is a considerable variety of 'whee-u' ('yoik') calls. Figure 7

TABLE 2. Black-headed gonolek (*L. erythrogaster*). Response times of duets between members of the same pair of birds, showing relation between the response time and distance between birds.

Distance between birds (yards)	Average response time of duet series (sec.)	Correction for distance (sec.)	Corrected reponse time (sec.)
25	Series 1–0·32	0·07	0·25
25	Series 2–0·32	0·07	0·25
15	0·24	0·04	0·20
10	Birds mutually concealed 0·31	0·03	0·28
10	Birds in visual contact 0·23	0·03	0·20
1	0·15	—	0·15
<1	Series 1–0·13	—	0·13
<1	Series 2–0·13	—	0·13
<1	Series 3–0·12	—	0·12

shows different calls produced by the (presumed) males of the six pairs. There is remarkable similarity between these calls and those of the gonolek (Figures 5 and 6).

Whenever possible, recordings were made of birds separated by a known distance. Apparent auditory response times have been calculated from sound spectrograms: from the material at present available it would appear that there is a discrepancy between the response time for birds duetting closer to each other than about ten yards, and times for birds separated at distances greater than this, even when allowance has been made for the time taken for the sound to reach bird B from A. Table 2 shows response times calculated from sound spectrograms of different duet series from one pair of black-headed gonoleks. Times when the birds were ten yards or more apart have been 'corrected' for the time taken for the sound to travel between them, using the correction of 1,130 ft/sec. as given by Thorpe (1963). The figures are not absolutely reliable as the recorded information is of field estimates, not controlled laboratory conditions.

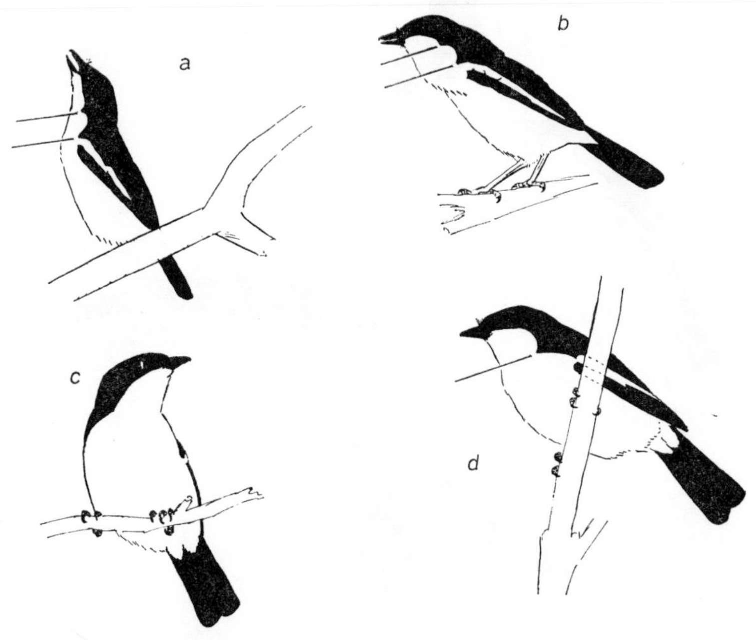

FIGURE 8. *Laniarius aethiopicus* (9 ins). Sexes similar.
a. Posture for the 'Snarl'—neck stretched; *b.* Posture for the 'bou-bou'—bobbing; *c, d.* Birds alert but not singing.
Contrasting contour between black and white is indicated by lines.

From observations on the black-headed gonolek and tropical bou-bou it seems likely that the more rapid response of birds close enough to be in visual contact is in fact a combination of auditory and visual signals. The cue would be the intention movements made by the initiator. Figures 8 and 9 show that in both the black-headed gonolek and the tropical bou-bou there is a marked contrast between upper and lower plumage—scarlet beneath in the former, white suffused with pink in the latter—the contrasting colours being separated by a clear-cut line of distinctive contour. When singing, the birds adopt characteristic postures which, in the tropical bou-bou, differ for the different calls (Figures 8*a* and *b*, 9*a*). Each call is preceded and accompanied by either stretching or bobbing movements, involving particularly the head and neck. That the answering bird responds to the movement and not the sound of the first note was confirmed on

FIGURE 9. *a. L. aethiopicus* (9 ins).
A pair singing the 'snarl/bou-bou' duet (Fig. 2, No. (2)). (Fig. 10*c*, No. (4)).
b. L. erythrogaster (8 ins). Sexes similar.
Alert. Note similarity of contrast pattern with that of *L. aethiopicus*. Underparts scarlet except for buff under tail coverts; iris bright yellow.
c. L. funebris (7 ins). Sexes similar.
Plumage: varying tones of glossy black and dull slate-grey with no marked contrast.

several occasions when it made intention movements before the first bird had produced a sound. This visual cue comes sufficiently in advance of the sound to increase the fore-period, or 'warning', for the selection and timing of a response. Considered as purely auditory reaction times, such cases are misleading as the apparent response time to the auditory signal is significantly shorter than the actual visual-plus-auditory signal-response interval. This could account for the differences shown in Table 2, where at distances between the birds

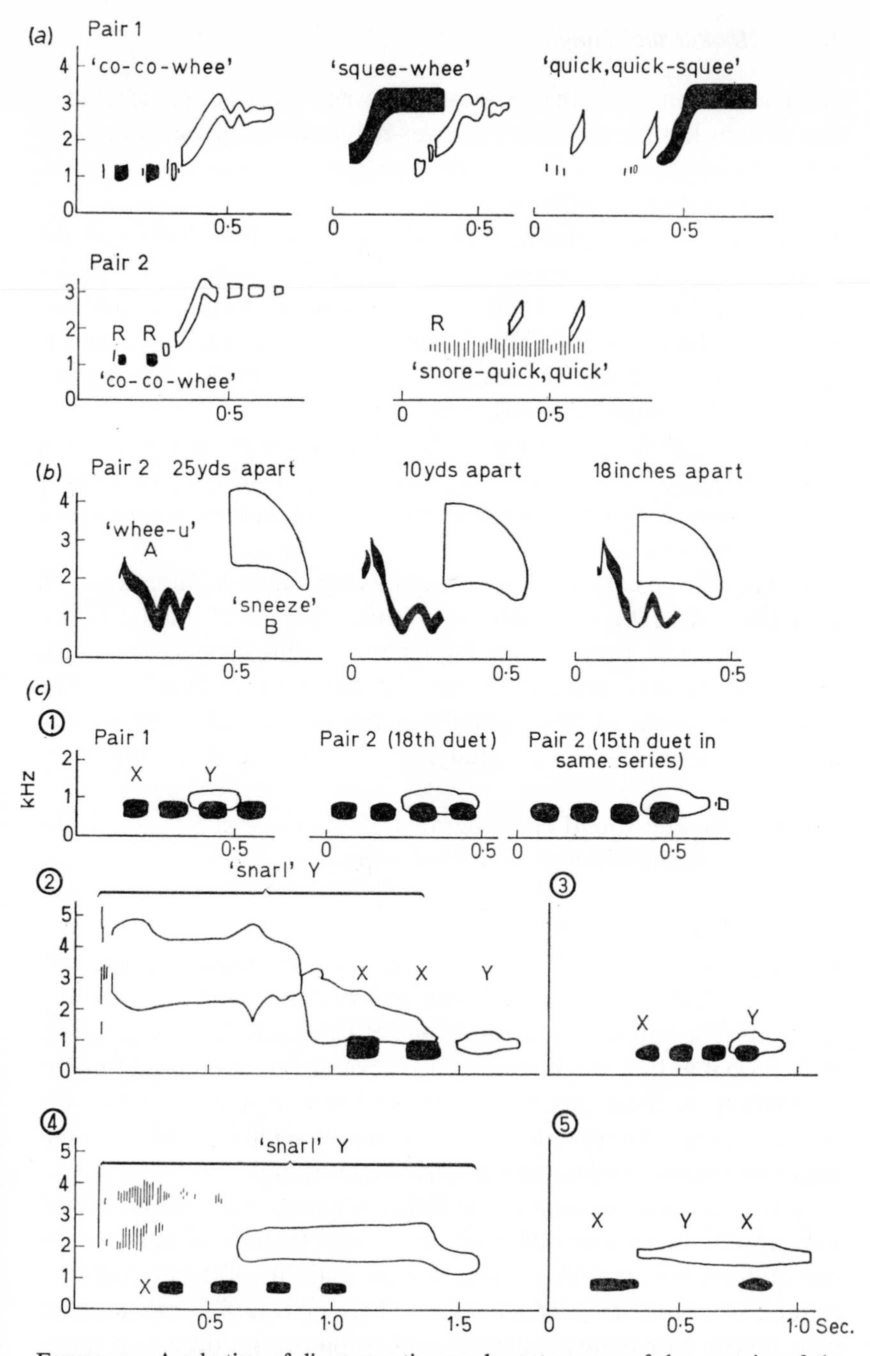

FIGURE. 10. A selection of diagrammatic sound spectrograms of three species of the genus *Laniarius*.

a. L. funebris; b. L. erythrogaster; c. (1 to 5): *L. aethiopicus.*

Pair 2 in (*a*), R indicates ringed bird;

X and Y in (*c*) represent presumed female and male respectively.

greater than ten yards, the response time was 0·20 sec. or longer, but at one yard or less the response time was reduced to 0·15 sec. or less.

SLATE-COLOURED BOU-BOU

In the same area as the black-headed gonolek there occurred the smaller slate-coloured bou-bou (*L. funebris*). The range of this species in East Africa is from Ethiopia and the Somali Republic, through Kenya, Uganda and Tanzania south to Lake Rukwa. Its habitat is again extremely dense vegetation, but generally in hot dry country and comprises thornbush and riverine forest.

The slate-coloured bou-bou tends to keep more to the lower scrub, unlike the black-headed gonolek which moves to the higher canopy when disturbed. From Figure 9*c* it can be seen that there is no marked plumage pattern.

The repertoire consists of a variety of very different patterns which lend themselves to onomatopoeic renderings rather than musical notation (Figure 10*a*). A first impression of this species is that the behaviour is very similar to that of the tropical bou-bou. The repertoires appear to be constant and sex-specific and either sex can initiate a duet series. We are studying the species further with aviary birds.

Diagrammatic sound spectrograms of the three species of shrike discussed are given in Figure 10 for comparison.

CONCLUSION

In the species of shrike described, duetting occurs throughout the year between members of an established pair occupying and defending a territory. The habitat is invariably one of extremely dense vegetation. It is probable that the birds pair for several seasons, if not for life.

Duetting in these species appears to function in maintaining the bond between family groups in dense vegetation and in joint aggressive vocal display during territorial disputes.

In tropical vegetation where visibility is greatly reduced, birds may have to rely much more on auditory signals than on visual cues for recognition. It is in such habitats that the majority of duetting species occurs. Duetting, in its wider sense, where male and female sing, or call, together or almost together and responsively, occurs in many families widely distributed throughout tropical and sub-tropical vegetation belts. To give but a few examples, duetting has been described in the marbled Guiana quail (Galliformes: Phasianidae) of Central American rain forests; some owls (Strigiformes: Strigidae)—the only case reported for Europe, but the birds are nocturnal so the same principles apply; the trogons (Trogoniformes: Trogonidae) widely distributed in tropical forests; the motmots (Caraciiformes: Momotidae) in forests of Mexico, south to northern Argentina; the barbets (Piciformes: Capitonidae) widely distributed in lush tropical forests, with the greatest development of the group in tropical Africa; the spinetails and ovenbirds (Passeriformes: Furnaridae) of southern South America; the tyrant flycatchers (Passeriformes: Tyrranidae) in the tropical lowlands of South America; the wrens (Passeriformes: Troglodytidae) of tropical Central America; the robin chats of the genus *Cossypha* (Passeriformes: Turdidae) of tropical African forests; the fantails or grass warblers of the genus *Cisticola* (Passeriformes: Sylvidae) very widely distributed, but particularly in Africa; and bush shrikes of the genus *Laniarius* (Passeriformes: Laniidae) of tropical Africa.

Duetting not only involves a long period of complex and accurate learning; but also, as has been shown in a few cases, the ability of one partner to mimic the other. This mimicry of antiphonal song patterns has been described for several bird species where the function would evidently seem to be the maintenance of a social bond. Waite (1903) described how two captive Australian magpies (*Gymnorhina tibican*) learned, as an antiphonal duet, a fifteen-note melody, each singing its own part. When one bird died, the survivor sang the complete melody which it had never before been heard to sing alone. Gwinner and Kneutgen (1962) described how, with pairs of captive raven (*Corvus corax*) and pairs of shama (*Copsychus malabaricus*), each member of a pair had certain more or less exclusive sounds or song elements not normally used by the mate. However, if the mate was absent the partner would use sounds normally reserved for the absent bird, with the result that the mate would return, as if called by name. This was found to be particularly effective if the male was absent, as nothing is more stimulating to a mated male than to hear its own repertoire repeated in its territory (Hinde, 1958).

In the shrikes we studied, playback of a complete duet resulted in both members of the pair approaching the tape recorder speaker and duetting aggressively from a conspicuous perch. Grimes (1966) also found this to be the case for the gonolek. On one occasion in the field, playback of the 'whee-u' call of a black-headed gonolek resulted in an immediate answering 'sneeze' from its mate, who apparently recognized his call from the tape recording. Captive tropical bou-bou shrikes can be made to respond to a whistled imitation of duet components. In all cases where response was obtained, it was by the appropriate bird singing the component required to complete the duet. The bird whose repertoire was being 'quoted' remained alert but silent (personal observations, 1967).

Thorpe and North (1966) quoted a case of a pair of captive tropical bou-bous which behaved in a similar way to the birds described by Waite and by Gwinner and Kneutgen. These two birds had developed a simple antiphonal duet pattern. One bird subsequently died, whereupon the other began to sing the complete pattern, continuing to do so for about a fortnight, whereas before it had never sung more than its own contribution to the pair's repertoire.

The relative importance, for recognition of mate or family group, of individual repertoire differences and of pair-specific response times is as yet far from clear. Further observations on captive birds under controlled conditions may throw more light on this fascinating phenomenon of duet singing in birds.

ACKNOWLEDGEMENTS

We would like to take this opportunity to thank Professor Thorpe who, with financial assistance from the Royal Society, made it possible for us to carry out field work in Kenya from September to December 1966. During this time Mr and Mrs T. J. Barnley did everything possible to facilitate the work and make the visit in every sense a memorable one.

REFERENCES

GRIMES, L. (1965) Antiphonal singing in *Laniarius barbarus* and the Auditory Reaction Time. *Ibis* **107,** 101–4.

(1966) Antiphonal singing and call notes of *Laniarius barbarus*. *Ibis* **108,** 122–6.

GWINNER, E. and KNEUTGEN, J. (1962) Über die biologische Bedeutung der 'Zweckdienlichen' Anwending erlernter Laute bei Vögln. *Z. Tierpsychol.* **19,** 692–6.

HINDE, R. A. (1958) Alternative motor-patterns in Chaffinch song. *Anim. Behav.* **6,** 211–18.

PUMPHREY, R. J. (1961) In A. J. Marshall, *Biology and Comparative Physiology of Birds*, vol. 2, p. 69. London.

SCHWARTZKOPFF, J. (1962) Vergleichende Physiologie das Gehors und der Lautausserungen. *Fortschr. Zool.* **15,** 214–336.

THORPE, W. H. (1963) Antiphonal singing in birds as evidence for avian auditory reaction time. *Nature* **197,** 774–6.

THORPE, W. H. and NORTH, M. E. W. (1965) Origin and significance of the power of vocal imitation with special reference to the antiphonal singing of birds. *Nature* **1208,** 219–22.

(1966) Vocal imitation in the Tropical Bou-bou shrike *Laniarius aethiopicus major* as a means of establishing and maintaining social bonds. *Ibis* **108,** 432–5.

WAITE, E. R. (1903) Sympathetic song in birds. *Nature* **68,** 322.

10. FUNCTIONS OF TERRITORIAL SONG IN THE WHITE-THROATED SPARROW

by J. BRUCE FALLS
Department of Zoology, University of Toronto

INTRODUCTION

Territorial bird-song has been the object of much detailed study (Thorpe, 1961; Armstrong, 1963). Variations in the sounds themselves, in the circumstances in which they are given, and in the response of different recipients have given rise to much speculation and some investigation concerning the information[1] communicated by song.

It is possible to infer that a particular message is encoded in a signal from the observation that the signal is normally associated with some condition or activity of the transmitter. Although such observation indicates the potential information content of a signal, the real test of effective communication lies in the response of a recipient. Similarly, syntactical studies of the signal may suggest which parameters are used to convey various kinds of information, but experiments in which these parameters are manipulated are necessary to determine which of the possible associations between message and the structure of the signal are biologically significant. Thus, to investigate the information communicated by bird-song a combination of observation and experiment is essential (Marler and Hamilton, 1966). Since the meaning of a signal to a recipient depends on the context in which it is received (Smith, 1965), it is particularly useful in studies of ecological function to carry out both observation and experimentation in the field.

I began to study song to elucidate territoriality, which I believe contributes to regulation of population density. In 1955, Judith Stenger and I attempted to map territories of ovenbirds (*Seiurus aurocapillus*) by playing back recorded songs in the field and

[1] In this essay information is used in its ordinary sense of knowledge or news rather than its more restricted, syntactical usage in communication theory.

discovered a stimulus-response system that lent itself to experimental manipulation (Weeden and Falls, 1959; Falls, 1963). Several studies followed with graduate students (R. J. Brooks, M. Lainevool, J. K. Lowther, W. E. Rees, H. B. Thorneycroft), undergraduates and my assistant, D. F. Robinson. At first we continued with ovenbirds but switched to the white-throated sparrow, *Zonotrichia albicollis*, which has a simpler song. The present essay is mainly a review of our research, both published and unpublished, with this latter species from 1957 to the present. Our aim has been to elucidate information communicated by the song of the white-throated sparrow, parameters of the song conveying different kinds of information, and the role of song in the territorial system of this species. The work is incomplete but illustrates an approach to these problems.

METHODS

We observe wild birds and experiment with them using recorded songs. In the field, various uncontrolled variables influence a bird's response to a recording and our experiments are designed to minimize their effects. Thus, because response varies among individuals, we play all the sounds to be compared to each of a sample of birds rather than play each sound to a different sample. To minimize the chance that a bird will become conditioned to the method of presentation or that the response to one sound will influence the next experiment, we leave a few days between trials. Experiments are confined to the breeding season and the daylight hours of the morning, when responses are normally strong. To ensure that each song is used throughout these periods, the order of presentation is random.

We have used two different experimental designs. The simpler one consists of comparing a bird's behaviour before, during and after a brief period (2, 3 or 5 min.) during which a song is played at normal, regular intervals (15 sec.). The first period serves as a control. Another control is to repeat the procedure using a sound (the song of another species) not biologically significant to the bird. When we have done this using unlike songs, we have never obtained a measurable response. This first method minimizes a bird's exposure to recorded sounds and we use it in most cases.

The second method, introduced by Brooks, minimizes variables associated with different times and days. The experiment consists of three or four consecutive fifteen-minute segments, each beginning with a three-minute period of playing as in the first method. A different sound is used in the middle segment (or the third if there are four). Comparisons are made between segments within the experiment and a control consists of playing the same sound in all segments. Results show no change in response from one segment to the next when the same song is played each time. We have used this design for comparing responses to very similar songs such as those of different individuals of the same species.

As far as possible, we have tried to use good quality recordings played at normal volume (except where volume is a variable in the experiment). Typical equipment and further details of the method are described elsewhere (Falls, 1963).

When an effective song is played through a loudspeaker located in an occupied territory, the territorial male typically approaches rapidly, moving after each rendition of the song. He may fly back and forth over the speaker, but usually stops moving after a few renditions of the recording and begins to sing nearby. However, song and movements may be intermixed. After the playing period is over, the bird may continue to sing from one or more positions. Calls or display postures sometimes occur early in the response. There is considerable individual variation in the mode of response, some individuals singing more than or sooner than others, while some are more prone to give calls or perform visible displays. Such differences are discussed below with respect to different plumage types in white-throated sparrows. However, the general description given above is typical of ovenbirds and several sparrows (mainly *Zonotrichia*) which we have studied.

We attempt to measure the strength of a response using its most obvious and universal features and relating them to the timed periods of the experiment. Thus, we note the bird's positions and vocalizations and determine the following: time of the bird's first song after playing begins, increase in number of songs (used most often), number of movements, and closest approach to the speaker. Comparisons can be made between periods in an experiment or between different presentations using 't' or χ^2 tests as applicable. Where possible, we try to compare data for the same bird.

Most of our studies are related to territorial behaviour. We have used songs of birds holding adjacent territories (neighbours) and others sufficiently separated that they do not normally hear each others' songs (non-neighbours or strangers). We carry out experiments on the regular (adjacent to the neighbour) or opposite boundary or near the centre of a bird's territory.

Our research has been conducted mainly in Algonquin Park, Ontario and near Fort Churchill, Manitoba.

THE WHITE-THROATED SPARROW AND ITS SONG

The white-throated sparrow is a common bunting breeding in open coniferous forest from Newfoundland to the Yukon Territory and as far south on the Appalachian ridges as West Virginia. It shows no taxonomic variation throughout its range but is polymorphic in plumage (Lowther, 1961) and karyotype (Thorneycroft, 1966). Two noticeable plumage types occur, one brightly marked especially in the region of the head (white-striped) and the other of similar pattern but much duller in colouration (tan-striped). The breeding plumage of the latter resembles the first basic plumage of most individuals. This polymorphism is not a function of age or sex. Birds of opposite plumage mate selectively. Among 110 mated pairs banded in Algonquin Park, 79 consisted of a white-striped male and a tan-striped female, 27 were the reverse, and only four consisted of two tan-striped birds (Lowther, 1961).

Lowther found several behavioural differences between birds of different plumage. For example, although males do most of the singing, white-striped females sometimes sing spontaneously early in the breeding season and often give a few songs in response to playback. Tan-striped females have not been heard singing in the field but Thorneycroft noticed a captive singing. Captives of both sexes and plumages have produced full song following injection of testosterone propionate. In the field, tan-striped males sing much less often than white-striped males, both spontaneously and when stimulated by playback. In general, tan-striped birds seem cryptic in behaviour as well as plumage.

The song, which is much admired for its purity, is easily distinguished from other vocalizations of this species and other natural sounds in the environment. It consists of pure, whistled notes without detectable harmonics, generally steady in pitch and arranged in a definite pattern. From a study of several hundred songs, Borror and Gunn (1965) listed the following commonly occurring types of notes: steady notes lasting more than a fifth of a second, similar notes beginning with a short upslur, and notes divided into three roughly equal parts (triplets). In their sample, the pitch of steady notes ranged from 2150 to 6500 Hz. Notes are separated by intervals of about 0·1 to 0·2 sec. Variations in the arrangement of different types and in the pitch of notes give rise to a variety of song patterns in a population.

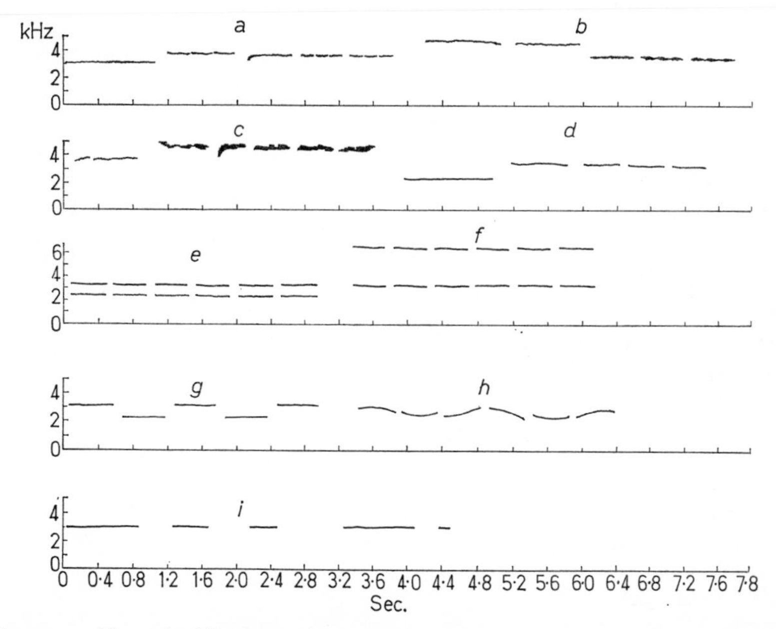

FIGURE 1. Normal white-throated sparrows' songs: ascending *a*, descending *b*, and female's song *c*. Artificial songs: normal *d*, two unrelated tones *e*, with first harmonic *f*, alternating pitch *g*, varying pitch *h*, and random timing *i*.

There are usually one or two changes in pitch in the course of a song. Borror and Gunn described 15 patterns but four made up over 96 per cent of their sample and two of them include the songs of nearly all the birds we studied (Figure 1*a* and *b*). The commonest pattern (ascending song) beings with a long steady note followed by a triplet, another steady note upslurred at the beginning, and a few more triplets; the second and later notes are of about the same pitch and higher than the first note. The second most abundant song (descending) begins with steady notes of about the same pitch (the second note may be slightly lower) followed by a third steady note considerably lower in pitch and a series of triplets at about the same pitch as the third note. Since either pattern can be rendered (by different birds) at any level of pitch within the normal range, the distinguishing features of the two patterns are confined to the first three notes; all the other notes are similar triplets.

TABLE 1. *Characteristics of white-throated sparrows' songs—occurrence and functions.*

Characteristics (with some common values and/or ranges)	Occurrence				Functions[2]		
	(1) Typical of species	(2) Stereotyped in individual	(3) Variable in individual	(4) Present in hand-raised[1]	(5) Species recognition	(6) Individual recognition	(7) Strength of motivation
Note types							
Long, pitch steady	+			+	+		
Others (1 or more types)	+			+[1]			
Triplets	+						
Pattern of notes		+					
Pitch							
Pure tones	+			+[1]	+		
Range (1500–6600 Hz)	+			+	+		
Changes (1 or more)	+			+	+		
Changes in direction (0–2)	+			+	+		
Absolute pitch		+				+	
Pattern of pitch		+				?	
Length							
Notes (0·1–1·7 sec.)	+			+	+		
Intervals (0·1–0·2 sec.)	+			+	+		
Song (3–4 sec, range 0·8–10 sec.)	+		+	+			+
Pattern of timing		+					
Number of notes							
Total (4–6, range 1–19)	+		+	+			+
Repeated (1–3, range 0–16)	+		+	+			+
Loudness							
Overall (up to typical level)	+		+	+			+
Pattern of loudness		+				?	
Singing rate							
Average (up to 4–8 songs/min)	+		+	?			+

[1] Besides at least one long note, pure and steady in pitch, hand-raised birds developed many abnormal sounds, a few of which were not pure tones. Their songs were stereotyped in the same respects as those of wild individuals. Based on research of H. B. Thorneycroft.

[2] Tentative lists based on experimentation.

There are no dialects in the usual sense in songs of white-throated sparrows, but the relative abundance of different patterns shows some geographic variation (Borror and Gunn, 1965). However, the two common patterns predominate both in Algonquin Park and at Fort Churchill.

The main features found in all songs are shown in Table 1 (column 1). From the results of Thorneycroft, I have also included in the table (column 4) those features that develop in birds taken as nestlings and raised in groups in sound isolation. Although their songs were normal in some respects, they contained many abnormal notes. It is particularly interesting that none of these birds developed triplets. Thorneycroft also found that young birds were capable of learning song patterns to which they were exposed from about 30 to 100 days of age. He postulated that they learn their adult pattern, possibly from their male parent, shortly after fledging.

Each adult male typically sings a single stereotyped pattern, but a few birds also have a second song of a different pattern, which they use less often. Songs of different birds vary in pattern, absolute pitch, pitch change between notes, length of notes, relative loudness of different notes, and the average number of notes. Since, apart from the number of notes, these features are relatively constant within the songs of any one bird, most individuals can readily be recognized by song. Repeated recordings of banded birds, studies of captives, and observations of wild birds with unusual patterns show that a bird's song remains essentially the same year after year. No characteristic features of songs of white-striped as opposed to tan-striped males have been discovered. Songs of females are often short, quavering, and variable in speed (Figure 1*c*) and in these respects resemble fall songs of males. That this may be a function of hormone level is suggested by the fact that Thorneycroft obtained songs like those of territorial males from injected females.

Songs of an individual vary in loudness, length, and rate of delivery. The usual length is 3 to 4 sec., the song consisting of five notes including two terminal triplets. While some individuals typically give longer or shorter songs, they all vary the length of their songs and we have heard a song with 16 terminal triplets. Occasionally, only one or two notes are given.

Thorpe (1961) has pointed out the tendency for territorial songs to include short notes of wide frequency range which aid in locating the singer, and the extreme rarity of acoustically pure tones. Yet here is a song composed almost entirely of pure tones. Perhaps location is aided by the repetitive elements and the relatively low frequency, for birds are able to determine its direction accurately and rapidly. The unusual character of the song makes it stand out against the acoustic background so it is presumbly an efficient signal. Only a few other birds in the white-throated sparrow's range produce whistles of constant pitch. The black-capped chickadee (*Parus atricapillus*) and the broad-winged hawk (*Buteo platypteris*) give whistles, but these are short and differ in quality from the white-throated sparrow's song. Harris's sparrow (*Zonotrichia querula*) occurs in the same general area as the white-throated sparrow near Churchill. Its song is a series of whistles without triplets but it also sometimes includes trills. Rees found that these two species, although very different in appearance, respond weakly to recordings of each others' songs. They usually occupy different habitats but, where they do occur together, their territories overlap and no strife is evident. We have found no character displacement in the white-throated sparrow's song where Harris's sparrow occurs.

During the breeding season in Algonquin Park, singing begins about 0330 h E.S.T., rises to a peak about 0400 h and then drops rapidly to a moderate level which is maintained through the morning. There is usually less song in the afternoon and a second peak occurs about 2000 h followed by a rapid decrease. Song finishes in the evening at a higher light intensity than that at which it starts in the morning. At Churchill the quiet period of the night is only about half as long as in Algonquin Park, with song commencing earlier and ending later; both events occur at higher light intensities in the north. In all cases, song begins and ends when light intensity starts to change rapidly. These observations suggest that the onset and ending of song is triggered by sudden changes in light intensity rather than any particular intensity. More data are needed to confirm this.

In Algonquin Park, temperatures below 50°F, which frequently occur early in the breeding season, seem to inhibit song. Similarly, there is little singing in periods of strong wind or heavy rain, but these conditions occur infrequently.

Like European buntings (Thorpe, 1961), the white-throated sparrow is strongly territorial during the breeding season. Its territory, which includes nearly all the activities of the pair, may be classified as Type A as discussed by Hinde (1956). Males arrive on the breeding ground about a week before the females. They soon commence chasing and fighting which continue until after the nest is begun but are rarely observed later. Although the female may accompany the male during skirmishes, she does not take part in territorial defence. We have seen no evidence of the exclusion of other species and territories of white-throated sparrows often overlap those of song sparrows (*Melospiza melodia*). The female builds the nest and incubates but the male shares in feeding the young. Lainevool found that the territory occupied by a pair is at first about an acre in extent but becomes small in the incubation and nestling periods. However, if a territorial bird is challenged either by a rival male or a recorded song played through a loudspeaker which is moved around, he will defend a larger area than he normally occupies. Thus, singing and feeding usually occur in only part of the area over which the male is dominant.

The fact that a recorded song elicits territorial defence in the same manner as an intruding male indicates that song plays an important role in the territorial behaviour of this species. Although singing begins during spring migration, frequent full song occurs only in the period when other males are driven from the territory. Chases and fights early in the season are often preceded and usually followed by song, and skirmishes along a boundary are normally interspersed with bouts of loud singing from one or both birds. However, most sustained singing occurs within the territory from two to four regularly used song posts. Once the boundaries are established, this advertisement of the presence of the male is apparently sufficient to maintain the integrity of his territory. Nevertheless, at any time during the season he will sing near the boundary, if challenged by a recorded song. However, late in the season responses are short, making it difficult to map a complete territory in this way. On a few occasions when Lainevool removed a male for a few days, allowing a new bird to take over the territory and female, and then returned the original territory-holder, there was a prolonged outbreak of fighting interspersed with song similar to agonistic encounters observed early in the season.

The singing of mated birds shows two peaks—in early June during egg-laying and incubation and again in early July after the young have left the nest. During the latter period, the young may have an opportunity to learn their song patterns.

In mid-July singing declines and females and young apparently disregard territorial boundaries. However, males often remain in their territories into August although defence is minimal. Previously unnoticed males sometimes establish territories and sing between the areas occupied by established birds near the end of the song season. They may return and establish territories in the same areas the following year. About half the established males and fewer females return. There is a brief recrudescence of song in September when migration has commenced.

What little evidence we have concerning the influence of song on females suggests that singing is important in locating mates and co-ordinating breeding activities. We have not found an early peak of song followed by a decrease after mating as noted in many other species (Armstrong, 1963). However low temperatures, which frequently occur early in the season, seem to inhibit song particularly in unmated birds. Later, when the song of mated birds fluctuates with the breeding cycle as described above, unmated males continue to sing strongly, averaging almost twice as many songs as those that are mated.

Lowther found that, when recorded songs are played in the field, both members of the pair may approach the loudspeaker, especially where the female is white-striped. Although both birds sing to some extent, the female responds chiefly by giving a precopulatory trill following each song from the loudspeaker or her mate. This often induces the tan-striped male to mount. Tan-striped females (with white-striped males) respond in the same way in the first few days following their arrival. Thorneycroft has observed that captive females may give precopulatory trills and assume a posture of solicitation when a male sings in a nearby cage.

SPECIES RECOGNITION

It is clear from field observations and playing back recordings that white-throated sparrows are able to distinguish songs of their own species from those of others. The stimulus-response situation provided by the playback technique makes it possible to investigate the parameters of song important for species recognition. Because of the simple character of the notes, a white-throated sparrow's song can be readily simulated with an audio-oscillator. If wild birds will respond normally to a recording of such an artificial song, it is possible to rule out any subtleties of the real song not reproduced in the copy as unnecessary for species recognition. In this way, an artificial song was prepared which accurately reproduced the pitch, timing and variations in loudness of an actual song but lacked the upslur of the third note and substituted continuous notes for triplets (Figure 1*d*). In a large number of field experiments, responses of territorial males to this song were indistinguishable from those elicited by the normal song which the artificial version copied (Falls, 1963). Thus, it appears that a song comprised of only one type of note is fully effective as far as our criteria allow us to judge the response. Bearing in mind species-specific features (Table 1, column 1), various altered versions of this artificial song were played to territorial males and the responses were judged using the sort of criteria outlined earlier. Some of these songs elicited responses similar in some or all respects to those obtained with the 'normal' copy; others could not be distinguished in their effects from songs of other species; still others were intermediate. In other words, some alterations reduced the response much more than others. The results can best be considered in relation to the properties of a normal song.

Continuous, pure tones are the most obvious feature of normal song. Making the notes variable in pitch (Figure 1*h*) greatly reduced the response. It may be relevant that females' songs, which are somewhat variable in pitch (Figure 1*c*), elicit weaker responses from males than do males' songs. The addition of an unrelated tone to the fundamental (Figure 1*e*) gave very weak responses and even the addition of a second harmonic (Figure 1*f*) had a similar effect.

The range of pitch within which normal songs fall was mentioned earlier and this proved to be rather critical. Birds did not respond to an otherwise normal copy two octaves higher, and one partly below the usual range received very weak responses. However, a song one octave higher and just inside the normal range elicited a fairly strong response. In natural songs, notes of different pitch are always present and the trend may change direction once or twice. A song all at one pitch received slightly less response than a similar version which included a change in pitch, while another song of similar timing but with four changes in pitch in alternating directions (Figure 1*g*) received still less response. Thus the pattern of pitch change is significant but, judging from the level of response, not as important as the character of the notes or absolute pitch.

The length of notes in a normal song is quite variable but the intervals between them are always rather short. Consistent with this, songs having notes of various lengths, including one that had only a single long note, gave roughly similar results. The only song in this series that resulted in very weak responses consisted of 15 notes of 0·1 sec. each. Although such notes sometimes occur in normal songs, they never comprise the entire song. In contrast to the effects of varying note length, a song in which the intervals between notes were up to four times the normal length and randomly arranged (Figure 1*i*) elicited only very weak responses. A song with equal intervals twice the normal length received intermediate responses. Thus, it seems to be important that timing should fall within certain fairly broad limits.

Eliminating the variation in loudness between notes had very little effect and the same can be said of varying song length. Both these features are rather variable in normal songs.

Without going into further detail, all the results of these experiments indicate that, to be effective, a white-throated sparrow's song must consist of unvarying, pure tones within a certain range of pitch. Less important is the presence of notes of different pitch arranged in a certain pattern. The notes should be of a certain minimum length and the intervals between them should not exceed a certain maximum (Falls, 1963).

Some features of the song seemed unimportant for species recognition in the context in which these experiments were carried out. Of these, some are quite variable features which may serve other functions. More puzzling is the case of the triplets which are invariably present in normal songs but could be replaced by continuous notes with no measurable effects. Possibly they provide species-specific information redundant under normal acoustic conditions. Although a certain amount of redundancy may be valuable in any message, the method used here rules out all but the essential content. It seems likely that such a constant feature of the song as triplets has some special function and this will be considered later.

INDIVIDUAL RECOGNITION

Since the songs of individual birds remain relatively constant from year to year and many of them are recognizable to our ears, it seemed likely that the birds themselves were capable of individual recognition by song. This was investigated by Brooks with several groups of territorial males using both the experimental designs mentioned earlier. By all criteria, birds responded more strongly to songs of strangers

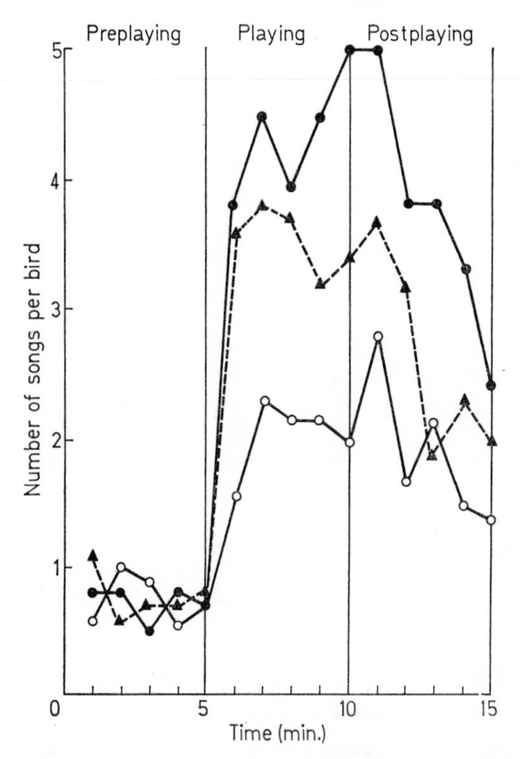

FIGURE 2. Number of songs given by territory holder in response to playback of stranger's (-●-), neighbour's (-○-), and the bird's own (---▲---) songs. Results of R. J. Brooks using first experimental design.

than to those of neighbours. The number of songs given by territory holders in sets of experiments using each experimental design are illustrated in Figures 2 and 3. The response to strangers was not only at a higher rate but also lasted longer so that, by the end of the observation period, the bird tested was still singing more than when a neighbour's song was played. However (Figure 3), this strong response did not carry over into the final segment of the experiment in which a neighbour's song was played. The sudden change in response between segments indicates that recognition is rapid, as does the observation that birds often responded after the first song of a stranger but only after several songs of a neighbour. On one occasion a bird ignored a recording of a neighbour's song to chase a stranger that had approached and sung in response to the broadcast. Responses to a bird's own song were intermediate in most respects (e.g. Figure 2).

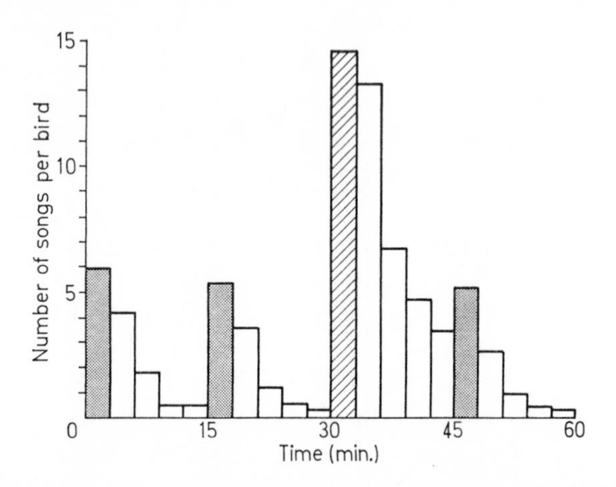

FIGURE 3. Number of songs given by territory holder during playback of neighbour's song (solid bar), stranger's song (hatched bar), and in periods following playback (open bars). Results of R. J. Brooks using second experimental design.

Clearcut differences in response to songs of strangers and neighbours were usually obtained if the experiment was performed on the territorial boundary adjacent to the neighbour whose song was used. Elsewhere, results were variable. Therefore, sets of identical experiments were repeated on the regular boundary, in the centre of the territory, and on the opposite boundary from the appropriate neighbour. On the regular boundary, responses were stronger to strangers than neighbours by all criteria, as already described. In the centre of the territory, responses tended to be stronger than at the boundary but strangers still elicited a stronger response than neighbours (significantly so with respect to number of songs and movements). However, on the opposite boundary, responses to neighbours were about as strong as in the centre and also about the same as responses to strangers by all criteria. Thus, when the neighbour was located in the wrong direction, it was treated like a stranger.

Another interesting result of Brooks's experiments was that

responses to neighbours' songs gradually decreased as the season progressed, while responses to strangers' songs remained strong. It seems unlikely that the small amount of experimental exposure to neighbours' songs was responsible for this result, especially as no decrement occurred when the same song was presented in four successive segments of an experiment.

Having demonstrated that territorial males can recognize each others' songs, the next question to consider is which features of song are important for this purpose. Songs of different birds differ in pattern, absolute pitch, pitch change between notes, length of notes, average loudness and the pattern of loudness, and the average length of songs (Table 1, column 2). Since absolute loudness was kept the same for all recorded songs in the experiments already described, differences in loudness are evidently not required for recognition. The pattern of loudness has not yet been investigated. Experiments were conducted to determine which of the other features were important.

First the major pattern was investigated, since that is the most obvious difference between individuals to our ears. Songs of strangers were played that were of the same pattern as or differed from that of the bird's neighbour. (Since there were only two patterns, one was similar to and the other unlike the bird's own song.) Descending and ascending songs of strangers elicited equally strong responses indicating that recognition of neighbours does not depend on the major patterns. Next, it was found that songs of strangers and neighbours could be distinguished when only the first three notes were played. That ruled out length or the number of terminal triplets which are rather variable features within an individual's songs.

In the remaining experiments of this series, neighbours' songs were altered in various ways. We considered that a feature important for individual recognition had been identified if an altered song evoked a stronger response than the original neighbour's song. Altered and original songs were played in different segments of the same experiment. First, we slowed the tape by about 15 per cent, altering both pitch and speed. Since these altered songs were treated like strangers', we had apparently changed an essential feature, but was it pitch or speed or both? Further experiments revealed that altering speed made no difference if pitch remained the same but that altering pitch alone or only the pitch of the first note rendered the song unrecognizable. Thus, absolute pitch and possibly pitch change between notes are important for individual recognition.

These results are consistent with what is known of natural songs since length of notes seems to be more variable than pitch in similar songs of the same bird. The long, steady notes of the white-throated sparrow are ideal for the analysis of pitch, and published accounts indicate that the pitch discrimination of birds is of the same order as our own (Thorpe, 1961).

Again, in studying individual recognition, several features of song were found to be non-essential. While some undoubtedly serve different purposes, it seems possible that others, although not actually necessary for the recognition of specific individuals, extend the possible number of signals that can be distinguished. For example, the existence of more than one pattern multiplies the number of songs of different pitch that can be fitted into the species-specific range of pitch. The minimum change in pitch required to eliminate individual recognition was found to be between 5 and 15 per cent. If we take 10 per cent as an approximation, about ten different songs of identical pattern could be recognized within the range of pitch found by Borror and Gunn. While a bird has fewer immediate neighbours than this, it can probably hear up to 15 or 20 birds from its territory. If there is any advantage in recognizing all their songs, the possibilities inherent in variation of pitch within songs of the same pattern and among different patterns should provide the necessary clues. Again, the possibility exists that several parameters provide redundancy thus ensuring individual recognition. We have no evidence for individual recognition of males by females but suppose that females can discriminate as well as males.

STRENGTH OF MOTIVATION

Three features of song which individual white-throated sparrows themselves vary remain to be considered. They are the number of notes (particularly the number of terminal triplets), loudness, and the rate of singing. With a few exceptions in the number of notes, these features were standardized in the recordings used to study species and individual recognition. In each case, a level was chosen typical of a bird singing steadily.

The first parameter to be considered is the number of songs given per unit time. By studying bouts of singing of several birds, we found the mean interval between the start of successive songs to be about 20 sec. with individual birds averaging 15 to 23 sec. There was a great deal of variation from song to song, intervals as short as 5 sec. occurring occasionally. The average rate of singing depends on frequency of bouts, bout length, and the average length of intervals in bouts. Tan-striped males are mostly deficient in all these respects, averaging about 0·25 songs/min. compared with about 0·8 songs/min. for white-striped males. Despite this apparent disadvantage, their territories are of normal size. Similarly, we have found no correlation between singing rate and territory size among white-striped birds.

Birds sing much more often following stimulation by a rival male or a recorded song. In playback experiments, average peak rates of singing for white-striped males are four to five songs/min. (about twice the rate for tan-striped birds). Individual birds often exceed this rate and eight songs/min. are sometimes given during the playback period. One bird sang 99 songs in 15 minutes. As described earlier, these periods of sustained singing are often associated with other manifestations of aggressive intent. For example, in an experiment, the bird approaches and moves about near the speaker. If an actual rival enters the territory, it is chased. In the afternoon, when spontaneous song is infrequent, birds often fail to respond or respond weakly to playback. From these observations one might postulate that rate of singing conveys information about the readiness of a bird to attack a rival.

In the experiments on species and individual recognition, we played four songs/min.—the normal rate for a stimulated bird. When strangers' songs were played at half this rate, responses were noticeably weaker. This demonstrates that infrequent singing constitutes a weak stimulus. It also strengthens the view, already put forward on the basis of observation, that variation in the rate of singing may convey quantitative information concerning the likelihood that the recipient will be attacked by the singer. Although two songs/min. elicited weak responses, playing eight songs/min. had little more effect than when the same song was played at a normal rate. Thus, beyond a certain rate, the effect on a recipient of delivering more songs is about the same.

In the song bouts already referred to, an average song consisted of five notes including two terminal triplets. Averages for different birds ranged from four to six notes. However, individual songs varied from one to ten notes and a majority of birds used most of this range at some time or other. Surely this great variation in song length must have some biological significance.

One possible explanation related to Hartshorne's (1956) anti-monotony principle, is that variation in length of an otherwise stereotyped song may facilitate frequent singing. If this were the case, one might expect successive songs to vary in length more often than would occur by chance. However, the rate of change appears to be random. Moreover, within a bout of sustained singing, there is no greater tendency to change length after a short interval between songs than after a longer one.

During the period when territories are maintained and song is frequent there seem to be no trends in song length through the day or the season. However, on a shorter time scale some trends emerge. After a pause of a minute or more, when a new bout begins, the first song is usually longer than those that follow. Perhaps this is why infrequent singers tend to give the longest songs. Also, the silent intervals between songs are longer after long songs than after short ones. Thus, there seems to be a tendency for song length to compensate for rate of singing.

The most obvious external factor influencing song length is the presence of danger in the form of a conspicuous observer. It is difficult to record a long song at close range and many of the songs given consist of only one, two, or three notes. Short songs are also sometimes given in response to playback, especially near the boundary of the territory and while the loudspeaker is actually playing. This may result partly from the tendency of a bird to stop singing when the playback occurs. Even when two birds are countersinging, they tend to sing alternately rather than overlap, which enables their songs to be heard with a minimum of interference. However, while short songs occur in some responses, normal or longer songs are given in others indicating that interference by the playback is not very important. It seems more likely that the occurrence of unusually short songs is associated with a conflict between the tendency for the singer to flee and the tendency to remain or approach.

Several experiments enable us to evaluate the effect of hearing short songs on territorial males. In experiments on individual recognition, when only three notes of a neighbour's song were played, responses were weaker than to a five-note version of the same song. In experiments with strangers' songs, a two-note song elicited weaker responses than the normal song from which it was made, even if the short song was played twice as often, but extra long songs were apparently not significantly more stimulating than songs of normal length.

Thus, short songs are weak stimuli compared with normal or long songs. This suggests that they communicate a weak tendency to attack on the part of the singer. Long songs, which often follow a long silent interval, may serve to attract the attention of potential hearers. They may also be given when the bird is not in any immediate danger and therefore when the motivation to attack may be high in a territorial species. This latter possibility should be investigated further.

Loudness of songs may be somewhat analogous to their length. Quiet songs are often given when an observer or a rival bird is present, and they are much shorter on the average than loud songs. They sometimes occur in response to playback, especially when the bird is near the loudspeaker, and often when the neighbour is nearby. Under these conditions, a bird may begin to sing quietly and increase the volume as time goes on. One might suppose that a quiet song was sufficient for communication at close quarters but the tendency for song to become louder argues against this interpretation and suggests that quiet songs more likely indicate weak or conflicting motivation. Also, playback experiments with quiet songs show that they are less stimulating to the hearer than loud songs. (These experiments started with two loud songs to attract the territory holder's attention.)

In this section, I have considered three parameters of song which are diminished in intensity by the birds themselves in circumstances suggestive of decreased motivation to attack or a conflict of motivation. The evidence for this interpretation is most convincing in the case of rate of singing. In each case, playback experiments with songs of considerably less than the typical intensity elicited reduced responses, indicating that the meaning communicated to rivals was less compelling. However, there was no corresponding increase in response when the intensity was supernormal. This apparent plateau of response may be real or only a result of limitation in the criteria used to measure response. Under natural conditions, a few exceptionally long songs are sometimes given when singing is resumed after a pause. They may serve to re-establish dominance. Perhaps this is also a function of the exceptionally sustained singing of early morning and evening.

DISCUSSION

It is generally agreed that bird-song embodies a variety of messages and several attempts have been made to list the items of information communicated (Marler, 1956; Welty, 1962; Armstrong, 1963). In the present studies, we have employed observation and experiment to investigate several of these possibilities with respect to territorial song and in a context where the recipient is a territorial male.

In view of its occurrence in boundary disputes and the way in which it seems to keep other males out of a territory, song evidently conveys to rivals the readiness of the owner to drive them from his territory. Only rarely is a singer challenged on his own ground. For his part, the territory-holder responds to song by approaching the source only when it occurs within or near the boundary of his territory. Thus there is convincing evidence that song carries a territorial message—a threat to rivals related to the area over which the singer holds dominance. However, if this is the principal message conveyed, it is at least qualified by a variety of other information.

Among related items of information communicated by the song of the white-throated sparrow are the actual location of the singer and his sex. A bird intending to approach can quickly and accurately locate the source, and the occasional songs of females, which seem to lack the precise control of pitch and timing characteristic of males' songs, elicit rather weak responses.

The song evidently identifies the species, since (weak) interspecific responses to recordings occurred in only one case (Harris's sparrow) and no evidence of interspecific territoriality was found. We have described the features of song common to all white-throated sparrows (Table 1, column 1) and shown, by determining those essential to evoke strong responses, which ones actually communicate species identity (Table 1, column 5). By contrast with these stereotyped features, there are several parameters—length, loudness and rate of singing, which the birds themselves vary, resulting in variation in the strength of response (Table 1, columns 3 and 7). Although there are typical intensities for all these parameters, they are not peculiar to the white-throated sparrow but are shared with other species including the ovenbird (*Seiurus aurocapillus*). The factors which have influenced their evolution are evidently common to many species—perhaps those with similar territorial systems. I have suggested that these parameters communicate the strength of the territorial message, i.e. the probability that the singer will attack a rival. Thus, they reflect the motivational state of the singer. However, unless combined with the more stereotyped species-specific characteristics they have no effect; they are necessary but insufficient to evoke a response. In a functional sense, characteristics communicating species identity may be regarded as the minimum group of additional features needed to elicit a normal response. Thus, although certain parameters alter the effectiveness of song quantitatively, I do not consider that they communicate a territorial message independent of species identity. The interspecific responses noted to songs of Harris's sparrow may be a case of mistaken identity where the signal has some of the characteristics of both species. I would similarly interpret the results recently reported by Brémond (1967) who found that an artificial song elicited weak responses from several species which have similar songs.

The need to keep territorial song species-specific may be more related to mate selection than territorial defence (Marler, 1960), but surely a significant conservation of energy is achieved by responding aggressively only to conspecific rivals.

There is no reason to suppose that any particular parameters of song will be the most important for species recognition in all species. In the white-throated sparrow's song, form of the notes, pitch, and to a lesser extent timing were important. In the ovenbird, which we studied in a similar fashion (Falls, 1963), the same parameters were used but timing seemed to be most critical. Similarly, Abs (1963) found pitch and timing to be most important in the song of the European nightjar (*Caprimulgus europaeus*). Using artificial songs, Brémond (1967) demonstrated convincingly that in the robin (*Erithacus rubecula*) organization of the song rather than the structure of notes is necessary to elicit a strong response. This song consists of sequences of a variety of notes with successive groupings of notes alternating between the upper and lower portions of the pitch range of the species. It seems likely that in each species the important characteristics are those that distinguish the song from its acoustic background. Thus Brémond laid particular stress on the variety of notes in the robin's song which distinguish it from similar songs of other species in the same environment.

While species-specific information is transmitted by stereotyped features, strength of motivation is apparently conveyed by variable features of song. Since in the white-throated sparrow quantitative variation in the appropriate parameters seems to parallel tendency to attack, the relationship of signal to message may be considered iconic (Marler, 1961). However, the relationship of the bird's motivation to its overt behaviour may be more complex.

In the white-throated sparrow's song, there is a close correspondence between characteristics typical of the species, those developed normally by hand-raised birds, and those important in species recognition (Table 1, columns 1, 4 and 5). The evidence suggests that these features are under more direct genetic control than those that characterize the songs of individuals (Table 1, column 2) which are learned during development. The triplets are an exception for, although they occur in all wild birds, they are not present in songs of hand-raised birds and are unnecessary for species recognition. Since they provide for a variable amount of repetition, I have suggested that they communicate quantitative information. Hand-raised birds often repeat other sounds which may serve the same function.

In 1961, Thorpe wrote 'The rival may recognize a given male not merely by the position of his territory but actually by his individual voice, independently of his territory'. These remarks seem at least partly justified in the light of recent work. We have experimentally demonstrated individual recognition of song by territorial males in ovenbirds (Weeden and Falls, 1959) and white-throated sparrows. Robinson's (1946–7, 1956) observations strongly suggest a similar phenomenon in the magpie-lark (*Grallina cyanobuca*) and the Western magpie (*Gymnorhina dorsalis*), and I was able to confirm these findings experimentally for *Gymnorhina tibicen* during a visit to Australia in 1964. In each case, it is clear that the song itself is recognized, since the response differs depending upon whether a known bird or a stranger sings from a location where the known bird usually sings. However, in the more detailed studies reported here, this difference disappears if the song originates from a location where the known bird does not normally sing. It is not clear whether location aids in recognition or there are other reasons associated with territorial maintenance for responding strongly to a neighbour that is out of place. More subtle criteria of response might show that the neighbour's song is recognized regardless of location, but we cannot conclude this from our data.

The parameters identified with individual recognition are some of those that vary among individuals but are stereotyped within the songs of each individual (Table 1, column 2). Considering the evolution of song, Marler (1960) has pointed out an apparent conflict between the requirements for stereotypy for species recognition and variation for individual recognition. His suggestion that these two functions might be relegated to different parameters of song is partly confirmed by our results (Table 1, columns 5 and 6). However, pitch seems to be important for both functions and a compromise exists whereby the range is critical for species recognition but, within that range, variation is used for individual recognition. A further dichotomy arises between the needs for stereotypy to serve both species and individual recognition and variation to express quantitative information. In this case, we have evidence that different parameters are employed (Table 1, columns 5, 6 and 7).

In our studies of individual recognition, we are evidently dealing with a stimulus specific decrement in response over a relatively long period, of the type known as habituation (Thorpe, 1956). Not only the song of a neighbour but also its general location may be aspects of the stimulus to which habituation occurs. This interpretation is supported by my observations on the response of Australian magpies to recorded songs played every few minutes in their territory. At first, the birds flew immediately to the speaker and sang but, after a few hours, the response had almost disappeared. When the speaker was moved to a new tree, the response was temporarily restored but waned even more rapidly than before.

From an ecological point of view, once territorial boundaries are established, a bird can conserve energy by not responding to a neighbour. As we have seen, a white-throated sparrow's response to neighbours declines as the season progresses but its response to strangers remains strong. This latter behaviour is also appropriate judging from results of experiments in which Lainevool removed territorial males. When the territory-holder was no longer advertising his presence, the vacated space and female were usually taken over by a male other than established neighbours. Similarly, there may be survival value in responding to a neighbour that is out of place. Thus learning to recognize individuals by song results in a restriction of actual fighting to cases where a genuine threat to a bird's territory may exist.

Learning may also be important in the interpretation of motivational information and possibly in species recognition as well. Falls and Szijj (1959) studied interspecific territoriality between two meadowlarks (*Sturnella magna* and *S. neglecta*) where one was ubiquitous and the other rare—occurring singly or in small groups among the common species. These species are similar in appearance and ecology but have different songs. Where both occurred together, they excluded one another from their territories and responded strongly to recordings of each others' songs. However, where the rarer species was absent, the common one did not respond to the other's song. Here we have a case where two competing species evidently learned to recognize each others' songs presumably by association of the song with disputes evoked by their similar plumages. One is tempted to speculate that the association of song with fighting, which occurs in many species, may be at least partly responsible for the effectiveness of song as an intraspecific threat. To a strange intruder, the song of a territorial male may be associated with eviction from the territory. To a territorial male, strange songs may be associated with fighting, while those of neighbours are associated with the status quo and a minimum of physical combat. Since territoriality may be regarded as a form of social dominance (Marler and Hamilton, 1966), it is reasonable to suppose that learning is involved in maintaining the system as in other social hierarchies.

Whatever the behavioural mechanisms involved, song appears to be a substitute for the more energetic and risky forms of agonistic behaviour, reducing physical combat to a minimum. During over 400 experiments with recordings, Brooks observed less than ten chases, even though two neighbours were often attracted to the speaker. Thus, although it may have aggressive motivation, song helps to promote a cooperative territorial system for the division of resources.

Song is particularly suited for maintaining dominance over a large area where vegetation obstructs vision. However, some individuals which sing very little still succeed in holding territories. This is particularly true of tan-striped white-throated sparrows. Other mechanisms may compensate for song in these cases. Although important, song is only part of the signalling system available to birds.

Finally, it is worth recalling that we have only investigated the responses to song of territorial males. It undoubtedly serves a variety of functions other than those discussed in this essay. Even in this restricted area of research, much remains to be investigated. Progress in studies of bird-song will continue to be stimulated by the work of Dr Thorpe and his associates.

ACKNOWLEDGEMENTS

I am indebted to those who took part in this research, to the National Research Council of Canada and the Ontario Department of University Affairs for support, and to the Ontario Department of Lands and Forests and the National Research Council for facilities in Algonquin Park and at Fort Churchill, Manitoba, respectively. My colleague, Dr David Dunham, read and discussed the manuscript.

REFERENCES

ABS, M. (1963) Field tests on the essential components of the European nightjar's song. *Proc. 13th Int. Orn. Congr.* 202–5.

ARMSTRONG, E. A. (1963) *A Study of Bird Song.* Oxford Univ. Press, London.

BORROR, D. J. and GUNN, W. W. H. (1965) Variation in white-throated sparrow songs. *Auk* **82,** 26–47.

BRÉMOND, J.-C. (1967) Reconnaissance de schémas réactogènes liés à l'information contenue dans le chant territorial du rouge-gorge. *Proc. 14th Int. Orn. Congr.* 217–29.

FALLS, J. B. (1963) Properties of bird song eliciting responses from territorial males. *Proc. 13th Int. Orn. Congr.* 259–71.

FALLS, J. B. and SZIJJ, L. J. (1959) Reactions of eastern and western meadowlarks in Ontario to each others' vocalizations. *Anat. Rec.* **134,** 560.

HARTSHORNE, C. (1956) The monotony threshold in singing birds. *Auk* **73,** 176–92.

HINDE, R. A. (1956) The biological significance of the territories of birds. *Ibis* **98,** 340–69.

LOWTHER, J. K. (1961) Polymorphism in the white-throated sparrow. *Can. J. Zool.* **39,** 281–92.

MARLER, P. (1956) The voice of the chaffinch and its function as language. *Ibis* **98,** 231–61.

(1960) Bird songs and mate selection. In *Animal Sounds and Communication,* Lanyon and Tavolga (eds.). Washington, D.C. pp. 348–67.

(1961) The logical analysis of animal communication. *J. Theoret. Biol.* **1,** 295–317

MARLER, P. and HAMILTON, W. J. III (1966) *Mechanisms of Animal Behavior.* New York.

ROBINSON, A. (1946–7) Magpie-larks: A study in behaviour. *Emu* **46,** 265–81, 282–91; **47,** 11–28, 147–53.

(1956) The annual reproductive cycle of the magpie, *Gymnorhina dorsalis* Campbell, in south-west Australia. *Emu* **56,** 233–6.

SMITH, W. J. (1965) Message, meaning, and context in ethology. *Am. Nat.* **99,** 405–9.

THORNEYCROFT, H. B. (1966) Chromosomal polymorphism in the white-throated sparrow, *Zonotrichia albicollis* (Gmelin). *Science* **154** (3756), 1571–2.

THORPE, W. H. (1956) *Learning and Instinct in Animals.* London.

(1961) *Bird Song: The Biology of Vocal Communication and Expression in Birds.* Cambridge.

WEEDEN, J. S. and FALLS, J. B. (1959) Differential responses of male ovenbirds to recorded songs of neighbouring and more distant individuals. *Auk* **76,** 343–51.

WELTY, J. C. (1962) *The Life of Birds.* Philadelphia.

11. EMBRYONIC COMMUNICATION, RESPIRATION AND THE SYNCHRONIZATION OF HATCHING

by **MARGARET A. VINCE**
Psychological Laboratory, Cambridge

THE EXCHANGE OF INFORMATION IN AVIAN EMBRYOS

In birds, the young begin to breathe through their lungs many hours, or even some days, before hatching; therefore it is not surprising that vocalizations are frequently reported in unhatched eggs. Such calls have been recorded and analysed in the chick embryo by Collias (1952) and Guyomarc'h (1966), and in duck embryos by Kear (1968). However, in dealing with embryonic communication and synchronization of hatching, we shall be concerned here not only with vocalizations but with other sounds, and in addition with some sub-audio frequencies produced by embryos. These sounds also become audible towards the end of the incubation period from about, or just before, the beginning of lung ventilation; thus they begin earlier than vocalizations.

Lung ventilation begins when, or just before, the embryo's beak pierces the inner shell membrane which separates it from the air space (Figure 1); the bird then breathes the air in the air space, which at this time has an extremely high carbon dioxide and extremely low oxygen content (Romijn and Roos, 1938, Roos and Romijn, 1941; Visschedijk, 1968*b*). After a few more hours, the egg 'pips': the outer shell membrane and the shell are perforated by the beak, and the embryo is now able to breathe fresh air from outside. The pip is visible where a small piece of shell is pushed outwards. In chick embryos studied by Romijn (1948), the inner shell membrane was pierced on day 19 (chickens hatch on day 20 or 21); for chickens also comparable figures given by Kuo (1932*a*) show that the majority of embryos examined had pierced the membrane on day 18, begun to breathe on day 19, and hatched on day 20. In isolated quail embryos our recordings have shown that this period between the beginning of lung ventilation and hatching may last as long as sixty-seven hours in the bobwhite (*Colinus virginianus*, incubation period twenty-three days) and up to forty-eight hours in the Japanese quail (*Coturnix coturnix japonica*, incubation period seventeen days). During this

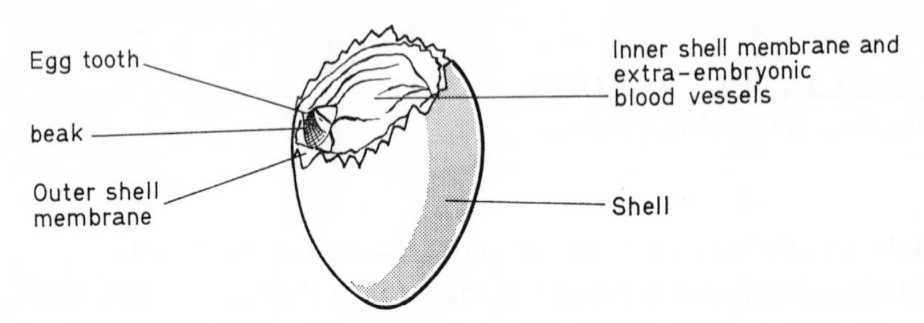

FIGURE 1. Egg opened over air space to show beak protruding through inner shell membrane.

period, foetal respiration by means of the allantoic vessels just inside the inner shell membrane is gradually superseded by lung ventilation, but the blood circulates in the allantoic vessels until just before hatching: if windows are made in the shell and the membrane rendered transparent the blood may be seen to drain out of the extra-embryonic blood vessels a matter of minutes before the embryo

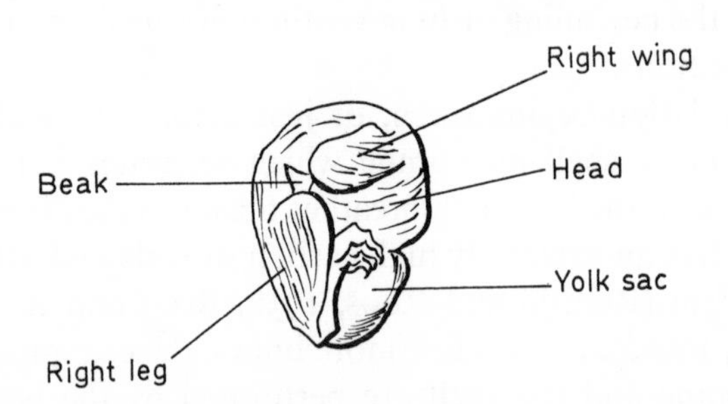

FIGURE 2. Pipped quail egg; shell, both shell membranes and extra-embryonic blood vessels have been removed.

begins to cut round the shell. According to Kuo (1932*a*) the yolk sac is completely withdrawn into the body cavity in 60 per cent of chicken eggs on day 18, but our observations suggest that this is completed later in quail, and in fact, in sixteen *pipped* bobwhite eggs which were opened, the yolk sac was only partially withdrawn in every case (Figure 2). Therefore, in listening for sounds from avian embryos, we are concerned mainly with the period between the piercing of the inner shell membrane and hatching, when gaseous exchange is gradually being taken over by the lungs and when embryonic development has still to come to completion.

During this period the embryo is already in behavioural contact with the outside world, responsive to external stimulation. Kuo (1932*a*) reported that chick embryos would respond to sound, vibration and sometimes to light after piercing the membranes, and since then there have been many reports of unhatched embryos calling, or responding in some way to stimulation from outside. For example, Goethe (1937, 1955) reported that hatching herring gulls (*Larus argentatus*) may stimulate each other to call and to make lively movements; that newly pipped embryos answered the calls of the disturbed colony overhead, and that shaking or sudden turning of the egg would produce vocalizations. Collias (1952), working with chick embryos, has reported that a pipped egg may give the appropriate call to warming or cooling, or can stop giving 'distress' calls at the sound of clucking, or will give 'pleasure notes' in response to a change in position when the egg is held in the hand. Lind (1961) mentions peeping, and an increase in activity at the beginning of cooling, in the black-tailed godwit (*Limosa limosa*). More recently Gottlieb (1968) has examined in detail vocal and other responses of unhatched domestic chicks (*Gallus gallus*) and ducklings (*Anas platyrhynchos*) to the maternal call of their own and other species, and in ducklings has demonstrated a selective responsiveness to different types of stimulation: an increase in oral activity to the maternal call of their own species, and an increase in heart rate in response to the maternal call of other species. This response to the maternal call is facilitated by the peeping of siblings from day 23 onwards (ducks hatch on day 28).

Some such discrimination clearly occurs if the embryos respond to specific stimuli, as in addition to sounds from outside the clutch, the eggs themselves produce a wide range of audio- and sub-audio frequencies. Some of these sounds and movements are easily identified, some are not; as there is evidence that some are used for communication between embryos, they will be considered now. Apart from vocalizations, the most characteristic sound is a regular loud click which is audible when the egg is held to an ear, and has long been recognized as indicating that hatching is imminent. In ducklings clicking is reported to begin soon after the onset of lung ventilation (Driver, 1967; Kear, 1968), but in chick embryos and in quail relatively silent lung ventilation can be recorded many hours (up to 46) before the beginning of regular clicking (Figure 3*a* and *b*); in

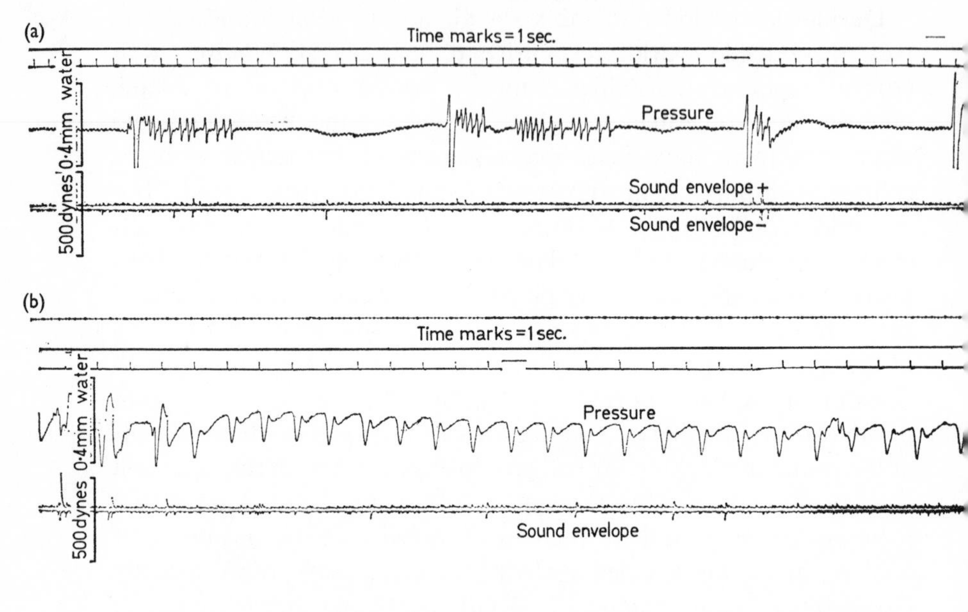

FIGURE 3*a*. Silent period: irregular breathing and beak clapping in the earliest stage of lung ventilation. Bobwhite, 1·5 h after first breath, and 66·25 h before hatching. In all recordings shown a deflection towards the time marks indicates a fall in pressure. *b*. Relatively silent period, and regular breathing before the beginning of regular loud clicking. Bobwhite 32·75 h. before hatching. For recording technique see Plate 1.

quail clicks, except certain soft clicks (Fig. 4*a*), are infrequent until about twelve to fifteen hours before hatching when regular clicking begins (Figure 4*b*) and they usually come to an end before the chick emerges from the shell (Figure 5) (Vince, 1964*b*, 1966*a*).

These clicks have sometimes been thought of as tapping against the shell but as long ago as 1911 Breed, working on chickens, observed embryos through holes made in the shell and noticed that the clicks coincided with the rhythmic pulsation of the bird, which he took to be breathing; the sound comes from inside the embryo. This has since been confirmed by Kirkman (1931) in the black-headed gull (*Larus argentatus*), by Lind (1961) in the black-tailed godwit

(*Limosa limosa*), and by Driver (1964) in a large number of ducks and other species. Driver found that the clicks could occur on inspiration, or expiration, or both (compare Figure 4*a* and *b*) and suggested (1967)

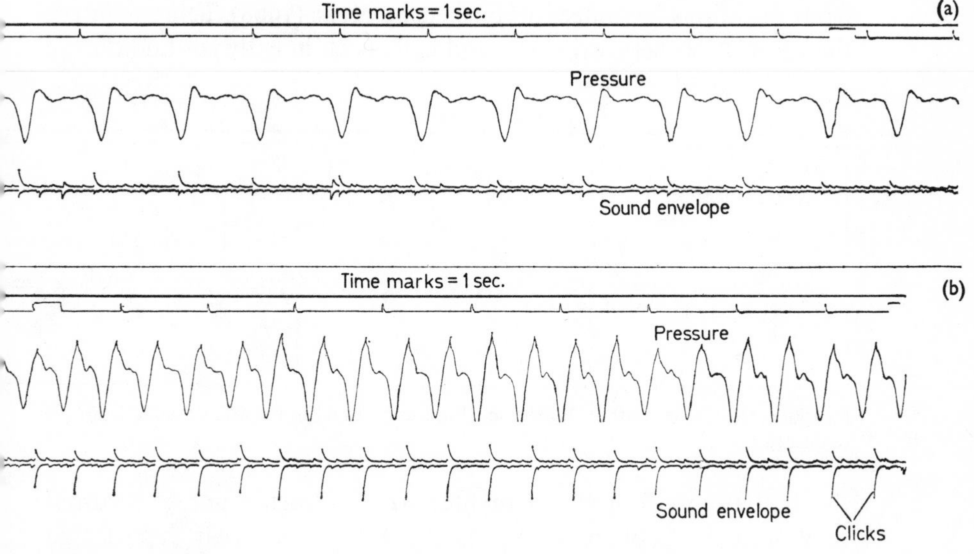

FIGURE 4*a*. Soft click, and regular breathing of medium amplitude. Bobwhite, 9·5 h before hatching.
FIGURE 4*b*. Regular loud click and rapid breathing of high amplitude. Bobwhite, 2·5 h before hatching.

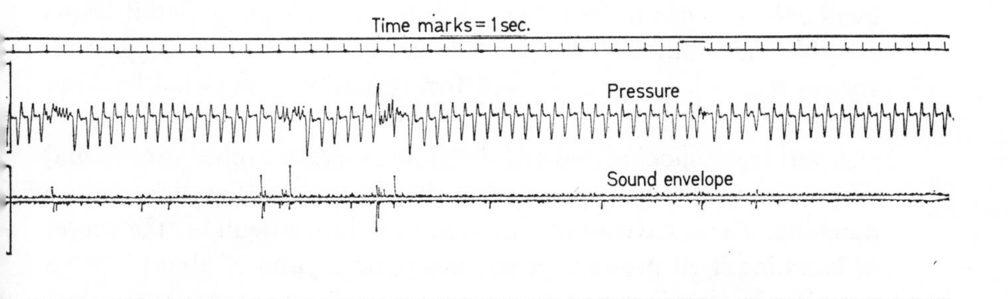

FIGURE 5. Relatively silent period after the end of regular loud clicking and before hatching; regular breathing of medium amplitude. Bobwhite, just before hatching.

that they are an accidental consequence of the respiratory system's early operation. Clicks have been recorded from two passerine species (thrush and sparrow) as well as game birds (Vince, 1966*a*), and it was then shown that each click consisted of a column of frequencies, some going up to about 8,000 Hz, and that some clicks consisted of multiple sounds (cf. Figure 17); also each click included both audio and sub-audio frequencies. Similarly, sonograms of clicks obtained from ducklings have been published by Kear (1968), demonstrating the association between clicks and calls both in embryos and during

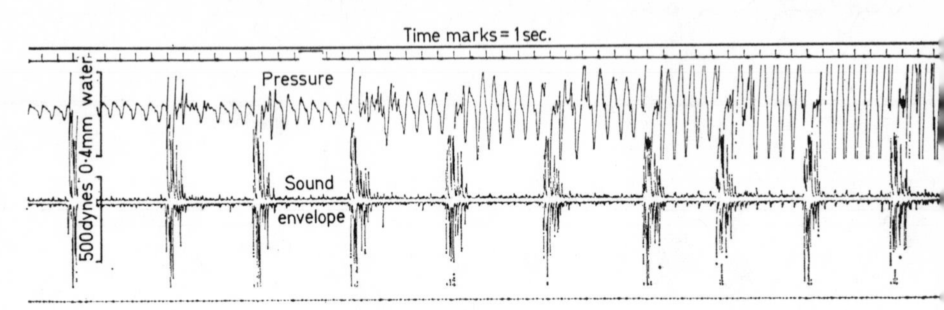

FIGURE 6. Hatching rhythm. Bobwhite. (Figures 3*a* to 6 are recordings made from the same egg.)

the first twenty-four hours of life: when 'pleasure notes' occurred they were in between clicks. Both Driver and Kear report that clicking may be heard in ducklings for some hours after emergence from the shell, and it is possible to hear quite loud clicks in nestling great tits some days after hatching, in particular if they are excited in some way (Vince, 1967). However, in recording newly hatched quail chicks, it has so far proved difficult to elicit clicks, and it seems possible that, although clicking has been heard from the egg in all species where it has been looked for, it may vary in detail between species.

Apart from clicking and vocalizations, there are other sounds and vibrations which can be recorded during the last few days before hatching. These have so far received very little attention; the sound of hatching itself produces a characteristic rhythm of about 8 to 10 signals/min. consisting of a regular series of blows against the shell and movements of the embryo as it turns within the egg before striking again (Figure 6). This cutting round takes about 10 to 30 minutes, and the whole process is often preceded and/or accompanied by rapid vocalizations. Very often the embryos begin to cheep repeatedly in the relatively silent period between the end of clicking and hatching. Rapid cheeps have been reported and analysed in the chicken at this stage by Guyomarc'h (1966) and by Collias (1952). Another characteristic sound of lower intensity and which occurs much earlier is a short burst of low frequencies (c. 20 to c. 40 Hz) lasting up to about one sec. These were not audible in our earlier recordings (Salter, 1966; Vince, 1966*a*) but were revealed by a new system produced more recently (Salter, unpublished; Vince and Salter, 1967). They are audible for many hours before the membranes

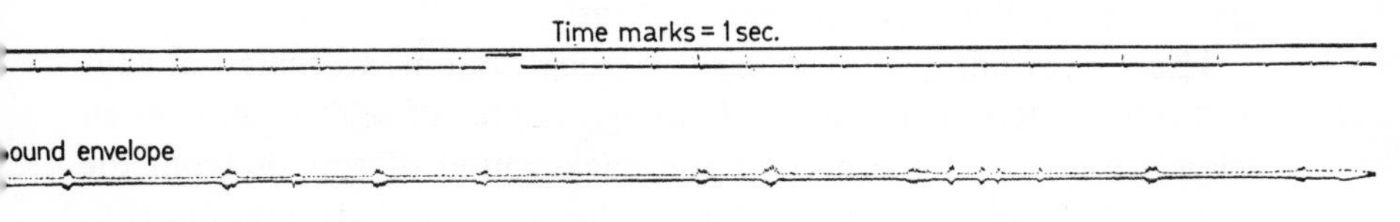

FIGURE 7. Short bursts of low frequency vibrations from anaesthetized quail chick, soon after hatching.

are pierced, occur at regular intervals, tend to be drowned by the sound of clicking, are audible again after that and may be recorded from hatched chicks (Figure 7). They may in fact be seen in newly hatched chicks as short bursts of vibrations. In the egg they can best

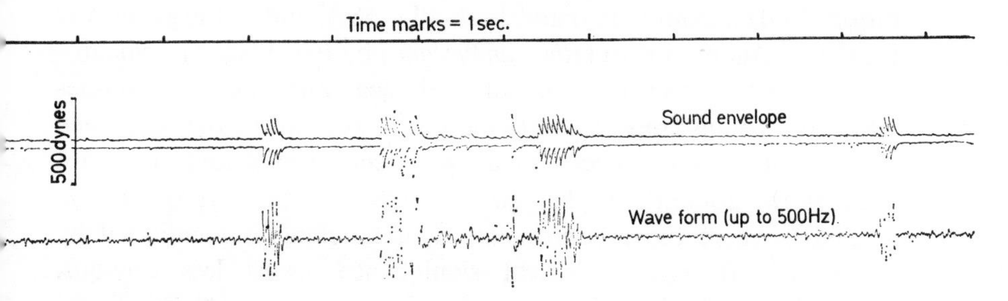

FIGURE 8. Low frequency sounds from opened egg (embryo had not pipped the shell!). Japanese quail.

be recorded when a large part of the shell over the air space is removed, presumably because this weakens the shell structure (Figure 8). It seems likely that they are produced by muscle tremor, and the forces involved are small.

Vocalizations, the sound of hatching, breathing, clicking and heart-beat and also these low frequencies are the most easily distinguishable of the sounds produced by quail embryos; however, other sounds occur intermittently and at various intensities and are produced by limb movements, head-lifting, beak-clapping, etc., and include both high and low frequencies; in fact, for some hours before the membranes are pierced, audio and sub-audio frequencies are produced at an increasing rate until the chick emerges. During this time, the amplitude of signals tends on the whole to increase, presumably as a result of growth and increasing strength in the embryos, the drying of the membranes, and, of course, the occurrence of regular loud clicking as well as the hatching rhythm.

Thus, discriminative responses to vocalizations, to maternal calls and other external stimuli imply a high degree of selectiveness when stimuli are considered against the background of sounds from the clutch which also form part of the embryo's immediate environment. In the second part of this paper, I shall consider the selective responsiveness of quail embryos to the 'background' itself.

THE SYNCHRONIZATION OF HATCHING

The synchronization of hatching has been studied in greatest detail in the quail, but it occurs in other species of ground-nesting birds, such as pheasants, partridges and grouse, and is described in detail in the mallard duck (*Anas platyrhynchos*) hatched under natural conditions by Heinroth (1959) and Bjärvall (1967), and in the artificially incubated common rhea (*Rhea americana*) by Faust (1960). Synchronization would appear to be of biological importance in species where the young are sufficiently mature to run out of the nest when dry, but it does not occur in all precocial species and does not occur in the domestic chicken (*Gallus gallus*). In the skylark (*Alauda arvensis*), where the synchronization of hatching is mentioned by Delius (1963), any biological significance seems less obvious. Normally, in a healthy clutch, quail embryos cut round the shell and emerge within an hour or two of each other, and may do so at the same time; in a tray of eggs in the incubator, hatching may begin at one side, and an hour or two later will have spread across the tray to the other side.

This synchronization of hatching occurs only when the eggs are in contact with each other; in eggs isolated about forty-eight hours before the expected time of hatching, embryos may emerge from the shell at different times over a day or two. This effect of contact suggested that the embryos stimulate each other, and this was demonstrated by placing eggs in clutches given twenty-four hours more incubation, when hatching still occurred more or less simultaneously; the spread of hatching within each clutch was never more than six hours (Vince, 1964*a* and *b*). It appeared therefore that the time of hatching could be accelerated by contact with more advanced embryos. It was explained earlier that the developmental stages leading to hatching must be completed before the chick emerges and are not, in fact, complete until just before the embryo begins to cut round the shell. Therefore, in considering the synchronization of hatching, we are faced with a number of different problems. For example: (i) does synchronization depends only on the acceleration of retarded embryos, or is the delayed egg really slowing down its more advanced siblings, or does the clutch work out a mean best time of hatching by bringing on the late embryos and holding back the most advanced? Or (ii) does the synchronization of hatching depend on some kind of rapid adjustment of the embryo at one time or on a gradual speeding up (or slowing down) of development, and if the latter, over what period of time? Also (iii) what is the stimulus, or what are the stimuli producing a possible slowing down, or speeding up, response? And (iv) what effect, if any, does the adjustment of hatching time have on the chick? These four main problems will now be considered in turn.

Possible methods of synchronization

A twenty-four-hour delayed egg hatching at the same time as neighbouring siblings does not tell us whether the delayed egg is being brought on or the siblings held back; it can, however, be shown that stimulation may produce actual acceleration by giving all the eggs an equal amount of incubation, then separating them, stimulating one artificially, and comparing its hatching time with the others. In one experiment of this type (Vince, 1966*b*) eggs were stimulated by sound or vibration at approximately the click rate (3/sec.). Almost all stimulated eggs hatched before the mean time of unstimulated siblings, the amount of acceleration varying up to a maximum of forty-nine hours. Hatching times of individual bobwhites stimulated by sound are shown in Figure 9. There is thus no doubt that hatching time may be accelerated.

If synchronization were achieved by acceleration alone, one might expect a shorter mean incubation period in groups than in isolates, and we do not find this to be the case. The eggs appear to work out a mean time of hatching for the whole clutch; the most advanced appear to be slowed down, presumably by stimulation produced by

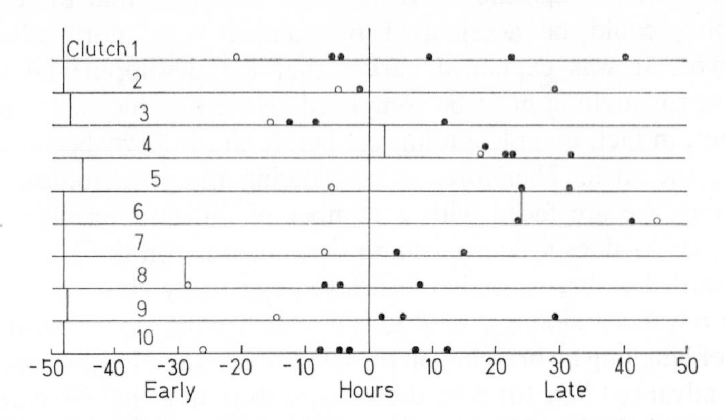

FIGURE 9. Hatching times of bobwhite eggs stimulated by sound, compared with unstimulated siblings. O h, expected time of hatching. ○, stimulated egg; ●, isolate; |, beginning of stimulation. Where there are two perpendicular lines, the second indicates the end of stimulation. (From Vince, 1966*b*, by courtesy of Baillière, Tindall and Cassell.)

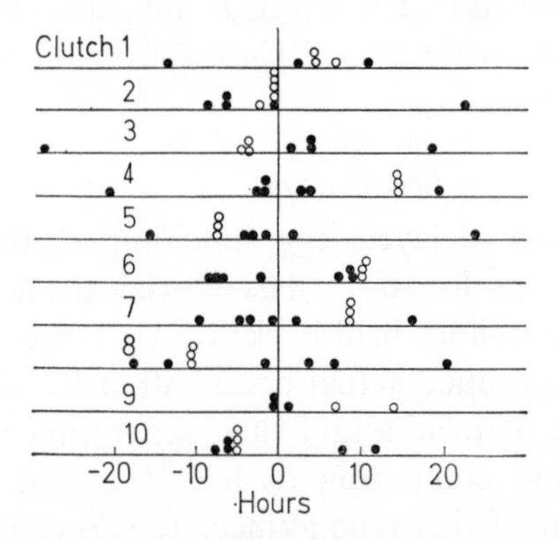

FIGURE 10. Bobwhite quail; hatching time of group compared with that of isolated siblings. O h, mean hatching time of isolates. ●, isolate, ○, egg in group. (From Vince, 1968*b*, by courtesy of Baillière, Tindall and Cassell.)

their more retarded siblings. The slowing down effect is not usually so dramatic as is the case with acceleration. It has, however, been shown that where a few eggs are taken from a clutch and hatched in contact and the remainder incubated in isolation (about 4 in. from each other), but in the same incubator, then the group will hatch at the same time and the hatching of isolates will be spread over quite a long period (up to about two days; Figure 10). Here the important point is that the time of hatching of the group is not consistently early or late in comparison with the hatching period of isolates, but rather towards the middle (Vince, 1968*b*). Again, a slowing down

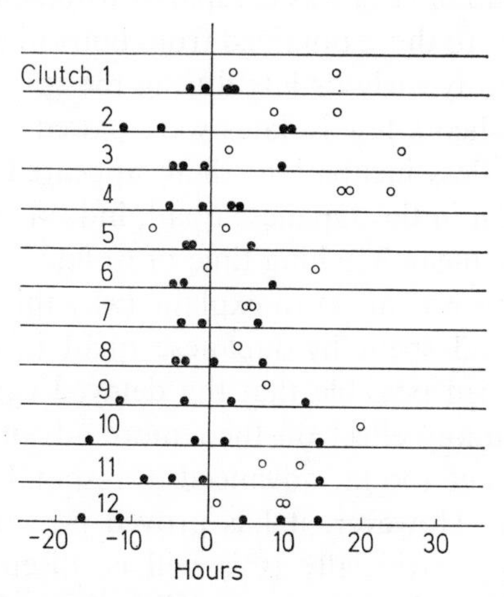

FIGURE 11. Bobwhite quail; hatching time of eggs paired with 24 h delayed eggs compared with that of isolated siblings. O h, mean hatching time of isolates; ●, isolated egg; ○, egg paired with one given 24 h less incubation. (From Vince, 1968*b*, by courtesy of Baillière, Tindall and Cassell.)

effect can be demonstrated in the bobwhite by separating the eggs in a clutch about forty-eight hours before they are due to hatch and pairing a few of them with eggs given twenty-four hours less incubation (Vince, 1968*b*). Results showed that in twelve clutches nearly all the paired eggs hatched after the mean of isolates, and in many cases well after all isolates had hatched (Figure 11). Hence, in the bobwhite we have a clear outline of the situation: synchronization is achieved by a two-way process of stimulation. The more advanced embryos bring on retarded ones and the more retarded hold back the very advanced. Thus experimental work using embryos of different ages to stimulate each other is likely to give results which are difficult to interpret; the time of hatching could be adjusted continuously between the two. A less ambiguous result must probably depend on the artificial stimulation of a single egg.

In the Japanese quail the situation is slightly different, and it seems possible that the two species use somewhat different methods for achieving the same result. There is evidence of acceleration when a single egg is stimulated artificially at approximately the normal click-rate, although the actual amounts of acceleration are slightly less than in the bobwhite: 11·6 hours, as compared with 19·6 hours before the mean time of hatching of siblings (Vince, 1966*b*). Again, where a small group of eggs was compared with isolated siblings, the results were similar to those obtained from bobwhites; the groups did not hatch consistently early or late during the spread of hatching of the isolates. But when a few isolates were paired with siblings given twenty-four hours less incubation, there appeared to be almost no effect of retardation in the Japanese quail, indeed some delayed eggs hatched before the mean hatching time of isolates (Vince, 1968*b*). At the time an attempt was made to explain away this result: considering the short period spent by Japanese quail in establishing lung ventilation, it seemed possible that the delayed eggs were relatively too young to have any effect on their more advanced siblings until the latter had become too far advanced to respond to stimulation of the retarding type. However, it has proved possible to slow down Japanese quail eggs artificially (this will be discussed later) under conditions which abolish the two-way effect. Recently, too, unfinished experiments have suggested that Japanese quail eggs can be retarded by contact with eggs given less incubation, if the intensity of the retarding stimulation is increased. It may, in fact, require contact with five, or six, delayed eggs to slow down one Japanese quail embryo.

The stage, or stages, when synchronization occurs

In an early experiment (Vince, 1964*b*) where clutches hatched simultaneously, it was found that the developmental stages of a retarded embryo and its older siblings approximated more and more to each other from about the last forty-eight hours before hatching; that is, the developmental stage of the retarded embryo was brought at successive stages more and more into line with the eggs with which it was in contact. In this experiment a bobwhite egg was placed in the incubator twenty-four hours later than its siblings, then was kept in contact with them until the time of hatching. At four-hourly intervals during the last forty-eight hours before hatching, each egg was examined and it was noted (i) whether it was pipped, (ii) whether it was clicking and (iii) whether it had hatched. The results showed that the time of pipping was very little, if at all, advanced in the delayed eggs, but the time interval between pipping and the beginning of regular loud clicking was very much shortened in those eggs put into the incubator late. This is where the largest amount of adjustment was to be found; the mean pip-click interval was 11·4 hours in the

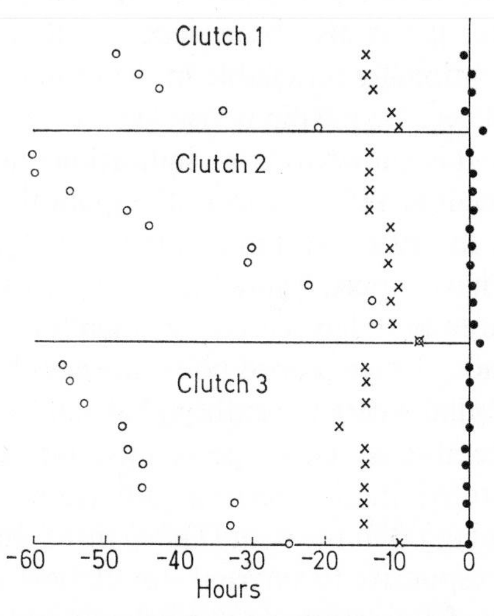

FIGURE 12. Bobwhite quail; shortening of pip-click and click-hatch intervals in eggs which pip late, compared with eggs which pip early in the same clutch. O h, approximate hatching time for clutch. For each egg ○, time of pipping; ×, beginning of clicking; ●, time of hatching. In each clutch the last egg to pip (bottom line) had been given 24 h less incubation.

delayed and 27·9 hours in their normal siblings. There was also a smaller, but still statistically significant difference between delayed and normal eggs in the click-hatch interval (11·8 compared with 13·6 hours) which suggests that the final stage of cutting round and emerging from the shell can also be brought forward. The gradual approximation of delayed eggs, eggs which pipped late, and eggs which pipped early is shown for three of these clutches in Figure 12, the figures for each egg being given separately. This illustrates a point made in this paper (Vince, 1964*b*) that those eggs which pipped first had the longest pip-click, or pip-hatching interval, this interval shortening progressively as the eggs, for whatever reasons, pip later; in fact, we find a correlation coefficient of −0·95 between the pip-hatch interval and the time of pipping when the latter is reckoned from the pipping time of the most advanced egg in the clutch.

Thus, late eggs will have a shorter pip-click interval, and a slightly shorter click-hatch interval, than their more advanced siblings; how much of this difference is achieved by acceleration and how much by slowing down, we do not yet know.

Subsequent work has tended on the whole to support the view that the time of pipping is not drastically altered by contact with more advanced eggs, but it has also been demonstrated that the time of pipping is an exceptionally unreliable measure in quail. Pipping was regarded as useful because the pip is visible on the outside of the shell, and in the chicken it is known to be an indication that lung ventilation has begun (Visschedijk, 1968*a*). But in the quail this has proved not necessarily to be the case: we have found eggs (and especially accelerated eggs) to have pipped *before* there has been any sign of breathing and before the embryo has pierced the membrane which separates it from the air space. Where pipped bobwhite eggs have been opened, some have been found where the embryo has not pierced the inner or the outer shell membrane. This type of false pip has been reported also by Drent (1967) in the herring gull (*Larus argentatus*) and also by Steinmetz (1932) in the coot (*Fulica atra*). In the quail, which appears to be so responsive to external stimulation, it seems tempting to regard this type of premature pip as a behavioural response to some as yet unknown external stimulus, whereas Visschedijk (1968*a*) has shown that pipping in the chicken occurs in response to chemical stimuli arising in the air space.

More recently the development of a new recording system (Salter unpublished; Vince and Salter, 1967) has made it possible to pinpoint the beginning of lung ventilation, so that we now have a more meaningful interval: the time between the beginning of lung ventilation and hatching, or the duration of lung ventilation in individual eggs. At the moment it is possible to record breathing only in one egg at a time, but a number of recordings have been made of (i) isolated eggs, (ii) an egg in contact with two others given the same amount of incubation and (iii) an egg in contact with one or two given between fifteen and twenty-four hours more incubation. The numbers are still small but figures for individual eggs given in Table 1 show the time intervals between the beginning of lung ventilation and clicking, and between the beginning of lung ventilation and hatching. Consideration of the means, together with the amount of delay in the delayed eggs, supports the view that there is little if any advancement in stimulated eggs at the time when the embryos begin to breathe through their lungs. In addition, these figures emphasize the dramatic shortening of the duration of lung ventilation in stimulated eggs.

Thus, under natural conditions it seems unlikely that pipping or the beginning of lung ventilation is advanced by stimulation provided

TABLE 1. *Time interval in hours between* (1) *the beginning of lung ventilation and clicking, and* (2) *the beginning of lung ventilation and hatching, in bobwhite and Japanese quail under* 3 *different conditions of incubation.*

	Lung ventilation-click		Lung ventilation-hatch		
	Bobwhite	Japanese quail	Bobwhite	Japanese quail	
Isolates	46·25	24	55·5	33	
	37	37	55	48·5	
	43	33	56·25	44·75	
	39·5	20·75	52·5	41	
	32	15·5	53·5	32·75	
Egg with eggs given same amount of incubation	23·75	24	36·75	36	
	30·5	13	41·5	discontinued	
	18	9·25	30·5	21·5	
	45·75	20·5 ± 2	59·75	33·5	
	39	17	52·5	26·5	
		23		36	
Egg with eggs given more incubation	18·25	5·5	32·5 19 h delay	15·5	24 h delay
	27·5	no record	41 16 h delay	14·25	
	17·25	13	32·25 16 h delay	23	
	17·25	no record	31·75 24 h delay	19	
		7		16	

by neighbouring embryos. On the other hand, an accidental heavy blow at an egg has been followed by hatching some fifteen hours later before siblings were pipped; also the amount of advancement in some eggs given artificial stimulation, as well as the false pip, encourage the view that we may sometimes get developmental stages brought on before the beginning of lung ventilation. It may be that the stage at which stimulation becomes effective as well as the rate of advancement depend on the intensity of the stimulus, a possibility which could open up quite a lengthy problem for research, considering the number of parameters which could be involved (frequency, amplitude and rate of occurrence of the stimulus, stage at which stimulation is provided and the possibility that different stimuli will be effective at different times).

However, considering the information we have so far concerning the stages of development affected by the synchronization of hatching, it could be argued that this is achieved in the main by bringing on, or holding back, the beginning of regular loud clicking, and, to a much lesser extent, by triggering off activities leading directly to hatching.

The effective stimuli for synchronization

We know that the embryos produce quite a variety of different frequencies at different times during the period between the beginning of lung ventilation and hatching: there are the means for communication. However, as it is possible not only that the synchronization of hatching is achieved by a two-way process, but also that it occurs in stages, the problem of which of these stimuli are the relevant ones and at what stages different ones operate could become quite complex. The answer may not be the same in bobwhites as it is in Japanese quail, and in neither species is it known; the discussion therefore will be provisional.

In experiments carried out so far, no attempt has been made to consider the effectiveness of stimuli at different stages, only an overall effect from about the time when lung ventilation begins. Isolated eggs have been stimulated continuously from about this time, and their hatching time has been compared with that of isolated siblings given the same amount of incubation, kept where possible in the same incubator and whose hatching times were recorded exactly (Salter and Vince, 1966). The results of one such experiment have already been described briefly; this showed that artificial clicks provided at a rate of about 3/sec. could accelerate hatching. It does not follow from this, of course, that the artificial stimulus must be a click, or that the rate is critical; simply that sounds like clicks can bring on development and hatching time, especially when accompanied by vibration. This is promising, but until the results have been compared with others, given different signals of similar intensity, we cannot judge the importance of clicks as such. Certainly the click rate can be varied over quite a wide range (Vince, 1968*a*) and still produce acceleration. In this experiment only Japanese quail have been used, and again the hatching time of the stimulated egg has been compared with that of unstimulated siblings. The click rate was varied from one every ten seconds to 1,000/sec. Results are shown in Figure 13. With rates between $1\frac{1}{2}$ and 60 clicks/sec. all stimulated eggs were speeded up in comparison with siblings. With rates above this up to 500 clicks/sec., all hatching times were slow in comparison with siblings (the numbers are small here), and outside these limits (above 500 and below $1\frac{1}{2}$ clicks/sec.) results were sometimes quite startling, but the direction of the effect varied, and a relatively large

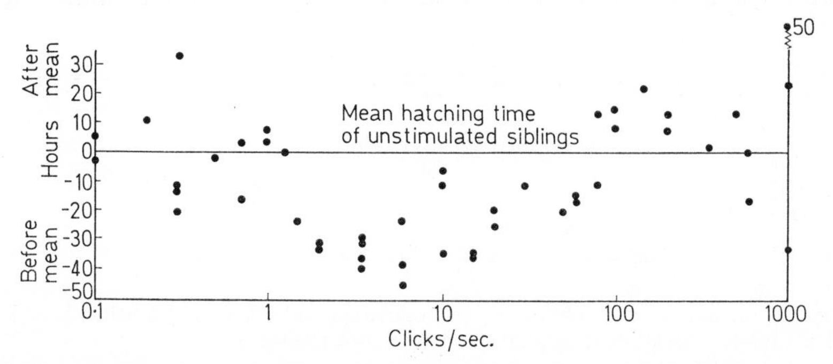

FIGURE 13. Effect of rate of stimulation on hatching time in Japanese quail; ●, hatching time of stimulated egg compared with mean hatching time of isolates in its own clutch. (From Vince, 1968*a*, by courtesy of Oliver and Boyd Ltd.)

number of embryos died before hatching. Thus the speeding up effect is not limited to click rates approximating to those produced by a single embryo. It is true, however, that under natural conditions each egg will be in contact with a number of others (up to six is a possibility) and as the eggs do not, in fact, click in time with each other, the actual rate of stimulation could be high.

The possibility that regular loud clicking is one of the more significant stimuli for the acceleration of development therefore remains open and more than likely. In fact it seems likely that regular loud clicking is a stage the beginning of which tends to be synchronized: that regular loud clicking encourages regular loud clicking in adjacent eggs.

It is tempting to link the evidence of retardation at the low audio frequencies shown in Figure 13 with the sounds of low frequency produced by the embryos and discussed in the first section. In Figure 13 evidence of retardation begins at 80 clicks/sec., that is 40 Hz, a level which is audible at any rate to the human ear at reasonable intensity. (Each click was produced by a transition in the stimulating wave form so that the frequency in Herz is half the click rate). The low frequency sounds discussed before were of this order, or lower. Again, the range over which the effect occurred included somewhat higher frequencies. The data given in Figure 13 provide the only evidence we have so far of artificial stimulation holding back development and hatching and it is interesting that this has been obtained in Japanese quail. Certainly these sounds of low frequency are most in evidence at the time when signals for retardation are

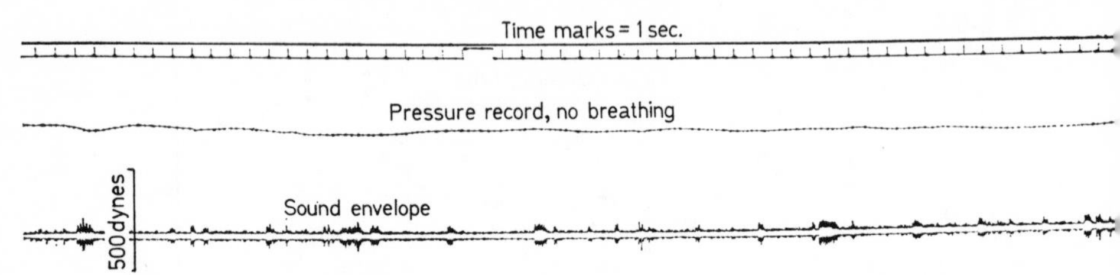

FIGURE. 14. Sounds of low frequency from three eggs in contact, 22½ h before hatching. (The sounds from all three eggs appear on the same channel.)

required if hatching is to be synchronized; they begin before the beginning of lung ventilation and become less audible when the other frequencies build up in intensity towards the time of hatching. A section of record obtained from three eggs in contact (signals from all three appear on the same channel) and 22½ hrs before hatching is shown in Fig. 14.

The possibility of a signal for hatching has been discussed earlier; there is no experimental work here, so far as I am aware. Listening in to a group of eggs in the immediately pre-hatching stage suggests a number of possibilities. Cutting round prior to emergence (which gives rise to the very considerable signals shown in Figure 6) could occur too late to provide the degree of synchronization which can occur, and in any case is the culmination of a whole series of developmental stages some of which could also serve as signals. In the quail, cutting round is usually preceded by relative or even total silence after the period of regular loud clicking, although some embryos go on clicking more softly all through hatching. Once more the intermittent sounds of low frequency may become audible. Very often also the embryos begin to cheep softly and rapidly in the pause between clicking and hatching. The rapid cheeps reported in the chicken at this stage by Guyomarc'h (1966) and by Collias (1952) have already been mentioned. In this paper Collias has made the interesting observation that a newly hatched chick still attached to the shell by the allantoic membranes will give 'contentment' calls whenever traction is placed on the membranes; it could therefore be illuminating to examine the relation between cheeps and the beginning of turning in the shell in the quail, with this idea in mind. It is important to note that since these preliminaries take some time, the earlier signals (the end of clicking, or beginning of cheeping) seem the most likely stimuli. However, this is mere speculation.

One difficulty peculiar to this investigation is that the response to stimulation is likely to be slow: on *a priori* grounds it seems unlikely that an embryo can take a sudden leap forward in its development. Thus a search for specific stimuli at different stages could be unrewarding. A different approach could be to look for a single explanation for the two-way effect. One such single explanation put forward (Vince, 1967) was that breathing is paced by the breathing of neighbouring eggs (it is known that the respiration rate rises towards the time of hatching). There are difficulties here: for one thing the embryos do not breathe in unison; for another the effective stimulation rates include some well above the highest respiration rate. Such an explanation would also fail to deal with the possible advancement of the beginning of lung ventilation or pipping, both of which normally occur when siblings are in the silent stage of breathing.

A brief attempt has been made to test a different type of theory in Japanese quail; that these embryos *require* a certain amount of stimulation, possibly nothing specific, to hatch at all. Eggs were hatched singly in an incubator kept for the final forty-eight hours of the incubation period in a soundproof room. However, the majority of these eggs hatched normally, if a little late, and the numbers were unfortunately too small to say definitely whether viability was reduced. Possibly an almost equally adequate experiment would be to compare hatchability over large numbers of groups and isolates.

In fact we know very little about the stimuli required for synchronization. In the future it is hoped to simulate different aspects of natural stimulation at different stages, and use hatching time as the criterion of success. Alternatively, or as an additional measure, it would be possible to analyse changes in the predominant frequencies occurring at different stages, using recordings of eggs hatched in isolation, then feed into other isolated eggs these natural recordings minus one or another of the predominant frequencies, and again, compare hatching times.

Effects of stimulation on the viability of the chick

Hatchability varies with a number of factors which have been summarized by Landauer (1961); hatchability (and the subsequent viability of the chick, if it survives hatching) goes down with length

TABLE 2. *Viability of chicks from stimulated, as compared with isolated eggs.*

(a) *Chicks which died before the 14th day, compared with other chicks.*

Time of hatching		Stimulated eggs Nos	Stimulated eggs %	Isolated eggs Nos	Isolated eggs %
Late i.e. after mean time of hatching	chicks which died	4	11·4	5	4·4
	chicks which lived	9	25·7	49	43·0
	Total chicks	13	37·1	54	47·4
Early i.e. before mean time of hatching	chicks which died	2	5·7	2	1·7
	chicks which lived	20	57·1	58	50·9
	Total chicks	22	62·8	60	52·6
Total	chicks which died	6	17·1	7	6·1
	chicks which lived	29	82·9	107	93·9
	Total chicks	35	100	114	100

(*b*) *Mean chick weights.*

	Stimulated eggs	Isolated eggs
hatching day	6·9	7·1
5th day of age	13·1	14·0
10th day of age	27·3	29·1
14th day of age	41·2	42·6

of preincubation storage, maternal age above a certain point, deviation from the optimum incubation temperature and so on. It seems possible that excessive shortening or lengthening of the incubation period resulting from the synchronization of hatching may be added to these factors. There is little work on this at the present time:

however, as pointed out in the previous section, a number of eggs given artificial stimulation failed to hatch. Examination of deaths in this experiment show that, although the numbers were small, there were twice as many in stimulated as compared with isolated eggs, and the viability of chicks from stimulated eggs also fell. The actual numbers considered, the percentage which died and the weights of chicks up to fourteen days of age are given here (Table 2). Results for accelerated and retarded eggs are shown separately. Thus the viability of stimulated chicks was lower than that of the isolates, and the larger part of this effect was due to the retarded chicks (where, curiously, the amount of adjustment was less). The weights of chicks which survived were very close to those which hatched without stimulation; this last effect could, however, have been biased by the deaths, as it is possible that stimulation killed off the weaker embryos.

The impression gained from handling the chicks in this experiment was that a small number of the stimulated chicks were, in fact, weaker than normal ones, but that the remainder were indistinguishable from isolates. The possibility that gross adjustments of hatching time can reduce viability emphasizes the biological importance of synchronization being effected by a *mean* hatching time as this will obviously reduce the amount of adjustment for each individual.

Possible reasons for the lowering of viability in stimulated embryos will be considered in the next section.

THE ESTABLISHMENT OF LUNG VENTILATION IN QUAIL EMBRYOS

Consideration of the synchronization of hatching brings us to the study of respiration, as the work summarized above has shown. We know that the adjustments in developmental rate take place in the main while lung ventilation is being established; we known also that the most characteristic signals produced by the embryos (clicks) are linked with respiration; again, we know that this stage of regular loud clicking, (one of the stages which appears to be brought close together in neighbouring eggs), is the most constant in duration; eggs appear to have to click for a certain length of time before they hatch. According to Kuo (1932*b*) lung ventilation must be well established at hatching to provide the embryo with sufficient strength to cut round and emerge from the shell; therefore, given the facts of synchronization and the importance of lung ventilation, we are faced with the problem of how the duration of the latter may be drastically curtailed in an accelerated egg (it may be halved) without *necessarily* harming the chick, and with the complementary problem of how the weak stimulated chicks differ from normal ones.

At the moment, it is not possible to answer these questions but *because we are dealing with species where the final stages of development including the establishment of lung ventilation are, to some*

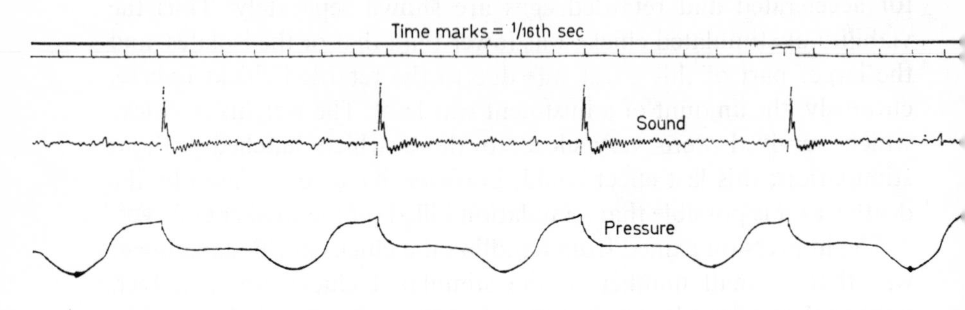

FIGURE 15. Very fast recording (16 cm/sec.) of breathing and clicking in Japanese quail. The sound channel here shows the actual wave form (not sound envelope).

extent, under external control, we are in a position to look into this. If we knew more about the stimuli required at different stages of development to ensure that hatching is synchronized, we should be able to regulate the rate at which lung ventilation is established, lengthening or restricting certain phases at will.

Recently, given the recording system developed for this purpose (Salter, unpublished; Vince and Salter, 1967), it has been possible to look rather more closely at some aspects of lung ventilation in different conditions of incubation, that is, with and without stimulation. The device consists of a pressure transducer providing a record of breathing which can be used at the same time as transducers for sound, and which thus give a measure of the more general activity of the embryo (Plate 1).

Results obtained so far confirm the finding of Romijn (1948) that the respiration rate rises towards the time of hatching, but in addition to this it has been possible to show that the most marked increase in rate occurs in the bobwhite at the beginning of regular loud clicking, and in the Japanese quail often one or two hours before. They show also that the respiration rate normally falls slightly immediately

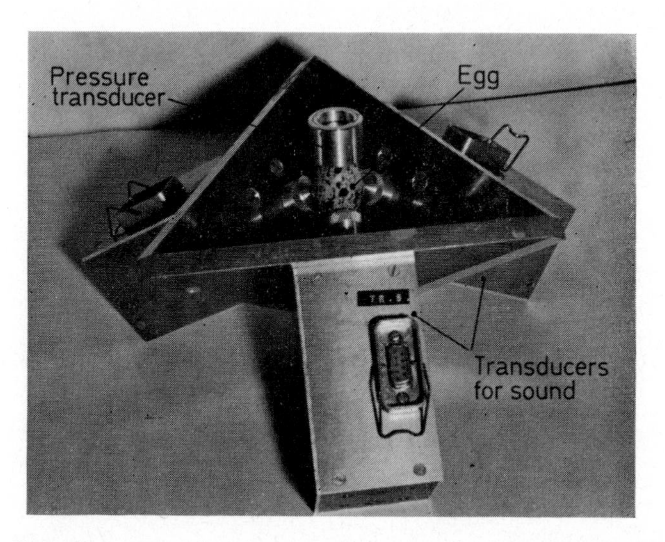

PLATE 1. Recording system; the egg rests on three mutually perpendicular piezo electric ceramic elements, which give a flat frequency response over the audio range. The pressure transducer seals a small hole in the shell over the air space and thus rests on top of the egg; it can measure to about 0·01 mm of water differential and provides a record of respiratory movements.

(*facing p. 254*)

before emergence (Vince, unpublished). A few accelerated eggs hatched without this fall in the respiration rate, presumably when their siblings produced signals for hatching, so it is possible that they emerged from the egg in a slightly immature state. Recordings also

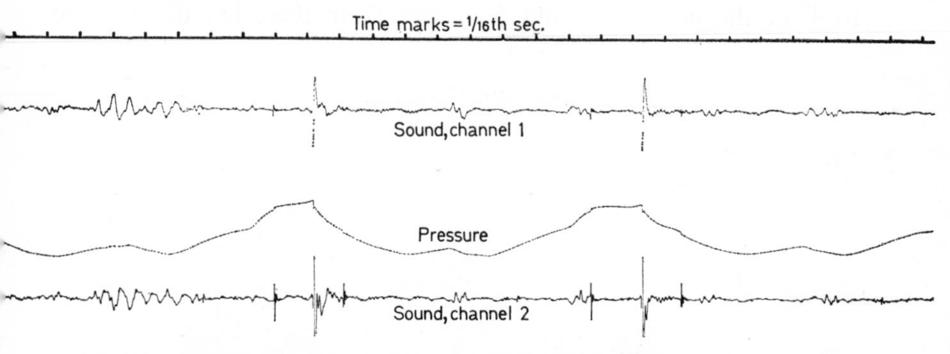

FIGURE 16. Very fast recording of clicks and breathing in duck embryo. Note multiple clicks, heart beats (at 260/min.) and also low frequency sound during 1st breath. Actual wave form, not sound envelope.

confirm the findings of Breed (1911) and Driver (1964) that clicks are in time with breathing (Figures 15 for Japanese quail and 16 for the duck); further, the linking of sound and pressure recording described

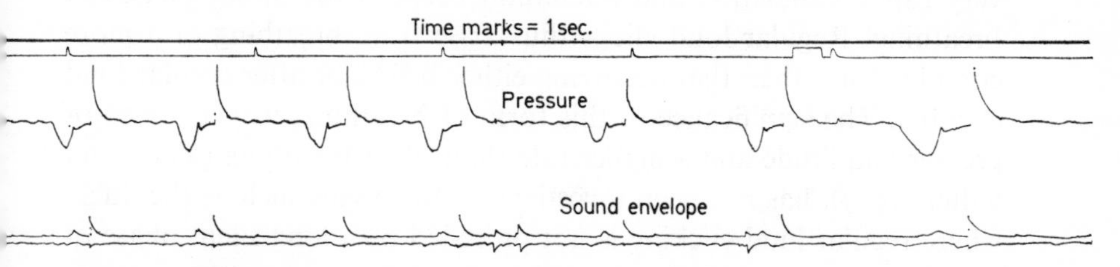

FIGURE 17. Bobwhite quail. Pressure recording made with beak and area below the inner shell membrane sealed off. N.B. Coincidence of rapid exhalation with click signal on sound envelope.

in detail elsewhere (Vince and Salter, 1967) show two types of click: the regular loud click which coincides with, or just precedes, a rapid rise in pressure between successive breaths, and a much softer click. Soft clicks (Figure 4*a*) occur earlier than the regular loud click, produce forces of different polarity and coincide with the in-out phase of breathing. The pressure changes shown in Figures 15 and 16 are produced partly by movements of the embryo's neck, etc., in to the air space, and partly by movements of the body below the membrane. The actual pattern of respiratory movement can best be seen if these are separated. This can be done by opening the shell over the pip and sealing off the remainder of the air space covering the back of the neck, etc., and recording from there (Figure 17); or

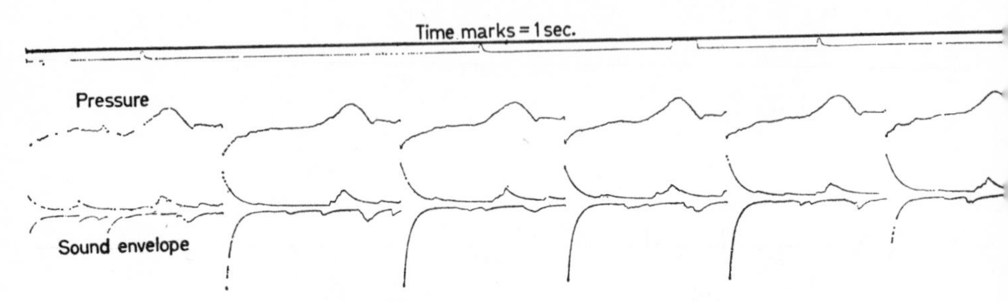

FIGURE 18. Bobwhite quail. Pressure recording made with neck area of air space sealed off, i.e. main pressure signals coming from below the membrane.

opening the shell over the neck, and sealing off for recording that part of the air space which includes the beak and the torn area of the inner shell membrane (Figure 18). These recordings are a mirror-image of each other, and show that the clicks coincide with an extra, very rapid, exhalation and inhalation between the in-out phases of breathing. Regular loud clicks thus arise from breathing of a more complex form than that occurring either before or after regular loud clicking. The significance of this form of breathing, which is also of greater amplitude and a higher rate than silent breathing (Vince and Salter, 1967), has not been investigated. In species such as the duck, where regular loud clicking is not preceded by a long silent phase of breathing, our recordings show that the respiratory pattern accompanying loud clicking is of the same pattern as that found in the quail. Also the multiple nature of the clicks, pointed out elsewhere (Vince, 1966*b*), is more evident in our duck recordings (Figure 16), possibly because of the larger size of the eggs. It may be of significance that the duck embryos from which these recordings were made (Aylesbury × Peking) do not synchronize their hatching, at any rate under conditions of artificial incubation, and in duck species where synchronization is reported (Heinroth, 1959; Bjärvall, 1967) this has occurred in clutches incubated by the parent.

In the quail it is possible to speculate about problems arising from the acceleration of development. The results described could suggest that the very characteristic manner of breathing occurring before hatching (click-breathing with a high respiration rate) is of some special importance to the organism, therefore its continuation for a minimum period is safeguarded even when embryos hatch early: the beginning of this stage can be drastically brought forward, and is roughly synchronized in neighbouring embryos. The postulated signal for hatching would be unlikely to cut it off much too early, although there are occasionally signs of embryos hatching in an immature state (i.e. with the yolk sac not fully withdrawn). It seems reasonable to consider that the considerable amplitude of these click-signals renders the embryo relatively safe from most competing signals except the cessation of the rhythm itself or the even louder signals arising from hatching. If this picture is at all correct, and if the sealing off of the allantoic blood vessels occurs simply as a preliminary to hatching (i.e. is triggered off by the same stimulus, thus accounting for the slow build up towards the beginning of actual cutting round), then slightly premature hatching in retarded chicks could be explained and weak chicks in the stimulated group would be those which clicked and breathed rapidly for the shorter periods. A brief pilot experiment measuring the oxygen uptake of stimulated and unstimulated quail chicks (Freeman and Vince, unpublished) has shown certain stimulated chicks to have an abnormal metabolic rate during the first ten hours after hatching (after this age the oxygen uptake increases with age in a regular fashion in chickens (Freeman, 1962)). In future work it is hoped to expand this finding by treating the foetal and hatched chick as a continuum, and obtaining records of rate and amplitude of respiration to cover this period in individual accelerated, retarded and unstimulated chicks.

Recordings of lung ventilation confirm the earlier findings that it is largely the silent period of breathing which is curtailed when development is accelerated. What evidence there is suggests that the opposite occurs when development is retarded: the beginning of clicking is held back, and the duration of clicking, or at any rate the interval between the beginning of clicking and hatching, may also be slightly lengthened. The fall in viability here could have quite different causes, as embryos which hatch late frequently become weak chicks.

It seems possible that synchronization involves mainly the advancement, or the holding back, of the beginning of regular loud clicking, as a vital developmental stage which cannot be curtailed very much; one obvious next step will be to compare the relative amounts of silent and click-breathing in species which do and do not synchronize their hatching.

Attempts have been made to show that avian embryos are unaffected by influences from neighbouring eggs (see, e.g. MacLaury and Insko, 1948, in chickens), but the evidence given above shows that this may be far from the case in a very large number of species: from the time when the membranes are pierced and the embryo begins to breathe through its lungs, it may respond differentially to signals from outside the egg, and where responses result in the adjustment of hatching time and the synchronization of hatching there must be considerable interchange of information. The actual mechanisms underlying this type of communication still remain obscure to some extent, but the general picture, at least in the bobwhite quail, is now becoming clear. As the adjustments to hatching time affect vital developmental stages including the establishment of lung ventilation, some quite intricate problems are raised by this specialized mode of embryonic activity.

ACKNOWLEDGEMENTS

The work on quail embryos has been carried out with the support of the Medical Research Council; in addition I am indebted to many colleagues who have helped to make the work possible, including S. H. Salter, who designed and built the equipment, Miss Violet Cane, of the Cambridge Statistical Laboratory who solved the statistical problems, and Dr B. M. Freeman and Miss D. Cooper of the Houghton Poultry Research Station who have given material aid, information, and encouragement.

REFERENCES

BJÄRVALL, A. (1967) The critical period and the interval between hatching and exodus in mallard ducklings. *Behaviour* **28**, 141–8.

BREED, F. S. (1911) The development of certain instincts and habits in chicks. *Behav. Mon.* **1**, no. 1, pp. iv+78.

COLLIAS, N. E. (1952) The development of social behaviour in birds. *Auk* **69**, 127–59.

DELIUS, J. D. (1963) Das Verhalten der Feldlerche. *Zeits. Tierpsychol.* **20,** 297–348.

DRIVER, P. M. (1964) 'Clicking' in the egg-young of nidifugous birds. *Nature* 206, 315.

(1967) Notes on the clicking of avian egg-young, with comments on its mechanism and function. *Ibis* **109,** 434–7.

DRENT, R. H. (1967) Functional aspects of incubation in the herring gull. Leiden, E. J. Brill, p. 132.

FAUST, R. (1960) Brutbiologie des Nandus (*Rhea americana*) in gefangenshaft. *Verhandlungen der Deutschen Zoologischen Gesellschaft in Bonn/Rhein.* **42,** 398–401.

FREEMAN, B. M. (1962) Gaseous metabolism in domestic chickens. II. Oxygen consumption in the full-term and hatching embryo, with a note on a possible cause for 'dead in shell'. *Brit. Poult. Sci.* **3,** 63–72.

GEOTHE, V F. (1937) Beobachtungen und Untersuchungen zur Biologie der Silbermöwe (*Larus a. argentatus*) auf der Vogelinsel Memmertsand. *J. für Ornithologie* **85,** 1–119.

(1955) Beobachtungen bei der Aufzucht junger Silbermöwen. *Z. Tierpsychol.* **12,** 402–33.

GOTTLIEB, G. (1968) Prenatal behaviour of birds. *Quart. Rev. Biol.* **43,** 148–74.

GUYOMARC'H, J-C (1966) Les emissions sonores du Poussin domestique, leur place dans le comportement normal. *Zeits. Tierpsychol.* **23,** 141–60.

HEINROTH, O. and K. (1959) *The Birds.* English edition. University of Michigan.

KEAR, J. (1968) The calls of very young anatidae. *Vogelwelt,* **1,** 93–113.

KIRKMAN, F. B. (1931) The birth of a black-headed gull. *Brit. Birds* **24,** 283–91.

KUO, Z. Y. (1932*a*) Ontogeny of embryonic behaviour in aves. I. The chronology and general nature of the behaviour of the chick embryo. *J. exp. Zool.* **61,** 395–430.

(1932*b*) Ontogeny of embryonic behaviour in aves. III. The structural and environmental factors in embryonic behaviour. *J. Comp. Psychol.* **13,** 245–71.

LANDAUER, W. (1961) The hatchability of chicken eggs as influenced by environment and heredity. *University of Connecticut Agricultural Experiment Station. Monograph* 1.

LIND, H. (1961) Studies on the behaviour of the black-tailed godwit (*Limosa limosa*). *Meddelelse fra Naturfredningsradets reservatudvalg nr.* 66. Munksgaard, Copenhagen, pp. 157.

MACLAURY, D. W. and INSKO, W. M. (1948) Effect of 'infertiles' and 'early deads' on embryo mortality in adjacent eggs. *Poult. Sci.* **27,** 127–35.

ROMIJN, C. (1948) Respiratory movements of the chicken during the parafoetal period. *Physiol. Comp. et Oecol.* **1,** 24–28.

ROMIJN, C. and ROOS, J. (1938) The air space of the hen's egg and its changes during the period of incubation. *J. Physiol.* **94,** 365–79.

ROOS, J. and ROMIJN, C. (1941) De gaskamer van het bebroede kippenei en haar beteekenis voor het kuiken. *Tijdschr. Diergeneesk.* **68,** 3–13.

SALTER, S. H. (1966) A note on the recording of egg activity. *Anim. Behav.* **14,** 41–3.

SALTER, S. H. and VINCE, M. A. (1966) The use of conducting paint for recording hatching times. *Med. and Biol. Engng.* **4,** 283–4.

STEINMETZ, H. (1932) Beobachtungen und Untersuchungen über den Schlüpfakt. *J. Ornith.* **80,** 123–8.

VINCE, M. A. (1964*a*) Synchronisation of hatching in American bobwhite quail (*Colinus virginianus*). *Nature* **203,** 1192–3.

(1964*b*) Social facilitation of hatching in the bobwhite quail. *Anim. Behav.* **12,** 531–4.

(1966*a*) Potential stimulation produced by avian embryos. *Anim. Behav.* **14,** 34–40.

(1966*b*) Artificial acceleration of hatching in quail embryos. *Anim. Behav.* **14,** 398–4.

(1967) Wie synchronisieren Wachteljunge im Ei den Schulpftermin? *Umschau,* 415–19.

(1968*a*) Effect of rate of stimulation on hatching time in Japanese quail. *Brit. Poult. Sci.* 9. 87–91.

(1968*b*) Retardation as a factor in the synchronisation of hatching. *Anim. Behav.* 16. 332–335.

VINCE, M. A. and SALTER, S. H. (1967) Respiration and clicking in quail embryos. *Nature* **216,** 582–83.

VISSCHEDIJK, A. H. J. (1968*a*) The air space and embryonic respiration. 2. The times of pipping and hatching as influenced by an artificially changed permeability of the shell over the air space. *Brit. Poult. Sci.* 9. 185–196.

(1968*b*) The air space and embryonic respiration. 3. The balance between oxygen and carbon dioxide in the air space of the incubating chicken egg and its role in stimulating pipping. *Brit. Poult. Sci.* 9. 197–210.

PART E
EVOLUTIONARY ASPECTS

INTRODUCTION

In the previous section the emphasis was on the extent to which particular vocalizations, or particular aspects of vocalizations, could be related to particular functional consequences. But the extent to which any particular character is beneficial inevitably depends on many other factors. All features of the structure, physiology and behaviour of a species form a complex adaptive to the environment in which it lives. Change in any feature of either the animal or the environment may have ramifying consequences throughout the whole. Two related methods have been used for studying this principle, both of which can be exemplified by the work of Tinbergen and his students on the reproductive behaviour of gulls (Laridae). The first, involving an intensive study of one species, is illustrated by their work on the black-headed gull (*Larus ridibundus*) (Kruuk, 1964; Tinbergen, 1967). The evidence indicates that the spatial distribution of broods in this species represents a compromise between the advantages of colonial nesting, which makes possible massed attack on predators, and spacing out, which decreases both intra- and interspecific predation on the young. The cryptic colouration of eggs and young, and the removal of broken egg-shells, also decrease predation. The colouration of adult gulls is a compromise between the requirements of crypticity, especially against fish living near the surface of the sea, and the requirements of intraspecific signalling. During the breeding season this species can 'afford' its conspicuous brown hood, which plays an important role in territorial aggression, since the gulls feed much on land and anti-fish camouflage is relatively unimportant. These few points represent only a fraction of the work on this species, but exemplify the manner in which the functions of a variety of characters, structural and behavioural, concerned with feeding, intraspecific signalling, and avoiding predation, all interlock to form an adaptive whole.

The second method, often used in conjunction with the first, depends on interspecies comparisons. While most gulls nest in low-lying areas or off-shore islands, the kittiwake (*Rissa tridactyla*) nests on cliff-ledges. Many of the differences between kittiwakes and other gulls, including the nature of their intraspecific fighting behaviour, nesting, copulatory and parental behaviour, and various aspects of the behaviour of the young, can be regarded as evolutionary consequences of cliff-nesting (Cullen, 1957). Certain other species of gulls (*Larus thayeri, glaucoides and hyperboreus*), which nest either on cliff-ledges or on level ground, are intermediate between the kittiwake and typical ground-nesting gulls, such as the herring gull (*L. argentatus*) (Smith, 1966). The gannet (*Sula bassana*), also a cliff-nester but quite unrelated to the gulls, resembles the kittiwake in many of the characters in which the latter differs from its ground-nesting relatives (Nelson 1967). The black-billed gull (*Larus bulleri*) shows a quite different set of adaptations, related to its habit of nesting in river beds subject to flooding (Beer, 1966).

Comparative studies of this sort provide an evolutionary perspective on the characters of the species, emphasizing the manner in which those characters form part of an interacting adaptive complex fitting the species to its niche. Another example is provided by Crook's (1964) detailed studies of the weaver-birds. In the following essay (chapter 12) he focuses on the vocalizations of these species, and in particular on the manner in which these form parts of adaptive complexes of characters related to the breeding habitats. That (so far as the evidence goes) the evolution of vocal signals appears to have been linked only loosely to visual ones, gives an added interest to his work.

Crook's chapter forms a link between the functional studies of Part D and two essays concerned with evolutionary aspects of vocalizations. About the course of evolution of avian songs and call-notes we know little. Although comparative studies have been a fruitful source of information about the evolution of other behavioural characters, they have so far yielded little data about that of vocalizations. The major reason for this is the adaptedness of vocal signals (see page 162), which has led sometimes to divergence between closely related species and sometimes to convergence between distantly related ones, in either case obscuring evidence of phylogenetic relationships between the signals. We are left with a few rather unreliable hints from ontogenetic studies concerning the ways in which call-notes may have been secondarily evolved into song (see Parts B and C).

Bird vocalizations are, however, sometimes of use as taxonomic characters. The need for characters additional to the traditional morphological ones is particularly acute in those phyletic groups which have undergone a relatively recent adaptive radiation, like the teleost fishes and birds, and a number of sources of information, biochemical, physiological and behavioural, have been tapped. Because of their adaptedness for specific distinctiveness, songs are of little value for the higher systematic categories, though quality of note and length of song are sometimes characteristic at the family level (e.g. Marler, 1955, 1957, 1959; Thorpe, 1961; Thorpe and Lade, 1961). As Lanyon shows in chapter 13, however, they can be of great value around the species level. Behavioural characters, like any others, cannot be used indiscriminately in taxonomy, and it is important to establish whether differences in a given call are of biological significance: the playback technique, already discussed by Falls (chapter 10) is of value here. Lanyon (chapter 13) discusses the precautions necessary for, and some of the conclusions obtained by, the use of vocal characters in avian systematics.

Another question concerns not the evolution of vocalizations, but the role of vocalizations in evolution. It is now generally agreed that the formation of new species depends on the prior occurrence of reproductive isolation between two related populations. Given such isolation, adaptive divergence is possible. Should the reproductive barriers break down, for instance by the re-meeting of two populations which have diverged in geographic isolation, selection can restore or promote reproductive isolation. Amongst birds, the characters principally involved are the auditory and visual signals used in pair-formation (e.g. Mayr, 1963; Dobzhansky, 1941): in many oscines, song seems likely to be of particular interest in this context. The nature and extent of the geographic variability of song characters within species is thus of considerable interest: the detailed studies by himself and others reviewed by Thielcke (chapter 14), show that the patterns of geographic variation found are in fact various and not always expected.

One line of evidence, relevant to the course of species formation sketched above, concerns characters important in pair formation or mating in species with partially overlapping ranges: these are usually more distinct in the sympatric zone than in the allopatric zones. Another is the disappearance of such signal characters in isolated populations where the danger of hybridization is absent. It is of particular interest that, as far as bird-song is concerned, Thielcke finds both types of evidence less firmly established than has often been supposed. It must be noted, however, that it may not always be song which forms the primary reproductive barrier in such cases: in the meadowlarks (*Sturnella*) a particular call-note used just before copulation seems to be more important (Szijj, 1966).

REFERENCES

BEER, C. G. (1966) Adaptations to nesting habitat in the reproductive behaviour of the Black-billed Gull *Larus bulleri*. *Ibis* **108,** 394–410.

CROOK, J. H. (1964) The evolution of social organisation and visual communication in the weaver birds (*Ploceinae*). *Behaviour, Supplement* X.

CULLEN, E. (1957) Adaptations in the Kittiwake to cliff-nesting. *Ibis* **99,** 275–302.

DOBZHANSKY, T. (1941) *Genetics and the origin of species*. Columbia University Press, New York.

KRUUK, H. (1964) Predators and anti-predator behaviour of the Black-headed Gull. Doctor's thesis, Oxford.

MARLER, P. (1955) Characteristics of some animal calls. *Nature, Lond.* **176,** 6–7.

(1957) Specific distinctiveness in the communication signals of birds. *Behaviour* **11,** 13–29.

(1959) Developments in the study of animal communication. In *Darwin's Biological Work: Some Aspects Reconsidered.* (P. R. Bell, ed.). Cambridge.

MAYR, E. (1963) *Animal Species and Evolution.* Harvard University Press, Cambridge, Mass.

NELSON, J. B. (1967) Colonial and cliff-nesting in the gannet. *Ardea* **55,** 60–90.

SMITH, N. G. (1966) Evolution of some Arctic gulls (Larus): an experimental study of isolating mechanisms. *Ornithological Monographs* No. 4, AOU.

SZIJJ, L. J. (1966) Hybridization and the nature of the isolating mechanism in sympatric populations of meadowlarks (*Sturnella*) in Ontario. *Z. Tierpsychol.* **23,** 677–90.

THORPE, W. H. (1961) *Bird Song*. Cambridge University Press, London.

THORPE, W. H. and LADE, B. I. (1961) The songs of some families of the order Passeriformes. I and II. *Ibis* **103a,** 231–45, 246–59.

TINBERGEN, N. (1967) Adaptive features of the Black-headed Gull. In *Proc. 14th. Int. Orn. Congr.* (D. W. Snow, ed.). Blackwell, Oxford, pp. 43–59.

12. FUNCTIONAL AND ECOLOGICAL ASPECTS OF VOCALIZATION IN WEAVER-BIRDS

by JOHN HURRELL CROOK
Department of Psychology, Bristol University

INTRODUCTION

The vocal repertoire of related birds comprising a taxonomic family normally consists of a collection of cries each emitted in a relatively precise context and with some determinable social function. There is little sign of randomization in the distribution of sounds either intra- or interspecifically, and it is commonly assumed that cries have evolved under natural selection in relation to precise survival values in communication between individuals. Avian vocalizations are often markedly affected by learning, but there are no reasons for supposing that the processes governing the phylogeny of visual and vocal signals differ in principle, even though the selection pressures may be different. For example, an especial enhancement of vocal signaling into duetting performances seems restricted to particular groups, and possibly to especial ecological circumstances (Thorpe and North 1966; see also Chapter 9). There have, however, been few studies concerned with the concomitant evolution of vocal and visual signalling systems. This paper discusses the functional relationship between the vocal repertoire and visual signalling of an avian sub-family in an attempt at a common interpretation of these two aspects of communication in the group.

The significance of a signal lies in its social context. While captivity observations are often of use in functional study, particularly in experiment, an essential background to this whole area of research is the field study of social systems of naturally-living wild populations. Without such background any statement regarding the function of a signal remains partial, biased and often wrong. This implies, however that the approach must be comprehensive in scope to a degree not easily attained. Long-term familiarity with many members of a family in the natural environment is required, and difficulties in interpretation of the collected data are often profound. For example, in listing the occurrence of a given cry in a variety of species the absence of evidence from a field study need not imply that the signal is never given; the right circumstances may not have been observed or the signal simply may be rare. Conclusions are thus apt to be tentative. In the making of evolutionary inferences the data used is cumulative, and periodic reviews in which the significance of newly acquired information is evaluated are essential.

The weaver-birds (Ploceidae, Ploceinae) have been studied in the field in Africa, India and the Seychelles Islands over a period of some six years and the data on the relations betwen the ecology of species, their social organization and patterns of visual communication analysed in some detail (Crook, 1964). They fall into four 'grades' of behavioural organization each representing specialization to particular modes of life through the co-adaptation of numerous morphological, behavioural and social characteristics:

(i) Insectivores of forest, savannah and secondary bush, often dispersed relatively solitarily through their habitats and showing sexual and seasonal plumage monomorphism. The birds are monogamous, and in pair formation the male's advances to a female are characterized by sexual chasing followed by close approach with song. He later leads her to his nest site.

(ii) Granivores of savannah and open country living in colonies or in neighbourhoods in grasslands. The species are mostly polygamous with marked sexual and seasonal plumage dimorphism and the males advertise nests with prominent displays by which females are attracted. Courtship and mating then follow.

(iii) Granivores of grasslands living in territories of varying size congregated into neighbourhoods. Males are commonly polygamous and sexual and seasonal dimorphism is usual. Males give aerial displays above territories in which nests are hidden. Females approach males and courtship sequences develop with approaches to the males' nest sites.

(iv) Granivores of forest fringe country. These few species feed on large seeds, move in itinerant family parties, and breed in compressed neighbourhoods relatively isolated from one another, commonly in tall grass patches. Sex-chasing with display at the nest has been described for one species. Sexual and seasonal dimorphism occurs.

The main axis of comparison concerns the extent to which species are gregarious and breed colonially or are relatively solitary. Species of the former kind are usually savannah dwellers feeding on seeds, while the latter are mostly forest insectivores. More solitary species show Pair Formation type I (Crook, 1964)—a highly motile courtship in which males pursue females a long way from the nest-site, alight near them in courtship and lead them back to the nest to which they are dramatically introduced by repeated entries and exits. In the colonial gregarious forms motility in courtship is often greatly reduced and displays advertising the nest, which is usually suspended from a tree, are given. These attract females to the site from afar. Courtship, which then follows, is restricted to the small territory in the packed colony (i.e. Pair Formation type II). Many of the behavioural features of courtship are homologous with those of forest dwellers but modified in orientation and position in time sequences. In grassland dwellers aerial display of the territory commonly replaces nest-display and precedes courtship (i.e. Pair Formation type III).

Adaptive radiation in the Afro-Asian biome types has been extensive, and it appears that most forest dwellers are derived from more open-country forms and have diverged from these in food habits and behaviour. While it remains unclear what type of social organization may be considered primitive in the family, it is certain that complex colonial life is well established and of ancient origin (Crook, 1963*a*,*b*, 1964).

In this paper we shall examine the available data on the vocal repertoire of the weavers, together with the contextual and functional aspects of the various cries, and try to establish whether there is a meaningful relationship between visual and vocal signals. Finally, we shall see whether an evolutionary approach covering both aspects of weaver communication is possible.

CONTEXT AND FUNCTION OF WEAVER-BIRD CALLS

In the Ploceinae some fourteen distinct calls have been identified and, since most of these are given in an easily identifiable social situation, they are here named contextually. This is a more convenient procedure than classification purely on a basis of the physical properties of the sounds and provides us with a background from which to discuss the social functions of each call. Certain calls, such as (5), (7), (8) and (9) below, show some variation in form depending on intensity of performance. The repertoire is as follows:

(1) Begging call.
(2) Juvenile location call.
(3) Social contact call.
(4) Flight call.
(5) Alarm and mobbing calls.
(6) Distress scream.
(7) Agonistic calls.
(8) The short courtship and territorial song.
(9) Long courtship and territorial song.
(10) Nest-advertisement calls.
(11) Copulation call of male.
(12) Solicitation call of female.
(13) Male call announcing female arrival at nests.
(14) Nest call of female.

The list covers our present knowledge of the range of cries given in the sub-family and may not yet be complete. In Table 1 the distribution of known calls of particular interest in a preliminary study, recorded from thirty-four closely related weaver species, is shown; all these calls have been described in the wild. In addition, Collias (1963) has described the vocalizations of *Ploceus cucullatus* kept in captivity in aviaries in California. He distinguishes fifteen different sounds. In this paper his low intensity and high intensity alarm and threat sounds are included under headings (5) and (7) while his 'scold chatter' comes also under (7). Two of his calls, listed above as (13) and (14), were not noted by me in the field. Collias did not record items (2), (9) and (12), of which (9) is certainly not part of the repertoire of his species.

Not all calls described have been heard in all species. In many cases this is doubtless because the circumstances in which they occur (e.g. mobbing) are relatively rarely observed or because the species has been less fully observed than others. However, special attention was paid to song in the field studies so that an absence of one type is less likely to be due to deficient observation. The study of vocalization in the field was largely incidental to a survey of social organization and visual communication, and no attempt at an exhaustive study of vocalization was made. Of the species in Table 1 only *Ploceus cucullatus*, *P. megarhynchus*, *P. nigerrrimus*, *P. phillipinus*, *P. manyar*, *P. benghalensis* and *Quelea quelea* were studied in detail in the field, observations on other species being largely restricted to observations of courtship and other reproductive sequences in colonies or at nests. This difference in the extent of observations is

TABLE I. *Calls described in the field from thirty-four species of Ploceine weaver.*

	Social contact call	Flight call	Alarm call with mobbing	Agonistic calls	Short song	Long song	Nest advertisement calls	Female solicitation cry
Amblyospiza								
unicolor					√	√		
superciliosus	√	√			√			
Malimbus								
malimbicus	√		√ √	√	√		√	
nitens	√				√		√	
rubriceps	√			√	√			
rubricollis					√		√	
scutatus					√			
Ploceus								
aurantius					√	√		
baglafecht	√				√	√		
benghalensis	√	√		√	√			
bicolor	√		√		√			
castanops					√	√		
cucullatus	√	√	√	√	√		√	√
heugleni					√	√	√	
insignis	√				√			
intermedius					√			
jacksoni	√	√	√	√	√	√	√	
luteolus					√			
manyar		√		√	√			√
megarhynchus	√	√	√ √	√	√		√	
melanocephalus	√	√	√ √	√	√		√	
melanogaster			√		√			
nigerrimus	√	√	√	√	√		√	√
nigricollis					√			
ocularis	√		√		√			
pelzeni	√			√	√			
philippinus	√	√	√	√	√		√	√
spekei				√	√	√		
tricolor					√			
weynsi	√	√						
xanthops	√				√			
vitellinus					√	√		
Quelea								
quelea	√	√	√	√	√			
cardinalis					√			

reflected in the large number of different cries recorded from *P. cucullatus*, *P. megarhynchus*, *P. nigerrimus*, *P. phillipinus* and *Q. quelea*.

Begging call (*1*) *and juvenile location call* (*2*)

Young weaver-birds in their nests emit frequent wheezy cries when hungry and the calling is reduced after a visit by a parent bringing food. The same cries are given on leaving the nest while the young are still dependent on parents for food. In *Quelea quelea* and *Malimbus rubricollis* females have occasionally been seen courtship begging to males and emitting similar begging cries. On leaving their colonies young *Ploceus phillipinus* scatter and may sit still for long periods waiting for a parent to come and feed them. They may emit a shrill 'location' call that appears to orient the parent. On the parents' approach they beg loudly. On leaving colonies the young of this species leave a protective environment and are subject to predation. Young *Quelea quelea* on leaving their nests in dense thorn bushes in Mauretania do not scatter but remain closely clumped for several days within the protective vegetation. They beg loudly but a location call was not recorded (Crook, 1960*a*,*b*). Collias (1963) describes the 'hunger distress chirp' of young *P. cucullatus*.

Social contact calls (*3*)

In groups of gregarious weavers a quiet call of short duration is commonly given at frequent intervals by all members while the party is active in foraging or hopping about. The call also occurs in flight but may then be replaced, especially at take-off and landing, by a louder 'flight call'. In emitting contact calls the birds assume no especial posture, the sounds simply occurring in the course of whatever activity the individual is performing. The cries vary little between species and may be rendered as 'chirt chirt' or 'chek chek' or, as in the small sized *P. pelzeni*, may be a higher pitched 'tik tik'. Minor variations between species were recorded mostly with respect to pitch. Among the more solitary insectivores, comparable cries are given between members of a pair or small party.

Flight calls (*4*)

Flight calls (Plate 5) are louder and more highly pitched than contact calls and have been described from some eleven species all of which are highly gregarious. No relatively solitary species has been heard calling in this way. The sounds are given together with preparatory movements for flight prior to take-off, during take-off, at certain times during flight, for example when a group changes direction, and prior to a group landing. Flocks of *Quelea quelea* on roosting flights crossing the River Senegal are silent until some 30 m from a tree on the bank in which they land. The cries appear to function in eliciting concerted preparation for flight and an integrated take-off by all members of a flock. They also appear to alert group members in some way prior to changes in flight direction and landing. Passing flocks may start calling together and were sometimes then seen to join up.

The emission of flight cries in aviaries was studied using captive groups of *Quelea quelea* and *Ploceus melanocephalus* in aviaries at Le Station Ornithologique of ORSTOM at Richard Toll in the Senegal valley. Caged birds of both species gave sudden outbursts of flight cries quite spontaneously. Often one bird would call and others immediately started. Frequently, however, outbursts of calling from caged birds were elicited by flocks of wild birds passing overhead, the cries or sight of which stimulated the reaction. After an outburst caged *Queleas* moved to the topmost perches of their aviaries and flew rapidly backwards and forwards under the wire roof. In aviaries at Cambridge captive gregarious weavers would likewise give flight calls during periods of intense flighting to and fro in large aviaries. These movements were especially common in the early morning (i.e. the period of dispersal from roosts in the wild) but also occurred repetitively following periods of rest throughout the day (Crook, 1961*b*). Some individuals would perch under the wire roof and show sustained wing-flicking, evidently produced by an inhibition of actual flight by the presence of the cage wire. Similar observations on flighting behaviour have been made on the spice finch (*Lonchura punctulata*) by Moynihan and Hall (1953).

The following details were obtained from caged *P. melanocephalus* at Richard Toll. There was a frequent temporal coincidence between flight cry emission in an open aviary (3×3 m), an uncovered cage (about 1×0·5 m) and free-flying flocks passing overhead. Of 64 recorded outbursts 41 occurred in all three conditions, six in cage and free flock only, two in aviary and flock only, eleven between cage and aviary with no overflying flock present and four occurred only in the cage. A number of birds kept in a covered cage, illuminated from within by electric light but not acoustically isolated from the exterior, showed no flight call responses during this same set of observations, but on four occasions *Quelea* kept in aviaries some distance away responded with flight cries to outbursts common to aviary, cage and flock. It follows that outbursts may occur in response to passing flocks, between cages and aviaries, in a cage without external stimulation, between unrelated species with similar cries, but not apparently in a covered cage.

The individuals in the covered cage were closely watched at the time of flight call outbursts external to the cage. During twenty outbursts in the neighbouring aviary no response occurred within the covered cage on nine occasions. Ten times a single bird—usually a particular female—sleeked its plumage on hearing the cries and gave one or two calls. Rarely, one or two others likewise sleeked the plumage—an intention movement of take-off into flight. Only on the remaining occasion did a small group show sleeking and emit cries together. Thus, while auditory stimulation yielded some responses, the blocking of the view around the cage inhibited the flight call outbursts typical of the species. Contact calls between members of the group and normal hopping about and feeding in the cage were not affected. As soon as the covering was removed the birds showed outbursts of flight calling in unison with another cage, and initiated outbursts by giving them spontaneously, even though the birds in the two cages were visually isolated from one another. Once the cage was uncovered the birds flew up onto the wire facing the courtyard when giving cries. It was further ascertained that the sound of flight calls alone was sufficient to elicit an outburst from a cage facing into a hedge when the birds inside were visually isolated from all others.

These limited field and captivity observations suggest that stimuli activating the flight system initiate both intention movements of flight and flight calling, but that once flight starts the calling ceases. Changes in stimuli determining the flight path, or which anticipate the termination of flight, elicit calling which ceases once the change of direction or landing has been effected. Calling has marked facilitatory effects on other individuals so that many birds are usually calling together. The calling appears to assist in integrating the activity of numerous individuals comprising the large flocks of gregarious ploceines.

Alarm, mobbing calls (5) *and distress scream* (6)

These cries vary from the high pitched scream (6) of alarm heard from *Quelea* when severely frightened or held in the hand to repetitive anxious calling reminiscent of flight cries but with a different pitch and emphasis. In *Quelea quelea* colonies, individuals were seen mobbing what was apparently a snake in the grass. The birds emitted high pitched 'pseep' cries while making intention movements of flight. *Quelea* were never seen to 'mob' avian predators perched in trees around the colony. Their nests in thorn scrub are largely protected from these by the spiny vegetation.

Male Finn's bayas (*Ploceus megarhynchus*) at colonies with eggs or young sometimes make mobbing attacks on human intruders. Mobbing of snakes was also seen in conjunction with other bird species of the locality, loud 'skeer' calls being emitted. Mobbing has also been reported from *Ploceus philipinus*, *P. melanocephalus*, and *Malimbus malimbicus*. Both *Foudia seychellarum* and *F. madagascariensis* mob cats or human intruders on Frigate Island, Seychelles (Crook, 1961*a*).

Agonistic cries

In agonistic encounters over food, nest-building materials or territorial space, weavers utter a variety of loud, harsh 'chak' calls and vibrating chatters (Plate 5). These vary considerably in pitch and loudness from species to species but have a general form and context instantly recognizable even at a distance. The sounds are typically accompanied by attack flights, supplanting or threat posturing, and are commonly interspersed with songs in territorial and courtship encounters.

Short (8) *and long songs* (9), *calls during nest-advertisement* (10)

In the breeding season male weavers develop a number of calls that are emitted in more or less precise circumstances during reproductive behaviour (Plates 1 to 4). The first type is the 'Short song'—a relatively brief phrase containing preliminary chattering or warbling sounds and usually ending in a harsh and sometimes prolonged wheeze. The second call, the 'Long song', consists either of a repetition of a modified short song in a linked series, or of elongated versions of the short song containing repetitions and modifications of phrases within the pattern. The third type of call is a simple repetitive phrase given in the territory, in pursuit of the female and/or (most commonly) at the nest during vigorous advertisement display. Only a few species give this special call in addition to either short or long song phrases or, rarely, both.

The occurrence and type of male song varies with the Pair Formation type (above). In Pair Formation type I singing is highly variable both between and within species. *Malimbus nitens*, a forest-adapted species closely allied to other large bodied *Ploceus* species, has two kinds of song—the short song given in many courtship and territorial situations and consisting of the phrase and wheeze pattern typical of the subfamily, and the 'winny', occurring near the nest, which appears to be the developed equivalent of the special advertisement calls of other species. In some other Malimbes the variety of calling is considerable (Crook, 1964) and females and males partake in vigorous vocal exchanges in contexts suggesting both that the sexes may occupy separate territories outside a breeding phase and that 'duetting' may be a feature of their vocal life (Thorpe and North, 1966). More detailed work on these obscure forest birds is needed.

Other forest or forest fringe weavers such as *P. melanogaster*, *P. nigricollis*, *P. ocularis*, and *P. tricolor* all give the short song phrase—each species having its own particular call. Peculiar songs are found in the forest dwelling *P. bicolor*, in which the pair perform a simple duet near the nest suspended high in a tree. The male's notes are noticeably flute-like, lacking the usual harsh wheezy component. The song, in consequence, probably carries much further through the forest vegetation than is the case with more typical songs. The forest dwelling island species *Foudia seychellarum* has a markedly reduced vocal repertoire correlating with the loss of distinctive plumage characteristics—a feature commonly found in isolated oceanic island endemics (Crook, 1961*a*).

The large number of species with Pair Formation type II all have typical short or long songs, and sometimes both. Among the *Euplectes* weavers of grasslands with PF type III, the development of striking aerial display patterns, performed over territories in which nests have been placed cryptically among the stems, correlates with the degradation of song into a series of machine-like repetitive sounds, buzzes or whirrs that accompany the visual display. Sizzling cries reminiscent of short songs are given in courtship in the grasses but these undistinguished cries have received little attention. In this genus there is a marked inverse correlation between the brilliance of the plumage and aerial display and the use of complex song patterns.

The distribution of advertisement and courtship vocalization patterns is complex and difficult to interpret (see below). The clearest relationship is between short-song production, the male's close approach to a female during courtship sex-chasing, and the male's approach to neighbouring males in territorial encounters. Song is usually given as part of the song-bow or stretch sequence in which the male approaches and bows, emitting the wheezy termination of the song as he does so. In motivational terms this sequence has been interpreted as conflict behaviour—posture components revealing tendencies to attack, flee and (in female company) to behave sexually (Crook, 1964).

Observations on captive *Quelea* suggest that song production varies annually in correlation with reproductive behaviour such as nest-building, and is probably likewise under testosterone control (Butterfield and Crook, 1968). Singing in agonistic behaviour relates to competition between males for sexually valent (territorial) space. However, song also occurs 'spontaneously' when males are sitting alone in territories and may be accompanied by threat posturing which, although undirected, is probably significant to other territorial males. These birds, being extremely colonial, always have their neighbours in view. Singing also occurs, moreover, when the birds are sitting in flocks in mid-day or dusk roosts prior to the occupation of the breeding colony and also, in aviaries, during the rest phases of the activity cycle (Crook, 1960*a*,*b*, 1961*b*). In this case it is not accompanied by posturing. Observations on wild and captive *Quelea quelea*, *Ploceus philippinus*, wild *P. cucullatus* and wild *P. capensis* (Skead, 1947) show that this singing occurs in synchronized bursts—a number of males singing more or less together. Each bout is followed by a silent period until a male starts the group off again by initiating song. Captive male *Quelea* kept in Hartshorn sound-proof chambers in small groups will sing in synchrony with bouts of song played to them on a loudspeaker. Such birds respond in the same way to bouts of white noise of the same duration, loudness and temporal patterning. Birds singing in a control group unexposed to bouts of either recorded song or white noise were found to be singing individually and more or less at random with respect to one another—the group apparently being too small to generate its own synchrony (unpublished experiment). Possibly these outbursts of synchronized singing, which are very impressive in the field, might aid in ensuring rapid seasonal maturation of the gonads so that the greater part of the male population is ready to breed as soon as environmental conditions (onset of rains, green grass etc.; Marshall and Disney, 1957) allow. The remarkable synchrony with which millions of *Quelea* start breeding in their immense colonies adds point to this suggestion, which remains to be tested experimentally.

Copulation (11) and solicitation calls (12)

Ploceus philippinus males give a high pitched 'titititti tt eeee' cry during copulatory mounting and may also make snipping sounds by clippering the mandibles when in sexual excitement. The copulation cry may also appear when the birds are sexually frustrated and displaying pseudo-female solicitation (Crook, 1960*b*). Females soliciting copulation give a variety of sibilant squeaking sounds (e.g. *Ploceus cucullatus*, *P. nigerrimus*, *P. manyar* and *P. philippinus*).

FUNCTION AND EVOLUTION OF WEAVER SONG

Song distribution

The variety of songs given by male weavers is considerable and poses questions that demand special attention.

The distribution of songs together with the known occurrence of the special call in advertisement is shown in Tables 2 and 3, which include thirty-four species. Of these at least seventeen and possibly twenty-three sing songs both at the nest and also during sex-chasing and courtship of females away from it. At least fifteen species are known to sing in territorial threat to rival males, while twenty are known to sing at the female when she has accepted a nest and enters it, but not before she does so. In five species song is given only at the nest and in four of these cases the song is a prime component of nest-advertisement display. In the fifth species, *Malimbus rubriceps*, a short song is given as the female enters the nest but it is not known whether the male also sings to her in courtship sequences away from the structure: probably he does so. In six species song is known only in courtship away from the nest but again only one of these species (*P. benghalensis*) has been studied in any detail (see below). Long songs are given by some nine species, three of which also sing short songs (*P. baglafecht*, *P. castanops*, *P. heugleni*). Four of them sing

TABLE 2. *The performance contexts of short songs (√), long songs (LS) and special cries (S) in thirty-four weaver birds, the Population Dispersion (PD) and Pair Formation (PF) types of which are shown. The population dispersal types (a) when not breeding and (b) when breeding are as follows: I(a) and (b) solitary individuals or pairs. II(a) Family party; (b) solitary pairs. III(a) Family party; (b) loose or tight colony or pairs. IV(a) Flock; (b) solitary pairs, harems or small colonies. V(a) Flock; (b) territories in neighbourhoods. VI(a) Flock; (b) territories in colonies. The main pair formation types are described on page 267: further details are given in Crook* (1964).

Species	PD type	PF type	Territorial encounters	Courtship far from nest	In advertisement display	On entry of female to nest	Courtship of female near nest
Amblyospiza unicolor	III	Ib			LS		
Amblyospiza superciliosus	III	Ib			LS		
Malimbus malimbicus	II	Ia		√			
Malimbus nitens	III	Ia	√	√	S?	√	√
Malimbus rubiceps	II	Ia				√	
Malimbus rubricollis	III	Ia	√	√	√		
Malimbus scutatus	III	Ia	√	√			
Ploceus aurantius	VI	IIb			LS	LS	
Ploceus baglafecht	II	Ia		√LS			
Ploceus benghalensis	VI	Ib	?	√			√
Ploceus bicolor	I	Ia		√			
Ploceus capensis	VI	IIb		√	S	?√	√
Ploceus castanops	VI	IIa	LS	S	S	LS	
Ploceus cucullatus	VI	IIb	√		S	√	

[continued

TABLE 2.—*continued*

Species	PD type	PF type	Territorial encounters	Courtship far from nest	In advertisement display	On entry of female to nest	Courtship of female near nest
Ploceus heugleni	IV	IIa	LS		S	√	
Ploceus insignis	I	Ia		√		√	
Ploceus intermedius	VI	IIb		√	S		
Ploceus jacksoni	VI	IIa	LS		S	LS	
Ploceus luteolus	I	Ib	√	√	√		
Ploceus manyar	VI	IIa	√		S	√	
Ploceus megarhynchus	VI	IIb			√		√
Ploceus melanocephalus	VI	IIa	√	√	S	√	
Ploceus melanogaster	I	Ia		√		√	
Ploceus nigerrimus	VI	IIb	√	√	S	√	
Ploceus nigricollis	I	Ia	√	√		√	
Ploceus ocularis	I	Ia		√		√	
Ploceus pelzeni	III	Ia		√		√	√
Ploceus philippinus	VI	IIa	√	?√	√S	√	
Ploceus spekei	VI	IIb			LS		
Ploceus tricolor	III	Ia		√			
Ploceus xanthops	I	Ib		√	√		
Ploceus vitellinus	IV	IIa	LS	LS	S		
Quelea quelea	VI	IIb	√			√	√
Quelea cardinalis	V or VI	IIIa	√	√			

the long song at the nest while performing advertisement display but the others sing them away from the nest and (three cases) when the female enters the nest. However, in only nine species altogether is song, whether short or long, a part of advertisement display. In twelve species the special call is given during advertisement although in two of these song may also be given in the same situation.

The variety of behavioural contexts in which weaver short and long songs, and the special advertisement cries, occur is not easily interpretable upon present evidence. For example, song is given in nest-advertisement by species that commonly nest solitarily (*Amblyospiza unicolor*, *Ploceus luteolus*, *P. xanthops*) as well as by species usually breeding in large colonies (*P. aurantius*, *P. megarhynchus*, *P. spekei*) and also by *Malimbus rubricollis* nesting either solitarily or in tightly packed nest-groups high in forest trees. Furthermore, while *P. aurantius*, *P. spekei* and the *Amblyospiza* sing long songs in display, the other species' songs are short. There is no reason to believe that any particular phyletic relationship between these various forms explains these facts (Crook, 1964 p. 146–157).

Long songs appear in some nine species of which some are close taxonomic relatives (i.e. *P. vitellinus*, *P. heugleni*, *P. spekei* and, separately, *P. castanops* and *P. jacksoni*). Three species are solitary nesters (*A. unicolor*, *A. pachyrhynchus*, *P. baglafecht*), but while the first two sing at the nest, the latter sings on approaching the female. This makes sense since, while the two former build conspicuous nests, usually in tall grass or reed beds that are not easily reached by terrestrial predators, the latter constructs cryptic nests in bushes. The remaining species are all colonial nesters but they do show great intraspecific variation in their nesting densities. *P. vitellinus* may nest quite solitarily and its song is then given mainly away from the nest. The location of singing when it nests in the colonies sometimes reported in the literature is not known. *P. castanops*, *P. heugleni* and *P. jacksoni* all sing away from their nests and when the female enters it, and use repetitive calls other than song during nest advertisement display. *P. castanops* and *P. heugleni* sing short songs in addition. By contrast *P. spekei* is a very dense colonial nester singing long songs in advertisement as well as to females entering nests. In four species long songs occur in territorial encounters and two of these may also sing short songs in the same situation. No species singing long song in advertisement also sings it in territorial encounters. It seems clear that long songs have developed separately several times in the family, but which contingencies of environment and social milieu are important to its appearance remain uncertain.

The distribution between species of special calls in advertisement is perhaps less puzzling. It occurs quite generally during nest-advertisement displays of colonial weavers but not in homologous displays by less social species. Generally, the most developed displays are associated with a strictly colonial breeding biology in which courtship

TABLE 3. *Numbers of species in categories of population dispersion and pair formation types showing song in differing circumstances* (*see legend Table* 2).

Dispersion and Pair Formation type	Approach to female away from territory	In advertisement or nest invitation	Special calls in advertisement	On entry of female to nest
Solitary breeding dispersion (P.D.I and II Crook 1964) 10 spp.	9	2	0	5
Small colonies, occasional solitary breeding. (P.D.III) 7 spp.	5	3	1	2
Gregarious, colonial breeders, rarely nest solitarily (P.D.IV and VI). 17 spp.	8	4	11	13
Totals. 34 spp.	22	9	12	20
Pair formation type I				
Ia. 13 spp.	12 } 15	1 } 5	1 } 1	7 } 7
Ib. 5 spp.	3	4	0	0
Pair formation type II				
IIa. 7 spp.	3 } 6	1 } 4	7 } 11	6 } 12
IIb. 8 spp.	3	3	4	6
IIIa. 1 sp.	1	1	0	0
Totals 34 spp.	22	10	12	19

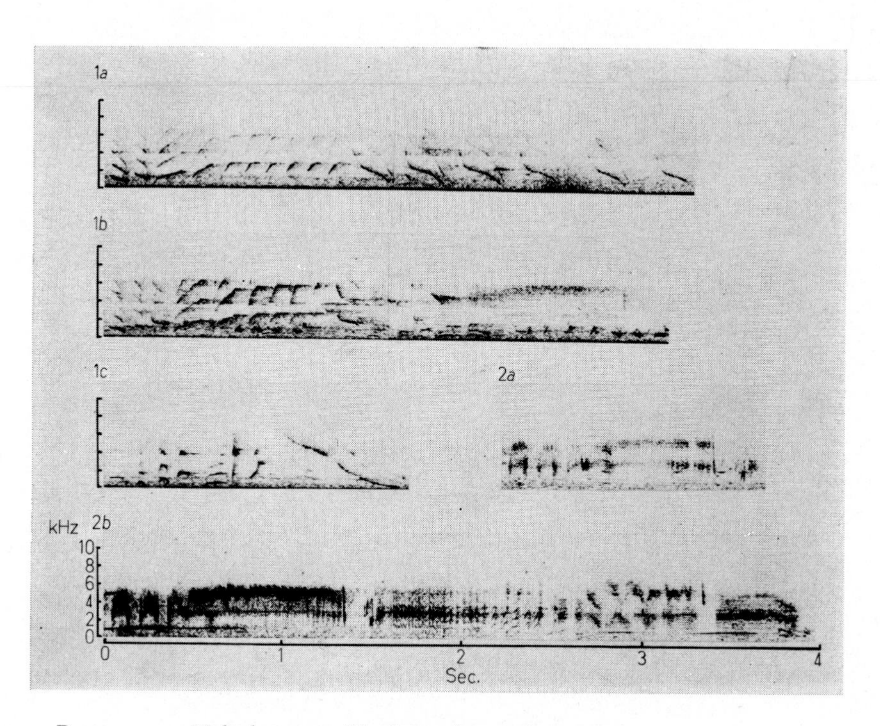

PLATE 1. 1*a*. *Malimbus nitens*. The 'winny' song phrase.
1*b*. *Malimbus nitens*. The 'winny' terminating in a wheeze.
1*c*. *Malimbus nitens*. The short song . . . 'wookitiki wheeze'.
2*a*. *Ploceus nigerrimus castaneofuscus*. Short song.
2*b*. *Ploceus nigerrimus castaneofuscus*. Short song with a second wheeze.

(*facing p.* 280

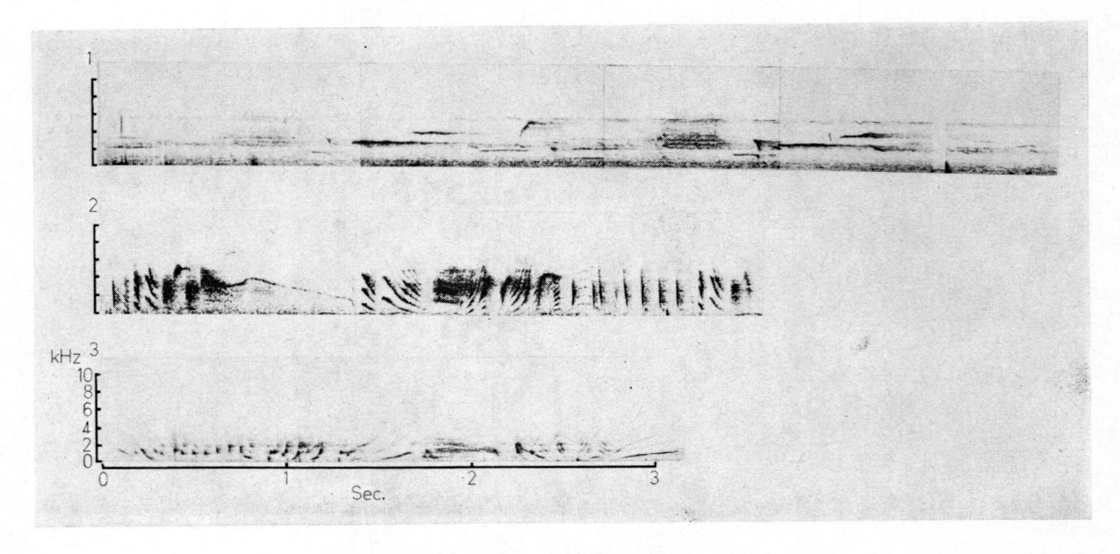

PLATE 2. 1. *Ploceus bicolor*. Short song.
2. *Ploceus luteolus*. Short song.
3. *Ploceus ocularis*. Short song.

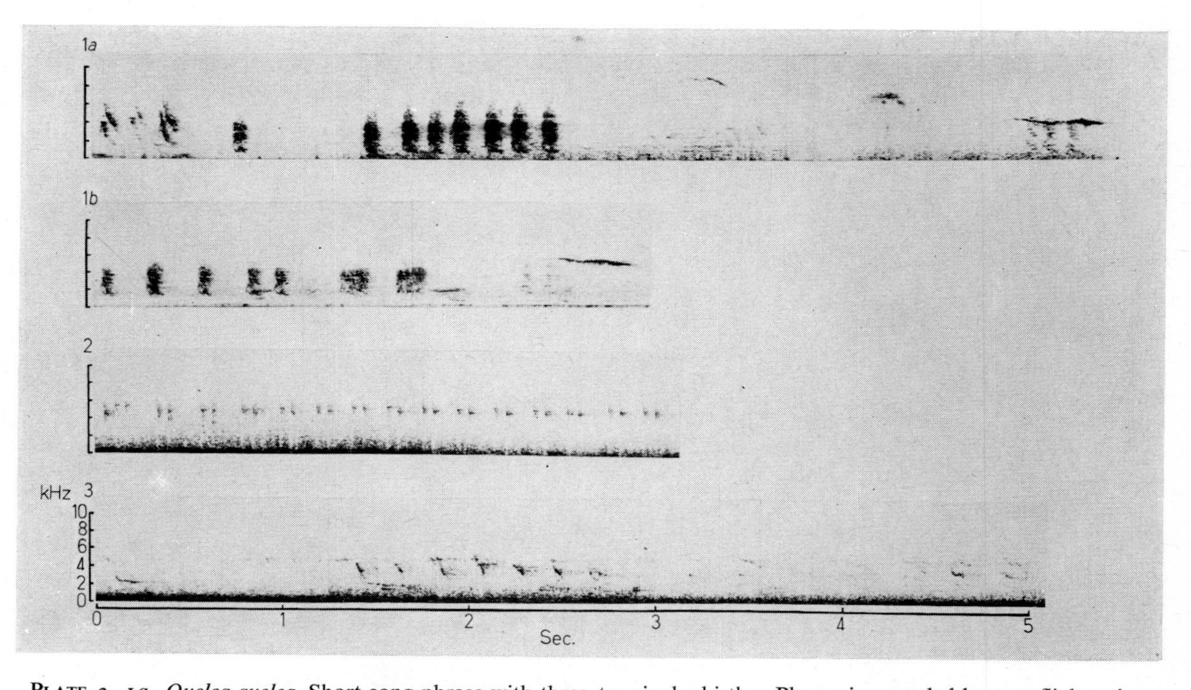

Plate 3. 1*a*. *Quelea quelea*. Short song phrase with three terminal whistles. Phrase is preceded by two flight cries.
1*b*. *Quelea quelea*. Short song.
2. *Foudia madagascariensis*. Trill.
3. *Foudia seychellarum*. 'Tsea tsea' cry.

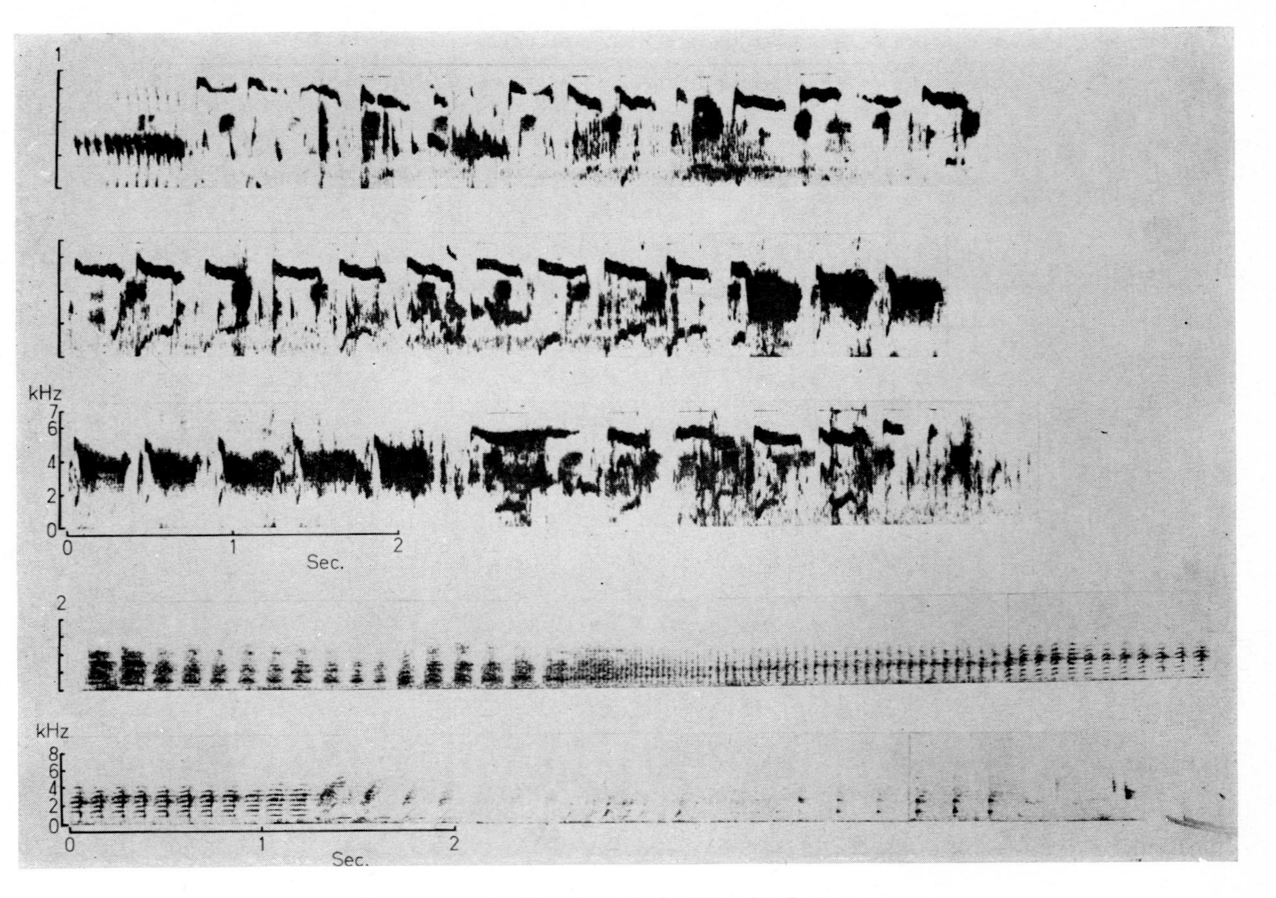

PLATE 4. 1. *Amblyospiza unicolor* (Nigeria). Long song.
2. *Ploceus spekei*. Long song.

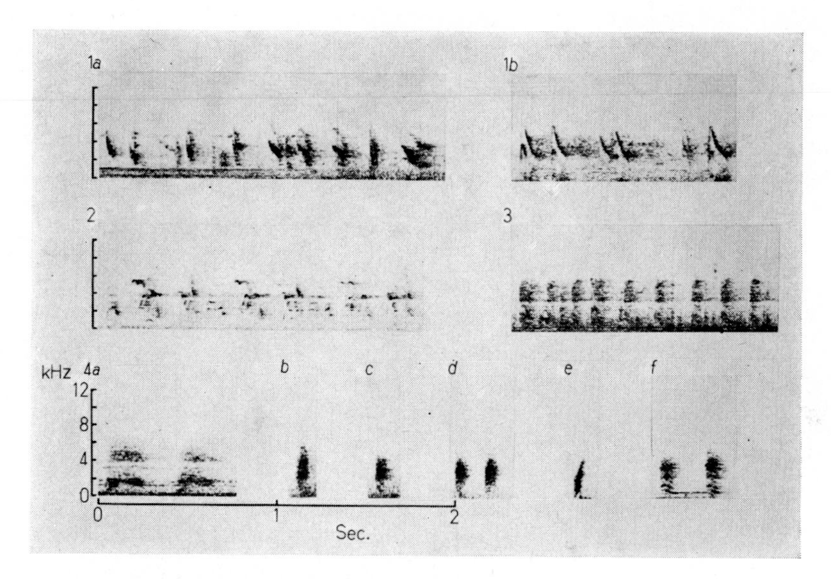

PLATE 5. 1*a*. Flight calls. *Ploceus melanocephalus*.
1*b*. Flight calls. *Quelea quelea*.
2. Advertisement calls. *Ploceus cucallatus*.
3. Agonistic chatter. *Ploceus cucullatus*.
4*a*. Agonistic 'chaks'. *Malimbus nitens*.
b. Agonistic 'chaks'. *Ploceus nigerrimus castaneofuscus*.
c. Agonistic 'chaks'. *Ploceus castanops*.
d. Agonistic 'chaks'. *Ploceus luteolus*.
e. Agonistic 'chaks'. *Ploceus megarhynchus*.
f. Agonistic 'chaks'. *Quelea quelea*.

is restricted to the territory to which the female has been attracted (PF type IIb, Crook, 1964). The advertisement displays are made more distinctive by the emergence of a specially characteristic cry differing from the songs given in agonistic sequences and courtship. It is not without significance here that, of the three colonial species that sing in advertisement, two of them have developed impressive long songs that likewise enhance the distinctiveness of the display. What factors have been significant in yielding display enhancement by song development in some cases and by special cries in others remains a mystery. A recourse to taxonomic relationship again does not appear to help.

Ecology and song

In spite of these difficulties a number of interspecies comparisons, both general and more specific, do provide suggestions as to the kind of explanation we may eventually expect. From the trend of the figures in Table 3 it is clear that songs are given in approach to females primarily in species where courtship involves sex-chasing outside the territories prior to courtship within them. Actual advertisement display of the familiar ritualized weaver pattern appears in simple form in birds that still retain extra-territorial sex-chasing. Song now correlates markedly with nest-advertisement and not with post-display courtship. In tightly colonial species all courtship sequences are preceded by advertisement at the nest by the male and

---

Points from Table 3.

(i) Singing in approach to females far from the nest is found in species of all population dispersion types although it has been lost in more than half of the more colonial birds.

(ii) Song in nest invitation or advertisement is never common but is generally more frequent when the female accepts and actually enters the nest.

(iii) The incidence of song following female acceptance of the nest is pronounced in the most colonial species. Here too song has been largely replaced by special calling during advertisement display.

(iv) Species the males of which are known to sing in approach to females are primarily those in which courtship precedes or coincides with nest invitation. Such song is less frequent when advertisement precedes courtship.

(v) Of species known to have special calls in advertisement display almost all show PF type II in which advertisement precedes courtship. Likewise such species are more likely to have special calls than songs in advertisement. However where courtship mobility is especially reduced (PF IIb) enhanced songs or special calls appear equally likely to associate with advertisement display.

(vi) Of species with males that sing to females as she enters the nest most have preceded courtship by advertisement and have not undergone behavioural interaction with her previously.

(vii) The single PF type III species included is *Q. cardinalis* (see text).

the song in display is now commonly replaced by a repetitive special call. Singing continues to occur in territorial and some courtship sequences, particularly when the female first enters the nest. Male advertisement is highly ritualized and the expression of conflicting tendencies to approach the female sexually or aggressively is inhibited until she actually moves to the nest. As she enters, display ceases, and changes in posture and behaviour sequence reveal a strong aggressive tendency conflicting with both sexual approach and fear. In some species at this time the display form changes and song appears, while in others song-bowing develops, the male moving to the nest entrance and singing loudly into it at his newly arrived mate. Occasionally the female may even be attacked. Song is thus very much associated with the conflict situations arising in weaver-bird courtship and territorial defence.

The occurrence of songs and special calls in varying behavioural contexts fits broadly into the already established pattern of correlation between visual communication by display in pair formation, population dispersion and species habitat requirements. For example relatively solitarily and cryptically sited nests are not advertised by song, presumably to reduce the risk of predation. By contrast colonies are commonly in protected sites and nests in them may be dramatically and loudly advertised over considerable distances. Species with prominent aerial displays over territories containing hidden nests have regressed songs. Island endemics with a history of isolation without competition from related forms also show reduction in both visual and vocal display.

The manner in which acoustic behaviour may radiate in evolution is illustrated by certain closely related species groups that have been observed in particular detail. The three Indian weaver species *Ploceus benghalensis*, *P. manyar* and *P. philippinus* are especially closely related. They show adaptive radiation into flood-plain grassland, swamp and forest fringe, and savannah respectively—the last habitat in India now being extensively modified by agriculture. *P. benghalensis* builds in tall grass in loose colonies and *P. manyar* mainly in colonies in reed beds usually over water (although in Java it builds in colonies in trees, the nests being pendant). *P. philippinus* builds pendant nests in large tree colonies. The sequences of reproductive behaviour differ in the three cases. *P. benghalensis* males leave their nests and territories on the approach of females, fly to them, perch near them and approach very close giving a dramatic 'upright wing-beating display' with song. Females near the colonies may be chased by one or more males far from the nest sites and the display approach repeated at a distance. Eventually the pair returns to the nest and the favoured male invites the female to enter it. In *P. manyar* the male gives an 'upright wing-beating display', clearly homologous with that of *P. benghalensis*, in his territory: only after the female's arrival does sex-chasing develop. *P. philippinus* males in trees display in advertisement upside down below the nest entrance—the posturing involved again being homologous to the displays of its nearest relatives. The female enters the nest on her arrival and only after this 'acceptance' do courtship sequences begin. Clearly the behaviour of *P. benghalensis* shows a marked convergence to that of the grassland euplectine weavers although it has not developed a fully aerial display. Its song in display, like that of the *Euplectes* spp, is markedly attenuated, resembling, as Salim Ali (*in litt.*) has written, 'the chirp of a cricket or the subdued short squeaks of an unoiled bicycle wheel'. It is emitted only during the close approaches to the female. Unlike *P. benghalensis* the songs of *P. manyar* and *P. philippinus* are given in territorial and courtship behaviour but not in the wing-beating displays, in which a special advertisement cry is given. In these three species then the visual display, although homologous, has different functions—courtship and attraction in *P. benghalensis*, territorial advertisement in *P. manyar* and nest-advertisement in *P. philippinus*. Display orientation likewise differs—upright to female, upright on stem and inverted below nest respectively. Song is retained as a display component in courtship in *P. benghalensis* but it is a very soft performance related to the intimate positioning of the birds. Song does not occur in the displays of the other two species but is retained in other courtship and territorial sequences. The significance of these contrasts clearly relate to the species' nest-site preferences. The cryptic grassland nests are not advertised while colonial nests are. *P. manyar* is intermediate (Crook, 1963*a*).

Differences in reproductive behaviour between two *Quelea* species provide another similar example of behavioural divergence. *Quelea quelea*, breeding in its immense colonies in dense thorn scrub or reeds, defends minute territories around the nests or nest entrances. Females approach the densely packed nests while males give upright raised-wing displays in advertisement near the entrance. In courtship sequences male and female show a variety of agonistic and submissive posturing during which the male sings a short song while posturing in the 'appeased threat posture', considered to be the homologue of the song-bow in *Ploceus*. Song is not given in the nest-advertisement display but may closely precede or follow it. *Quelea cardinalis* by contrast nests in loose neighbourhoods in grassland in which the nest itself is well hidden. The small territories are advertised by patrolling, and a prominent upright wing-beating display performed on the top of tall grass stems. Females enter the territories and males approach them again performing this display and at the same time emitting a simple short song phrase, much quieter than but similar to that of *Q. quelea* and ending likewise in a soft whistle. Clearly the songs and displays of the two species are homologous but while the colonial species displays at the nest without song and sings only in courtship sequences, the grassland species sings in display which is used as part of the courtship sequence. Adaptive radiation into grassland niches has thus imposed changes in nest-site selection and breeding dispersion in three Ploceine genera, *Ploceus*, *Quelea* and *Euplectes*, with closely comparable effects upon display and song in each case.

The two *Foudia* species in the Seychelles Islands reveal contrasts in song production that relate clearly to the extent of their isolation on oceanic islands. The genus is related to *Euplectes* and is basically a savannah stock. The open country 'Cardinal' (*F. madagascariensis*) of Madagascar has been introduced by man into most of the Indian Ocean islands, on which in several cases an endemic *Foudia* already lived. The cardinal is essentially a flocking seed-eating bird that nests territorially in the breeding season. It produces a trilling sound in territorial advertisement and this appears to attract females. It also has a quiet wheezy song, homologous apparently with the short song phrase of *Ploceus* spp, which it emits when moving about with the female and which appears to be significant in courtship. In the Seychelles, at least, this song is somewhat variable in form. The endemic Seychelles weaver, the toq toq (*F. seychellarum*), by contrast, has completely lost the wheezy song, although a colourless trill may sometimes be heard during courtship wing-beating displays given on approaching females in territories. There is a monotonous 'Tsea tsea' call which appears to act in maintaining social cohesion. The toq toq (so named from its alarm cry) is insectivorous and relatively territorial throughout the year, living in forest or plantation undergrowth. The poor quality of its vocalization is considered to be a degenerative effect of isolated island life in the absence of selection pressures maintaining species distinctiveness. Regression evidently occurs to different extents on different islands, however, because Gill (1967) reports that on Rodriguez *F. flavicans* retains a variable song in both sexes and that duetting may occur in forests.

Social selection

The songs of weaver-birds are highly characteristic of species and, together with contrasts in the patterns of visual display, must play an important role in reproductive isolation—particularly in the rich savannah of Central Africa where many species live in broad sympatry, some even breeding in large joint colonies. Weaver-bird wing-beating displays, courtship posturings and songs are all basically homologous and derived from a common behavioural source. It seems probable that the basic courtship pattern involved sex approach, either away from or close to the male's nest site, and that this involved conflicting tendencies that produced a complex posture that soon underwent some degree of ritualization into a display. The action patterns of song-bowing appear to represent partial fixation of movements still expressing motivation in conflict. In relatively solitary insectivores behaviour became specialized in terms of the sex-chase that precedes nest invitation, whereas in colonial seed-eaters courtship centred increasingly on the nest. A close comparison of the posture composition and dynamic components of advertisement display and other courtship postures suggests strongly that the former are derived from courtship song-bowing given on the nest itself, although elements of the sex-chase (wing-beating) became incorporated, thus increasing the effect of the display: similar incorporation is found too in a number of approach displays not necessarily given at the nest. This basic posture, increasingly ritualized, emerged as a nest-advertisement display attracting females prior to the initiation of courtship. In advanced colonial species the display eventually replaced the male's excursion from the nest completely. During the ritualization process the song of the original song-bow was either lost, being replaced by simpler loud and repetitive calling, or, as in the case of certain long songs, maintained or extended as a longer and more forceful performance. In either case song is retained in its original context in song-bowing in courtship-territorial conflict behaviour.

The selection pressures modifying the orientation and differentiation of the song phrase throughout the family remain obscure. Many are certainly related to ecological adaptations, as we have seen, while others involve intra- and intersexual selection, and selection for increased distinctiveness, Distinctiveness is important in two ways: first to distinguish a signal from others in the repertoire of the species and, second, to distinguish a signal from those of other species with the same function and based upon homologous components. Such signals enhance reproductive isolation between related forms to the extent that they are species-specific and unambiguous.

It seems probable that a simple wheezy cry originated as the behavioural component of a stress response to situations involving approach/withdrawal conflict in territorial and courtship behaviour. These cries then associated rapidly with other calls, primarily from the agonistic repertoire, and were selected, together with postural components, as a relatively stereotyped intraspecific signal providing motivational information. These simple song-bows or stretches would subsequently have responded to any selection pressures favouring differentiation. It is suggested that selection against hybridization operated to increase song diversity under favourable ecological conditions which imposed local sympatry between members of rapidly subspeciating ploceine stocks. Indeed, African palaeo-ecological history provides ample evidence for such periods. A rapid diversification of the song patterns involved in intimate courtship sequences would have greatly reduced interspecific signal ambiguity and allowed an increased reproductive isolation between stocks. Many contrasts between broadly sympatric weavers in Central Africa today are doubtless maintained by such selection. Such change in song pattern could account for the observed contrasts in species song patterns, especially with regard to their length; and also for the fact that this diversity appears to have developed largely independently of the visual displays by colonial weavers in advertisement, whose diversity has been attributed mainly to allopatric differentiation involving intrasexual selection in relation to contrasting ecology (Crook, 1964). Indeed, as we have seen, the precise contexts in which songs are given, (as distinct from their forms), in well-studied contemporary species-groups are considered to be a range of adaptations to the social requirements for successful reproduction in differing ecological niches. Clearly the enhancement of the song-bow into the stereotyped nest-advertisement sequence of tight colony birds did not always require the concomitant vocal component, which was then replaced by the simpler repetitive cries of so many species (Table 2). It seems probable, however, that birds evolving from stocks already possessing long songs utilized these fine performances in advertisement even though this involved a lesser degree of distinctiveness between intra-specific signals. This interpretation, panglossian as it seems at present, is largely subject to confirmation or rejection by close examination of the relevant correlations between species attributes, but this depends on the collection of more data from the birds in the field. It is also important to devise Tinbergen-type experiments whereby the functional significance of song in weavers' lives may be demonstrated rather than merely surmised. For example—do deaf weavers miscegenate?

Although direct tests for the possible involvement of learning in the patterning of weaver songs have yet to be made, the available evidence does not suggest this type of ontogenetic control. Comparisons of song phrases show that the main differences between species involve contrasts in song amplitude (loud or soft singing), extent of emphasis on the terminal wheeze, elongation of phrase through repetition of phrase components or of the complete phrase, the addition of linking notes through modification of existing syllables, omission of components and regression of part or whole songs. All these changes are of the type generally found in the modification of displays under the influence of selection for communicatory functions (Blest, 1961). Song performances are however by no means invariable. Recordings of some twenty-eight songs produced at a window pane by a single male *Ploceus brachypterus* at Georgetown, Sierra Leone, reveal a considerable range of variants involving omission of certain syllables, emphasis on differing syllables, contrasts in syllable sequence, presence or absence of terminal wheeze, and differences in phrase length. Similar but simpler contrasts are found for example in agonistic calling where the intensity of 'chaks' and their association into a rapid chatter vary with the apparent intensity of aggression. The contrasts between individual performances of songs raise important questions regarding the control of song production because, while all songs are recognizable as species characteristic, their ritualization does not extend to the production of an invariant signal. Indeed the lability of composition and phrasing may well carry information both to the recipient of a song and its performer. Thus minor shifts in motivation with social context may be reflected in each song pattern. Further investigation of these intriguing and complicated cries should prove of great value to our understanding of bird song. The little we now know suggests a rich area for future exploration.

CONCLUSIONS

(i) The vocal repertoire of the weaver-birds (Ploceidae, Ploceinae) consists of some fourteen distinct types of call.

(ii) Among gregarious species flight calls appear important in the close integration of flocking movements, and observations of calling in relation to flight in captivity are described.

(iii) Reproductive vocalizations, short and long songs and advertisement calling, occur in a wide variety of behavioural contexts in differing species. This distribution remains very difficult to interpret with the information at present available. Nevertheless, both general and more detailed comparisons between species reveal a meaningful association between the occurrence of song in displays and the contexts in which the displays are given. These can generally be shown to have survival value in relation to their signal functions in contrasting habitats.

(iv) The evolution of song patterns in the subfamily is only loosely linked to that of patterns of visual communication within the contrasting types of social organization found in the different Afro-asian biome types. An attempt at explanation suggests that the diversity of songs evolved prior to and independently of that of visual advertisement displays. The fine detail of this vocal radiation ultimately requires an analysis that must be based upon further detailed comparative and experimental work.

ACKNOWLEDGEMENTS

This study, and other more extensive investigations of social organization in weavers, would certainly never have happened had it not been for the stimulation and encouragement of Professor W. H. Thorpe FRS in my first years as a postgraduate at Cambridge. It was he who first drew my attention to the exceptional promise of this group for comparative studies, and he who first sent me off into

Africa to seek an ornithological fortune. For several years, and in letters to many lands, his advice, assistance and friendship lay as a secure foundation to my work. My gratitude to him for so many happy times in so many richly rewarding and remote places cannot possibly be expressed in these few words. This small article is intended as a tribute and a thank-you.

The research here presented was supported by the Science Research Council. Assistance from Mr John Deag in the preparation of sound spectrographs and illustrations is most gratefully acknowledged. Dr Gerard Morel assisted me greatly with facilities and hospitality during visits to Le Station Ornithologique, Senegal, in 1955 and 1956.

REFERENCES

BLEST, D. (1961) The Concept of Ritualisation. In *Current problems in Animal Behaviour* (W. H. Thorpe and O. L. Zangwill). Cambridge University Press.

BUTTERFIELD, P. A. and CROOK, J. H. (1968) Annual cycle of nest building and agonistic behaviour in captive *Quelea quelea* with reference to endocrine factors. *Anim. Behav.* **16**, 308–317.

COLLIAS, N. E. (1963) A spectrographic analysis of the vocal repertoire of the African Village Weaver bird. *Condor* **65**, 517–27.

CROOK, J. H. (1960*a*) Studies on the social behaviour of *Quelea quelea* (Linn) in French West Africa. *Behaviour* **16**, 1–55.

(1960*b*) Studies on the reproductive behaviour of the Baya Weaver (*Ploceus philippinus*). *J. Bombay nat. Hist. Soc.* **60**, 1–48.

(1961*a*) The Fodies of the Seychelles Islands. *Ibis* **103a**, 517–48.

(1961*b*) The Basis of Flock Organisation in birds. In *Current Problems in Animal Behaviour* (W. H. Thorpe and O. L. Zangwill, eds.). Cambridge University Press.

(1963*a*) The Asian Weaver birds: Problems of coexistence and evolution with particular reference to behaviour. *J. Bombay nat. Hist. Soc.* **60**, 1–48.

(1963*b*) Comparative studies on the reproductive behaviour of two closely related weaver bird species (*Ploceus cucullatus* and *P. nigerrimus*) and their races. *Behaviour* **21**, 177–232.

(1964) The evolution of social organisation and visual communication in the weaver birds (Ploceinae) *Behaviour*, *Supplement*, **10**.

GILL, F. B. (1967) Birds of Rodriguez Island (Indian Ocean). *Ibis* **109**, 383–90.

MARSHALL, A. J. and DISNEY, H. J. de S. (1957) Experimental induction of the breeding season in a xerophilous bird. *Nature Lond.* **180**, 647–9.

MOYNIHAN, M. and HALL, F. (1953) Hostile, sexual and other social behaviour patterns of the Spice finch (*Lonchura punctulata*) in captivity. *Behaviour* **7**, 33–77.

SKEAD, C. J. (1947) A study of the Cape Weaver. *Ostrich* **18**, 1–42.

THORPE, W. H. and NORTH, M. E. W. (1966) Vocal imitation in the tropical Bou-Bou Shrike *Laniarius aethiopicus major* as a means of establishing and maintaining social bonds. *Ibis* **108**, 432–5.

13. VOCAL CHARACTERS AND AVIAN SYSTEMATICS

by **W. E. LANYON**
Department of Ornithology, The American Museum of Natural History, New York

INTRODUCTION

For practical reasons, classifications of organisms have traditionally been based on morphological characters, i.e. those attributes that are readily observable in museum specimens. It is the taxonomist's good fortune that this classical approach is sufficient to discern relationships among the vast majority of the taxa upon which he works. But in those instances where the evidence furnished by morphological characters is inadequate or equivocal, the modern taxonomist can turn to supplementary data derived from studies in behaviour, ecology, physiology, and biochemistry. There is no longer serious doubt as to the advisability of using such non-morphological data in systematics. The principle concern is when and how they should be used, and hence there is a need for specialists to formulate guidelines as the answers to these practical questions become apparent.

Among the behavioural data receiving increasing attention by taxonomists in recent years are the sounds made by animals. It is not surprising that the first serious consideration given to the use of animal sounds in classification was by systematists working in those particular groups in which the sound environment plays an especially significant role, e.g. in the orthopteran insects (Alexander, 1962), anuran amphibians (Bogert, 1960; Blair, 1962, 1964), and birds (Mayr, 1956; Löhrl, 1963; Thielcke, 1964). In this chapter I will examine the use of vocal characters in avian systematics, offer procedural guidelines for the procurement and analysis of vocal characters, and attempt to assess the present state of the art.

SOME GENERAL CONSIDERATIONS

Field observers have long been cognizant of the usefulness of the voices of birds in field identification. Indeed, in a few cases, specific identification is completely dependent upon some degree of vocal divergence. In view of this history of a general awareness of the utility of such vocalizations, why have avian systematists been reluctant to make use of vocal characters until comparatively recent times? There appear to be two explanations, one conceptual and the other technological. General acceptance of such non-morphological characters in systematics had to await the shift from a morphological to a biological concept of the species. Avian taxonomists, I might add, were among the most effective proponents of this shift. But even the most progressive among them were understandably reluctant to use characters that could not be treated with the same degree of objectivity as was the case with morphological characters. The real impetus did not come until the early 1950's, when advancements in technology at last provided biologists with equipment that enabled them to gather samples of vocal characters on magnetic tape, analyse these characters spectrographically, and reproduce them pictorially in order that other workers might have an opportunity to evaluate them.

Vocal characters are influenced by the same factors that govern sampling procedure and interpretation of results when using more conventional taxonomic characters. The sounds made by birds exhibit the same types of variation as, for example, do wing-length and plumage colouration: genetic and nongenetic, intrapopulational, geographic, species-specific, and polyphyletic variation. The degree of objectivity associated with such characters is directly related to the operational skill and objectivity of the personnel involved, but the same can be said for even the simplest of conventional taxonomic procedures, e.g. the measurement of wing-length. The subject of geographical variation in avian vocalizations (Borror, 1961) is discussed in greater depth by Thielcke in another chapter of this volume.

Poor sampling procedure can completely invalidate what would otherwise be a most thorough and detailed study, hence attention should be given to the composition of the sample (sex, age, geography) and the adequacy of its size, which is in turn related to the variability of the vocal characters being sampled. Identification of the particular bird from which recordings were made is a critical problem in the field. I know of some workers who habitually obtain recordings from birds unseen, and consequently their tapes are of little or no use for systematic studies. The actual collection and preservation of the bird as a scientific specimen is imperative where sibling species or semispecies are involved, or where field identification is otherwise equivocal.

Recording equipment should be selected for uniform and broad frequency response, good signal-to-noise ratio, and low distortion. One must be aware of the limitations of this equipment and the variations inherent in each stage of the recording and analytical process. As one example, consider the many ways in which the frequency spectrum, one of the important parameters of vocal characters, may be distorted: poor microphone response, attenuation of low frequencies by an undersized parabolic reflector, changes in tape speed during 'record' and 'playback', deformation of the tape due to stretching, failure to check the frequency scale against the standard tone of the original tape, etc. Fortunately there are safeguards and precautionary measures for avoiding these difficulties (Kellogg, 1960).

Original field recordings should not be edited or filtered and should bear complete data with respect to date, locality, identification of bird, and collector's number in those instances where the bird is preserved. A signal of known frequency (e.g. a 440 Hz tone from a pitch pipe) should appear periodically on all original recordings to serve as a standard for determining the frequency scale in subsequent analyses. All recordings should be catalogued and preserved for subsequent analysis and reexamination.

The sound spectrograph has become a popular tool for the analysis and portrayal of complex avian sounds, and several authors have published on its use for this purpose (Borror and Reese, 1953; Borror, 1960; Davis, 1964; Thielcke, 1966). These sounds have generally been described in terms of their qualitative appearance after spectrographic analysis, with little attempt at a more detailed physical description of the nature of the sound energy involved. Efforts to quantify such parameters as frequency, time, and amplitude have, in some studies, implied a degree of precision not yet attainable with present techniques and equipment. Workers should be cognizant of the fact that our characterization of avian sounds is still primitively qualitative and that we are still dealing with gross aspects of vocal behaviour, in this case, just as avian anatomists are working with only the gross aspects of anatomy. Familiarity with the manner in which the analytic equipment portrays experimental signals of known composition will aid in the discrimination between artifacts generated by the equipment and the actual sound energy generated by the birds themselves (Davis, 1964; Stein, 1968; Watkins, 1968).

The comparative analysis of vocal characters in systematic studies ultimately leads to value judgments by the investigator as to the biological significance of a difference in voice whenever such divergence can be demonstrated. When differences in vocal characters are apparent in sympatric forms, and these differences prevail even on a common breeding ground, one needs only minimal supporting evidence from morphological data to confirm the specific integrity of the forms. If there is reason to believe that vocalizations play a role in species discrimination by the birds themselves, as in forms exhibiting little morphological divergence, valuable additional information can often be gained through the use of standardized playback experiments to test the ability of territorial males to discriminate between their own vocal repertoire and that of the sympatric form (Dilger, 1956; Stein, 1958; Thielcke, 1962; Lanyon, 1963; Gill and Lanyon, 1964). Such playback experiments are especially informative and critical in studies involving allopatric and insular populations where the investigator is denied the test of reproductive isolation (Thielcke, 1962; Lanyon, 1965, 1967). The ability of territorial males to discriminate between their own vocal repertoires and those of congeners from other populations, when sound recordings are played back simultaneously to stimulate a condition of 'sympatry', is in many cases the best index of reproductive isolation available for allopatric forms. The entire vocal repertoire can be used if data are lacking as to which vocal patterns might be involved in species discrimination. Such experiments, when carefully controlled and executed, provide safeguards against drawing erroneous taxonomic conclusions based on comparisons of non-homologous vocalizations and eliminate the necessity of speculating as to which patterns and which parameters have the greatest stimulus value to the birds themselves.

To be useful in assessing relationships, vocal characters must be species-constant and distinctive, yet recognizably related to the homologous characters in other species. There is a danger, of course, that similar vocalizations shared by different populations may be polyphyletic in origin. Highly adaptive vocalizations, such as alarm notes, often show convergence among quite unrelated birds (Marler, 1955, 1957), and hence have no taxonomic value.

It is not necessary that the systematist become involved in the semantic argument as to whether such behaviour is 'innate' or 'learned'. As Amadon (1959) has indicated, evolutionary biologists are prepared to accept the concept that 'learning is always a part of what may too facilely assumed to be "instinct",' but at the same time are cognizant of what to them, at least, is the more critical aspect of the ontogeny of behaviour, 'that although learning may be involved, there is a genetic basis insuring that under normal life conditions a certain pattern of behaviour will develop with reasonable consistency'. In this regard, data on the ontogenetic development of vocalizations of young birds, hatched and hand-reared in experimentally controlled sound environments (Lanyon, 1960*a*) may have significance in the use of certain vocal characters in systematics. At issue, of course, is the extent to which exposure to experienced individuals of the same species is necessary for the full maturation of a characteristic vocal repertoire, and the possible influence of exposure to the voices of sibling or closely-related forms on the development of the same repertoire.

The avian systematist is aware that a given taxonomic character may be useful at one level of evolution and not at another; i.e. those characters that are meaningful at the specific or generic level may not be helpful at the familial or ordinal level, and vice versa. This axiom almost certainly applies to the use of vocal characters as well. Several authors, in grouping genera at the subfamily or tribe level, have evoked differences in vocalizations to support their views, but the vocal characters involved in these studies have been only loosely defined and are of relatively limited taxonomic value. Delacour (1943), in classifying the Estrildine finches, pointed out that all waxbills (Estrildae) have 'high-pitched, chirping or sweet calls and song', while the grass-finches (Erythrurae) 'have an unmelodious voice, clucking, mournful, trumpeting, metallic or low, never sweet, chirping and finchlike'. Likewise, the division of the Hawaiian honey-creepers (Drepaniidae) into two subfamilies has been based, in part, on differences in voice (Amadon, 1950). Marshall (1966) has used voice in helping to determine the relationship of the bay owl,

Phodilus badius, with barn owls (Tytonidae) and all remaining owls (Strigidae). But in general, I would agree with Thielcke (1964) and Thorpe (1961) that vocalizations appear to have little value for discerning relationships above the generic level. The very attribute of vocal characters that tells us most about avian relationships is their specific distinctiveness. Selection in favour of such divergence has tended to obliterate relationships at the higher taxonomic levels.

DIAGNOSTIC VALUE AT THE SPECIES OR SEMI-SPECIES LEVEL

It is doubtful whether vocal characters will gain wide usage among practising taxonomists even at the species level. For the most part they are not needed. Morphological characters are usually adequate to demonstrate specific limits and are less difficult to measure and express. In certain 'problem groups', however, vocal characters may permit discrimination between taxa in which there has been comparatively little selection for morphological differentiation. In these instances the systematist at once looks to those vocalizations that may conceivably serve as cues or attractants in species recognition, mate selection, and courtship, for it is differentiation in these characters that may help to prevent or restrict hybridization and hence maintain the integrity of species. This would appear to be the most promising role for vocal characters in avian systematics.

Taxonomists have classified birds into two groups of nearly equal numbers on the basis of the degree of development of the syrinx or vocal apparatus: (i) the non-passerines, typically without intrinsic syringeal muscles, and (ii) the passerines, having up to seven pairs of intrinsic syringeal muscles. As one might predict as a result of this anatomical distinction, the vocalizations of the non-passerines, as a group, are less diversified and complex than those of the passerines. With less variability in their vocal repertoires and a comparatively limited capacity for vocal differentiation as a part of the process of speciation, the non-passerines generally have developed visually oriented mechanisms for species recognition.

Non-passerines

The 'problem groups', as defined above, among the non-passerines are principally those birds whose courtship is performed under conditions of low light intensity or in habitats that restrict visibility.

Among crepuscular and nocturnal birds whose voices have been studied with modern techniques, the owls (Strigiformes) and goatsuckers (Caprimulgiformes) are noteworthy. Marshall (1966, 1967), for example, has relied heavily on vocal characters for determining specific limits within owls of the genus *Otus*, and it seems certain that the classification of other genera of owls will be revised when founded on similar studies. According to Marshall (1967), four almost completely allopatric forms ('incipient species'), having different vocal characters but 'no really good morphological characters to distinguish them', comprise a single North American species, *Otus asio*. He was influenced in this decision by striking similarities in certain aspects of the vocal repertoires, by the lack of appreciable morphological divergence, and by evidence of interbreeding in the single locality where two of the forms are known to overlap. His value judgment of what constitutes a biologically significant difference in *Otus* voices was made without benefit of data from playback experiments. One wonders whether or not territorial birds would be able to differentiate between the simultaneous playback of their own repertoire and that of one of the other forms? From the occasional breakdown of isolating mechanisms in one locality, particularly at the range periphery, we need not necessarily infer that these same mechanisms are ineffective in all areas of sympatry. Marshall (1966) has made another value judgment, but with a different conclusion, in his remarks concerning *Otus scops* of the Old World and *Otus flammeolus* of western North America, closely related ecological equivalents that have been regarded as conspecific by some taxonomists: 'their songs, however, are as different as owl voices can be . . . it is inconceivable that a female of *scops* would recognize the singing male *flammeolus* as a potential mate and *vice versa*. They cannot be in the same species'.

Davis (1962) attempted to show relationships within the goatsucker genus *Caprimulgus*, but his inattention to morphological and zoogeographical data as well as his unorthodox nomenclatural procedures make his study of more value for the voice descriptions than for the taxonomic conclusions. He has spectrographic evidence to suggest a slight divergence in homologous vocal patterns of the whip-poor-wills of eastern (*vociferus*) and south-western (*arizonae*) North America, heretofore considered to be representative forms of one species, but the difference is so slight that his recommendation that *arizonae* be given full specific status cannot be adjudged without additional, more detailed, analyses.

The chachalacas and their allies (Cracidae) are prime examples of those non-passerines typically found in dense habitats and having acoustically-oriented means of specific recognition. Of particular interest in this discussion is the role played by vocal characters in a recent redetermination of the specific limits of the chachalacas (*Ortalis*) of Mexico. Colouration of the underparts and general size had heretofore been given great weight in the classification of the four most well-marked and largely allopatric Mexican populations. The three smaller forms, with whitish or ochraceous buff abdomens (*vetula*, *poliocephala*, and *leucogastra*), were considered conspecific (*O. vetula*) by most workers, and the larger, chestnut-bellied *wagleri* was given specific status. Davis (in a paper presented at the annual meeting of the American Ornithologists' Union in 1952, and published in 1965) presented evidence that neither *leucogastra* nor *poliocephala* could be considered conspecific with *vetula*, because of pronounced differences in vocal characters, and that *O. poliocephala* and *O. wagleri* might best be considered sibling species because of similarities in their voices, regardless of their morphological differences. Moore and Medina (1957) and Vaurie (1965) re-examined the morphological and zoogeographical evidence, as well as vocal characters, and agreed with Davis with respect to the specific status of *O. leucogastra* and *O. poliocephala*. But the discovery of a narrow zone of intergradation in Jalisco led these authors to regard *wagleri* as conspecific with *O. poliocephala*. The exact relationship of *wagleri* and *poliocephala* in the zone of overlap must await a thorough field study, though the presence of hybrid specimens suggests that the slight differences in voice reported by Davis (1965) are not adequate to maintain specific integrity when these two well-marked forms occur together. Additional data are needed on the alleged differences in vocalizations, and much could be learned from playback experiments as the basis for comparing the territorial responses of males in the overlap zone and in areas of allopatry.

Passerines

The process of speciation is a dynamic one and we must be prepared to find evidence of all of its evolutionary stages among contemporary birds, i.e. from populations that exhibit only slight reduction in reproductive compatibility to those that hybridize only rarely and under restrictive circumstances. Since the ultimate criterion for the attainment of specific status is reproductive isolation, the relationships of closely related populations having allopatric distribution are extremely difficult to determine. It is in such cases that the systematist must make critical value judgments of the sort discussed earlier in this chapter. An analysis of the vocal characters of disjunct populations, as well as considerations of their morphology, will not necessarily ensure the correct interpretation but does broaden the base for comparison of the forms involved, in order to determine how far the process of speciation has progressed.

The relationship of the red-breasted nuthatch (*Sitta canadensis*), a widespread Nearctic species, to such geographically restricted populations as *Sitta whiteheadi*, endemic to the island of Corsica, is an example of this type of problem where vocal characters can be helpful. Löhrl (1961, 1963) has called attention to consistent differences in the location notes of these birds, which develop at an early age and retain their distinctive characteristics even when the young are reared by foster parents. He believes that the very different vocal repertoires of the adults develop from these location notes, and he has some evidence from captive birds that the adults can discriminate between their respective repertoires. Presumably the Corsican population is a relic from an earlier, more widespread distribution that included not only *canadensis* but some of the Asiatic forms as well.

The two Mexican populations of crows (*Corvus*) are allopatric and differ only slightly in their morphology. Davis (1958) called attention to their recognizably distinct voices and recommended that they be considered specifically distinct ('*C. sinaloae*' and *C. imparatus*). But Johnston (1961) has argued against this treatment until there is more evidence as to the biological significance of these vocal differences. His study of the American crows illustrates the difficulty of evaluating such vocal differentiation when it can be demonstrated that there are apparent inconsistencies in the biological significance of such differences throughout the genus. A divergence in vocal characters apparently contributes to the complete reproductive isolation between the fish crow, *C. ossifragus*, and the common crow, *C. brachyrhynchos*, in the southeastern United States. But, according to Johnston (1961), a notable difference in voice between *C. brachyrhynchos* and the northwestern crow, *C. caurinus*, is inadequate to keep these forms from interbreeding in areas of overlap in British Columbia and Washington. Quite clearly, data on the behavioural responses of the birds themselves to playback of their respective vocal repertoires would help to resolve these differences in value judgments of the relationships of allopatric forms.

There is now experimental evidence of the important role played by vocal characters as isolating mechanisms in a variety of passerine genera: *Myiarchus* (Lanyon, 1963,1965,1967); *Empidonax* (Stein, 1958, 1963; Johnson, 1963); *Parus* (Thönen, 1962); *Certhia* (Thielcke, 1962); *Phylloscopus* (Thielcke and Linsenmair, 1963); and *Catharus* and *Hylocichla* (Dilger, 1956). These studies lend substantial support to the use of standardized playback experiments wherever there is reason to believe that voice may function in this capacity and especially when allopatric forms are involved.

In my own work with *Myiarchus* flycatchers, the specific limits of allopatric populations have been determined principally by the responses of territorial males to the playback of their own vocal repertoires and those of congeners (Lanyon, 1965, 1967). Two examples will suffice to illustrate this point. *Myiarchus barbirostris*, the smallest of three forms of *Myiarchus* found on Jamaica, has no representative form elsewhere in the West Indies. Close morphological relationship to *M. tuberculifer*, a successful and polytypic species of Middle and South America, has been acknowledged by many workers and the Jamaican form sometimes has been treated as an insular race of that continental species. Similarities in portions of their respective vocal repertoires also support the morphological evidence of a close affinity. But the Jamaican birds do not have the prolonged whistled note that is given so frequently and characteristically by *M. tuberculifer* throughout its extensive range. The presence or absence of this diagnostic vocal character, perhaps in conjunction with more subtle differences in other vocal patterns, provides a sufficient basis for *barbirostris* and mainland *M. tuberculifer* to discriminate between their respective repertoires. In the simulated conditions of 'sympatry' afforded by the simultaneous playback of both vocal repertoires, these forms behaved in the manner of sympatric species of *Myiarchus* that have been tested elsewhere (Lanyon, 1963). I have recommended that *M. barbirostris* be considered an endemic Jamaican species.

The five differentiated populations of *Myiarchus* in the Lesser Antilles have all been treated as conspecific at one time or another by virtue of their morphological similarity, and generally have been treated as insular races of *M. tyrannulus*, a successful and widespread flycatcher of Middle and South America. When the inventories of their respective vocal repertoires were completed and analysed spectrographically, it was obvious that the four northern populations (*M. oberi berlepschii, M. o. oberi. M. o. sclateri*, and *M. o. sanctae-luciae*) share the same vocal repertoire, in which the most diagnostic character is a prolonged, plaintive whistle. It appeared questionable, then, that they could be recent derivatives of *M. tyrannulus* of Trinidad and the South American continent, as previously supposed, for that species has a repertoire devoid of any pure, unmodulated whistled notes. Playback experiments subsequently demonstrated that territorial males of the various insular populations of *M. oberi* do not, in fact, respond to the vocalizations of *M. tyrannulus*. The southernmost Lesser Antillean form, *M. nugator*, does appear to have evolved rather recently from *M. tyrannulus*, for its vocal repertoire shows great similarity to that of the South American species and likewise has no whistled notes. That *nugator* merits specific status, however, is strongly suggested by vocal differences established spectrographically and by the lack of response of territorial males to playback of *tyrannulus* vocalizations. This vocal divergence is also in concordance with a difference in colour of the mouth lining of these forms, a character which is not known to vary intraspecifically within the genus.

The relationship of the brown creeper (*Certhia americana*) of the New World to the two sibling creepers of the Old World (*C. brachydactyla* and *C. familiaris*) is an excellent example of the inadequacy of morphological characters alone for determining specific limits in some allopatric forms. Thielcke (1962) has reported on field experiments in which the European birds were able to discriminate between their own vocal characters and those of the American birds, and showed little response to the vocalizations of the latter. The result of the playback experiments together with a spectrographic analysis of the differences in the vocal characters led to his conclusion that *C. americana* should be regarded as specifically distinct from the European populations.

Vocal characters have also proved helpful in studies involving forms that are suspected of having contiguous or possibly continuous breeding ranges, but whose relationships have been questioned or are unclear. Until 1945 the house wren, *Troglodytes aedon*, and the brown-throated wren, *Troglodytes brunneicollis*, were thought to be allopatric species, the former representing the breeding form of the United States and Canada and the latter the resident form of the mountains of the Mexican plateau. In that year, a new subspecies of the Mexican form was described from the extreme southwestern part of the United States, implying sympatry of the two species in that region. This initiated a reexamination of populations of *Troglodytes* along the United States-Mexican border, with particular attention to an alleged difference in song between the house wren and brown-throated wren (Lanyon, 1960*b*). A cline of morphological characters between the two forms was demonstrated, but the gradient is steep due to discontinuities in distribution of preferred habitat and a resulting restriction of gene flow. Spectrographic analyses of vocalizations revealed a greater similarity between Mexican and western North American populations than between the eastern and western races of the North American species. A strong response of Mexican wrens to playback of the vocal repertoire recorded from house wrens outside of the range of *T. brunneicollis* confirmed the view that *T. aedon* and *T. brunneicollis* should be considered conspecific.

Two well-marked populations of the redwinged blackbird, *Agelaius phoeniceus grandis* and *A. p. gubernator*, are reported to be in secondary contact in two areas of central Mexico (Hardy and Dickerman, 1965). In the older of these contact zones, the two forms hybridize freely and the songs of the males show intermediacy between the songs of the two forms where they occur allopatrically. In an area of recent contact, the two forms are within sight of one another but are ecologically separated and do not interbreed, or do so rarely. In this region vocal characters show no evidence of intergradation and hence provide useful supplementary data for the solution of the problem.

There is an interesting geographical hiatus in southern Central America in the range of the flycatcher *Myiarchus tyrannulus*. One population (*brachyurus*) within the area of this hiatus has been treated as a distinct species by some workers and by others as a race of *M. tyrannulus*. Spectrographic analysis revealed that all Middle American populations of *M. tyrannulus*, as well as *brachyurus*, have the same vocal characters (Lanyon, 1960*c*). With these data and a series of specimens of known vocalizations, it became possible to demonstrate unequivocally that *brachyurus* is conspecific with *M. tyrannulus*. The several geographical races of *M. tyrannulus* have undergone some morphological divergence, such as marked difference in size and extent of cinnamon colouration in the rectrices, but have retained a remarkable uniformity in vocal characters, an observation that seems to apply to most of the species in this difficult flycatcher genus. A knowledge of the relative plasticity or conservativeness of vocal characters, as well as morphological characters, is of fundamental importance in avian systematics.

Still another example of a study involving forms with contiguous or continuous breeding ranges is that by Smith (1966), who recommends separation of the breeding kingbirds (*Tyrannus*) of north-eastern Mexico (*couchii*) as specifically distinct from *T. melancholicus*, the tropical kingbird, with which they have been associated by all workers in this century. The recommendation for specific status rests exclusively on an alleged difference in voice, for there is no morphological evidence to suggest any relationship other than the usual type of intergradation between a larger, partially migratory race in the north with a smaller, resident race immediately to the south. It is unfortunate that Smith did not have more data on geographical variation in the vocal repertoire of the wide-ranging *T. melancholicus*, for any significance attached to a difference in the voice of one segment of a species' population should be predicated on a thorough analysis of variation throughout the range of the species. Furthermore, a taxonomic decision of this kind should not be made without an exhaustive field study in the critical area of contact, in this case the central eastern lowlands of Mexico, involving the collecting of birds of known vocalizations. It is true, we may properly anticipate that minor or incipient divergence in the vocal characters of different races of birds may foreshadow differences between species, but this only accentuates the need for determining the progress of speciation in the critical area of contiguity or of overlap. In view of his ultimate decision to admit *couchii* as a full species, Smith's observations implying a certain amount of combining of the features of the vocal repertoires of both *couchii* and *T. melancholicus* in the vocalizations of individuals within the critical contact zone is especially disconcerting. The possibility of such intergradation is a question that must be answered satisfactorily before taxonomists can take Smith's recommendation seriously.

The chiffchaff (*Phylloscopus collybita*) of the Old World exhibits little geographical variation in morphology, other than the usual Eurasian clines in size and colouration, but the populations in southwestern Europe ('Spanish' chiffchaffs) are reported to have a song that differs just as markedly from that of the 'normal' chiffchaffs elsewhere within the range of the species as the song of that species differs from that of the sibling willow warbler, *P. trochilus*. Thielcke and Linsenmair (1963) used vocal characters effectively to study the relationship of the 'Spanish' chiffchaff and the 'normal' chiffchaff in a narrow zone of secondary contact in southwestern France. Spectrographic analyses of geographical variation in song throughout the range of the species provided them with a more objective basis for evaluating the alleged divergence in voice than had been possible heretofore, and disclosed intermediacy in song in some individuals within the contact zone. Males of the 'normal' chiffchaff responded to the playback of the song of the 'Spanish' chiffchaff, but less vigorously than when exposed to playback of their own song. The authors hypothesize that the 'normal' and 'Spanish' chiffchaffs differentiated vocally when isolated during the last glacial period, but the vocal divergence apparently was not great enough to effect complete reproductive isolation upon secondary contact. Because of the equivocal nature of the morphological evidence involved, resolution of this problem would have been impossible without a consideration of vocal characters.

According to Thönen (1962) a similar relationship exists between two song forms of the willow tit, *Parus montanus*, in central Europe. The birds of the Alps ('Alpine' song form) have a recognizably different territorial song from that of willow tits in the lowland regions ('normal' song form) and, as in the case of the chiffchaffs, the birds of one song form show much less response to playback of the territorial song of the other song form than they exhibit when exposed to their own song. Spectrographic analyses demonstrate that the territorial songs of both forms are remarkably constant throughout the ranges of the two populations, even though willow tits have evolved considerable differentiation in both size and colour. The two song forms largely represent each other geographically and the zone of overlap, within which some hybridization occurs, is quite restricted. Here again, 'hybrid' or intermediate vocal characters were of great importance in establishing relationships within the area of sympatry.

Thönen believes that the situation is not adequately explained by a development of a different song dialect *in situ*, as suggested by some workers, but rather by a secondary contact of two forms, following a lengthy period of isolation. The two forms are still not differentiated enough ecologically to permit their ranges to overlap to any great extent, and should properly be considered races of one species.

In no area of passerine systematics has the application of vocal characters been more appreciated and so eminently successful as in the establishment of specific limits for sibling species, i.e. those forms that are sympatric, at least in portions of their respective ranges, and whose identification through morphological characters is a troublesome task even to so-called experts. One need only be reminded of some of the outstanding instances where differences in voice provided the first or principal clue that certain sibling forms had reached the species level. In the Old World there is the classical case of the chiffchaff and the willow warbler, recently investigated with modern techniques by Thielcke and Linsenmair (1963). Playback experiments demonstrated that males responded positively to their own songs and seldom if at all to songs of the sibling form, as discussed more fully on page 304. In North America, Stein (1958, 1963) has presented convincing evidence for regarding the alder flycatcher, *Empidonax traillii*, and the willow flycatcher, *E. brewsteri*, as distinct species, and my own studies (Lanyon, 1957, 1962, 1966) and those of Szijj (1963, 1966) have confirmed the specific status of the eastern meadowlark, *Sturnella magna*, and western meadowlark, *S. neglecta*. Indeed, there may be few exceptions if any to Mayr's (1956) observation that there are no species-pairs among European and American birds that do not differ in their voices. In addition to the species mentioned above, vocal characters have been used effectively in studies of the following sibling species: *Empidonax hammondii*, *E. oberholseri*, and *E. wrightii* (Johnson, 1963); *Myiarchus cinerascens* and *M. nuttingi* (Lanyon, 1961); *Certhia familiaris* and *C. brachydactyla* (Thielcke, 1961, 1962); North American thrushes of the genus *Catharus* (Dilger, 1956); *Cassidix major* and *C. mexicanus* (Selander and Giller, 1961).

The techniques and procedures used in most of these studies of sibling species have followed a similar pattern. Spectrographic analyses of geographical variation have provided a basis for determining which vocalizations are most diagnostic and most trenchant. Specific limits have then been formulated through a concordance of morphological and vocal characters, as well as a consideration of ecological and behavioural differences. With the establishment of specific limits, which in many cases differ substantially from those of contemporary workers who used morphological characters exclusively, these investigators have then been able to hypothesize on the evolution of the forms involved. Playback experiments have generally confirmed speculation by earlier workers that reproductive isolation in these sibling species is dependent primarily upon the evolution of species-specific vocalizations. Females appear to have the selective role in pair formation, and are capable of differentiating between males of their own species and sibling males on the basis of diagnostic vocal characters. In some cases, the males do not differentiate between their own vocal repertoire and that of sibling males, nor do they show obvious preference for females of their own species. This lack of specific discrimination results in mutually exclusive territories and interspecific territorialism (e.g. *Sturnella*, *Cassidix*, and *Empidonax*). In other instances, the males do discriminate between their own vocal repertoire and those of sibling species, and hence they frequently have overlapping territories in areas of sympatry (e.g. *Myiarchus*, *Catharus*).

ASSOCIATIVE VALUE AT THE LEVEL OF THE SUPERSPECIES

In addition to their diagnostic value, as indicated above, vocal characters may also have an associative value, i.e. they may reveal close relationship between two or more species that might not be evident on the basis of morphology alone. Similarities in vocal repertoires can be most useful in studies directed toward the zoogeography and evolution of entire genera, enabling the recognition of groups of allopatric forms that comprise a superspecies (Mayr, 1963). Detailed studies of the vocal repertoires of all the forms of certain polytypic genera which show relatively little morphological variation offer excellent opportunities for improving our understanding of species formation and patterns of evolution. They may also provide some insight into the intriguing question of differential rates of evolution of the various components of a vocal repertoire. Unfortunately the use of vocal characters for their associative value has been seriously impeded because of a dearth of broad-scale comparative studies.

It is to the more conservative part of the vocal repertoire, i.e. the part more resistant to modification with time, that one looks for evidence of relationships among species within a given genus. In the flycatcher genus *Myiarchus* there is evidence that the so-called 'dawn song', i.e. that assemblage of certain vocal patterns normally used during the daylight hours but rendered in a particular and characteristic sequence just prior to daybreak, is more conservative than any of the component calls which are rendered individually during the day (Lanyon, 1967). For example, there is a monotypic species (*Myiarchus barbirostris*) endemic to the island of Jamaica that is, as stated above, unquestionably an island derivative of *M. tuberculifer*, a successful and polytypic species of Middle and South America. The Jamaican birds have a whistled component in their 'dawn song', but no whistled or unmodulated note in their daytime repertoire of calls, which is so characteristic of *M. tuberculifer* throughout its extensive range. I have hypothesized (Lanyon, 1967) that *barbirostris* may well have had whistled notes in its daytime repertoire at one time, but subsequently lost them except for those retained within the 'dawn song' assemblage. Though I have argued that *M. barbirostris* has reached specific status, as revealed by the response to playback experiments, I recommend that the close relationship with continental *M. tuberculifer* be indicated by including both species under the superspecies *barbirostris* (the older name).

In another example from West Indian *Myriarchus*, a remarkable similarity between the 'dawn song' of the polytypic *M. oberi* of the Lesser Antilles and that of *M. antillarum* of Puerto Rico is the only present-day clue to the close relationship of these species, now obscured by considerable evolution in size and plumage colouration and pattern. Moreover, the 'dawn song' of *M. antillarum* appears to be intermediate in its configuration between *M. stolidus* and *M. sagrae* of the Greater Antilles and *M. oberi* of the Lesser Antilles, suggesting a closer relationship of all of these species than had heretofore been suggested in the literature. Consequently, I have recommended that all of these species be placed in the superspecies *M. stolidus* (the oldest name) in order not to veil their relationship, which is anything but obvious on morphological characters (Lanyon, 1967).

CONCLUSION

The objective analysis of the voiced sounds of birds for systematic purposes is a relatively new research procedure, scarcely twenty years of age. The development of the technique, with able assistance from technology, has not been without growing pains, but it appears to have reached a desirable degree of maturity and to have found a significant role in avian systematics. For the solution to many problems of relationships among birds, particularly at the levels of the superspecies, species, and semispecies, we can no longer be content with the classical approach of the morphologist. Vocal characters, when sampled and analysed in accord with those principles governing the handling of more classical characters, and when interpreted in relation to pertinent morphological and zoogeographical data, hold much promise for the student of avian systematics and evolutionary biology. It is hoped that this cursory examination of the state of the art will have been a revelation to those previously unfamiliar with the field, and will serve as a catalyst for further investigations.

REFERENCES

ALEXANDER, R. D. (1962) The role of behavioral study in cricket classification. *Sys. Zool.* **11**, 53–72.

AMADON, D. (1950) The Hawaiian honeycreepers (Aves, Drepaniidae). *Bull. Am. Mus. Nat. Hist.* **95**, 151–262.

(1959) Behavior and classification: some reflections. *Vjschr. naturf. Ges. Zürich* **104**, 73–8.

BLAIR, W. F. (1962) Non-morphological data in anuran classification. *Syst. Zool.* **11**, 72–84.

(1964) Isolating mechanisms and interspecies interactions in anuran amphibians. *Quart. Rev. Biol.* **39**, 334–44.

BOGERT, C. M. (1960) The influence of sound on the behavior of amphibians and reptiles. *In* W. E. Lanyon and W. N. Tavolga (eds), *Animal Sounds and Communication*, Am. Inst. Biol. Sci. Publ. 7, Washington, D.C., 137–320.

BORROR, D. J. (1960) The analysis of animal sounds. *In* W. E. Lanyon and W. N. Tavolga (eds), *Animal Sounds and Communication*, Am. Inst. Biol. Sci. Publ. 7, Washington, D.C., 26–37.

(1961) Intraspecific variation in passerine bird songs. *Wilson Bull.* **73**, 57–78.

BORROR, D. J. and REESE, C. R. (1953) The analysis of bird songs by means of a Vibralyzer. *Wilson Bull.* **65**, 271–303.

DAVIS, L. I. (1958) Acoustic evidence of relationship in North American crows. *Wilson Bull.* **70**, 151–67.

(1962) Acoustic evidence of relationship in *Caprimulgus*. *Tex. J. Sci.* **14**, 72–106.

(1964) Biological acoustics and the use of the sound spectrograph. *S. West. Nat.* **9**, 118–45.

(1965) Acoustic evidence of relationship in *Ortalis* (*Cracidae*). *S. West. Nat.* **10**, 288–301.

DELACOUR, J. (1943) A revision of the subfamily Estrildinae of the family Ploceidae. *Zoologica* **28**, 69–86.

DILGER, W. C. (1956) Hostile behavior and reproductive isolating mechanisms in the avian genera *Catharus* and *Hylocichla*. *Auk* **73,** 313–53.

FISH, W. R. (1953) A method for the objective study of bird song and its application to the analysis of Bewick Wren songs. *Condor* **55,** 250–57.

GILL, F. B. and LANYON, W. E. (1964) Experiments on species discrimination in Blue-winged Warblers. *Auk* **81,** 53–64.

HARDY, J. W. and DICKERMAN, R. W. (1965) Relationships between two forms of the Red-winged Blackbird in Mexico. *The Living Bird*, **4,** 107–29.

JOHNSON, N. K. (1963) Biosystematics of sibling species of flycatchers in the *Empidonax hammondii-oberholseri-wrightii* complex. *Univ. Calif. Publ. Zool.* **66,** 79–238.

JOHNSTON, D. W. (1961) The biosystematics of American crows. *Univ. Washington Press, Seattle*, 119 pp.

KELLOGG, P. P. (1960) Considerations and techniques in recording sound for bio-acoustics studies. *In* W. E. Lanyon and W. N. Tavolga (eds.), *Animal Sounds and Communication*, Am. Inst. Biol. Sci. Publ. 7, Washington, D.C., 1–25.

LANYON, W. E. (1957) The comparative biology of the meadowlarks (*Sturnella*) in Wisconsin. *Publs. Nuttall Orn. Club*, no. 1, 1–67.

(1960*a*) The ontogeny of vocalizations in birds. *In* W. E. Lanyon and W. N. Tavolga (eds.), *Animal Sounds and Communication*, Am. Inst. Biol. Sci. Publ. no. 7, Washington, D.C., 321–47.

(1960*b*) Relationship of the house wren (*Troglodytes aedon*) of North America and the Brown-throated Wren (*Troglodytes brunneicollis*) of Mexico. *Proc. 12th Inter. Orn. Congr. Helsinki*, 1958, 450–8.

(1960*c*) The Middle American populations of the crested flycatcher *Myiarchus tyrannulus*. *Condor* **62,** 341–50.

(1961) Specific limits and distribution of ash-throated and Nutting flycatchers. *Condor* **63,** 421–49.

(1962) Specific limits and distribution of meadowlarks of the desert grasslands. *Auk* **79,** 183–207.

(1963) Experiments on species discrimination in *Myiarchus* flycatchers. *Am. Mus. Novit.*, no. 2126, 1–16.

(1965) Specific limits of the Yucatan Flycatcher, *Myiarchus yucatanensis* *Am. Mus. Novit.* no. 2229, 1–12.

(1966) Hybridization in meadowlarks. *Bull. Am. Mus. Nat. Hist.* **134,** 1–25.

(1967) Revision and probable evolution of the *Myiarchus* flycatchers of the West Indies. *Bull. Am. Mus. Nat. Hist.* **136,** 331–370.

LÖHRL, H. (1961) Vergleichende Studien über Brutbiologie und Verhalten der Kleiber *Sitta whiteheadi* Sharpe und *Sitta canadensis* L. Teil II. *J. Ornith.* **102,** 111–32.

(1963) The use of bird calls to clarify taxonomic relationships. *Proc. 13th Int. Orn. Congr. Ithaca*, 1962, 544–52.

MARLER, P. (1955) Characteristics of some animal calls. *Nature, Lond.* **176,** 6–7.

(1957) Specific distinctiveness in the communication signals of birds. *Behaviour* **11,** 13–29.

MARSHALL, J. T. Jr. (1966) Relationships of certain owls around the Pacific. *Nat. Hist. Bull. Siam Soc.* **21,** 235–42.

(1967) Parallel variation in North and Middle American Screech-Owls. *Monographs of the Western Foundation of Vertebrate Zoology*, no. 1, 1–72.
MAYR, E. (1956) Gesang und Systematik. *Beiträge zur Vogelkunde*, **5**, 112–17.
(1963) *Animal species and evolution.* Belknap Press (Harvard Univ.), Cambridge, Mass.
MOORE, R. T. and MEDINA, D. R. (1957) The status of the chachalacas of western Mexico. *Condor* **59**, 230–4.
SELANDER, R. K. and GILLER, D. R. (1961) Analysis of sympatry of great-tailed and boat-tailed grackles. *Condor* **63**, 29–86.
SMITH, W. J. (1966) Communication and relationships in the genus *Tyrannus*. Allen Press, Lawrence, Kansas. *Nuttall Ornith. Club, publ.* no. 6, 1–250.
STEIN, R. C. (1958) The behavioral, ecological and morphological characteristics of two populations of the Alder Flycatcher, *Empidonax traillii* (Audubon). *Bull. N.Y. St. Mus. Sci. Serv.* **371**, 1–63.
(1963) Isolating mechanisms between populations of Traill's Flycatchers. *Proc. Am. Philos. Soc.* **107**, 21–50.
(1968). Modulation in bird sounds. *Auk*, **85**, 229–243.
SZIJJ, L. J. (1963) Morphological analysis of the sympatric populations of meadowlarks in Ontario. *Proc. 13th Orn. Congr. Ithaca*, 1962, **1**, 176–88.
(1966) Hybridization and the nature of the isolating mechanism in sympatric populations of meadowlarks (*Sturnella*) in Ontario. *Z. Tierpsychol.* **6**, 677–90.
THIELCKE, G. (1961) Stammesgeschichte und geographische Variation des Gesanges unserer Baumläufer (*Certhia familiaris* L. und *Certhia brachydactyla* Brehm). *Z. Tierpsychol.* **18**, 188–204.
(1962) Versuche mit Klangattrappen zur Klärung der Verwandtschaft der Baumläufer *Certhia familiaris* L., *C. brachydactyla* Brehm und *C. americana* Bonaparte. *J. Ornith.* **103**, 266–71.
(1964) Lautäusserungen der Vögel in ihrer Bedeutung für die Taxonomie. *J. Ornith.* **105**, 78–84.
(1966) Die Auswertung von Vogelstimmen nach Tonbandaufnahmen. *Die Vogelwelt*, **87**, 1–14.
THIELCKE, G. and LINSENMAIR, K. (1963) Zur geographischen Variation des Gesanges des Zilpzalps, *Phylloscopus collybita*, in Mittel- und Südwesteuropa mit einem Vergleich des Gesanges des Fitis, *Phylloscopus trochilus. J. Ornith.* **104**, 372–402.
THÖNEN, W. (1962) Stimmgeographische, ökologische und verbreitungsgeschichtliche Studien über die Mönchsmeise (*Parus montanus* Conrad). *Orn. Beob.* **59**, 101–72.
THORPE, W. H. (1961) *Bird Song. The biology of vocal communication and expression in birds.* Cambridge University Press; Cambridge Monographs in Experimental Biology, no. 12.
VAURIE, C. (1965) Systematic notes on the bird family Cracidae. No. 2. Relationships and geographical variation of *Ortalis vetula*, *Ortalis poliocephala*, and *Ortalis leucogastra*. *Am. Mus. Novit.* no. 2222, 1–36.
WATKINS, W. A. (1967) The harmonic interval: fact or artifact in spectral analysis of pulse trains. In *Marine Biacoustics II.* W. N. Tavolga (ed.) Pergamon Press, Oxford, England, 15–43.

14. GEOGRAPHIC VARIATION IN BIRD VOCALIZATIONS

by **GERHARD THIELCKE**
Max-Planck-Institut für Verhaltensphysiologie, Vogelwarte, Radolfzell

INTRODUCTION

Geographic separation is always a precondition for the genesis of a new species (Mayr, 1963; but see Thoday and Boam, 1959). The first aim of any study in this context must therefore be to recognize intraspecific geographic variation. Stereotyped vocalizations are particularly appropriate for such studies because of their probable role as isolating mechanisms, because they are qualitatively independent from inanimate environmental factors, and because they lend themselves to objective results. The present review is limited to studies describing the variation of several populations.

Calls and notes are defined here as continuous sounds not divided by pauses: they differ in function. A note is the smallest temporal unit of a song (example: A, B, C, D, e, E, and F in Figure 4). A song consists of a group of notes separated from another group of notes by a pause longer than the pauses between the notes themselves (example: Figure 4 shows seven notes of one song). In German, a distinction is also made between Strophe (song) and Gesang (general song pattern). Examples of songs of the same song-type are *a* and *b*, and *c* and *d*, in Figure 1; notes of the same type are shown in Figure 4*c*. The term 'phrase' is used in this paper as defined by Thorpe (1958). The term 'dialect' is used only for vocalization variants with a mosaic distribution.

VARIATION IN A COMPACT RANGE

A compact range is defined in this paper as a geographic area inhabited by a species without excessive gaps. 'Gap' is of course a relative term: the decisive consideration here is its effect on the bird species concerned, i.e. whether or not the gap hinders or even prevents birds crossing to settle or resettle on the other side. Since exchange between two populations is hindered if they are separated by large geographic areas offering scanty or no survival possibilities for the species concerned, it is of interest to study the effects of extremely compact and of extremely fragmented ranges on their populations. It should be borne in mind, however, that our knowledge of the actual interchange of individuals between two populations is very limited.

Mosaic-type variation

Chaffinch (Fringilla coelebs) songs. All songs are based on the same basic pattern. Each male has one to six (on the average two) song-types (Marler, 1956*b*): songs of the same type are highly stereotyped. After a series of songs of one type, a male sings a series of another type, and so on. The frequency of each song-type is increased if the bird hears a tape recording of that song-type (Hinde, 1958). The males learn the fine points of their song in their first summer, up to the thirteenth month of age (Thorpe, 1958; see also p. 19–22 of this volume).

Promptoff (1930) and Marler (1952) have studied in detail the geographic variation of chaffinch songs. We still, however, lack a sound spectrographic analysis which is nowadays considered essential for studies of geographic song variation. According to the above two authors, local dialects can only be demonstrated statistically. On the other hand Conrads (1966) found a strong 'concentration' of a song-type in one area.

The extent of the similarity between the same type of songs sung by different males is illustrated in Figure 1*a* to *d*. The terminal flourish and the entire song vary geographically. Very similar terminal flourishes may be found in widely separated areas (Figure 1*c*, *d*, *e*).

In many areas of Europe, but not in all, chaffinches conclude a complete song with a short note sounding like 'kit' (Figure 1). The map in Figure 2 is only intended to illustrate the wide separation of areas in which this kit occurs. The map is grossly incomplete with respect to the actual frequency of occurrence of this version; in many regions the dots should be spaced so densely that whole areas would be black. We do not yet know how high the proportion of kit-singers is in different populations, nor whether there are males who

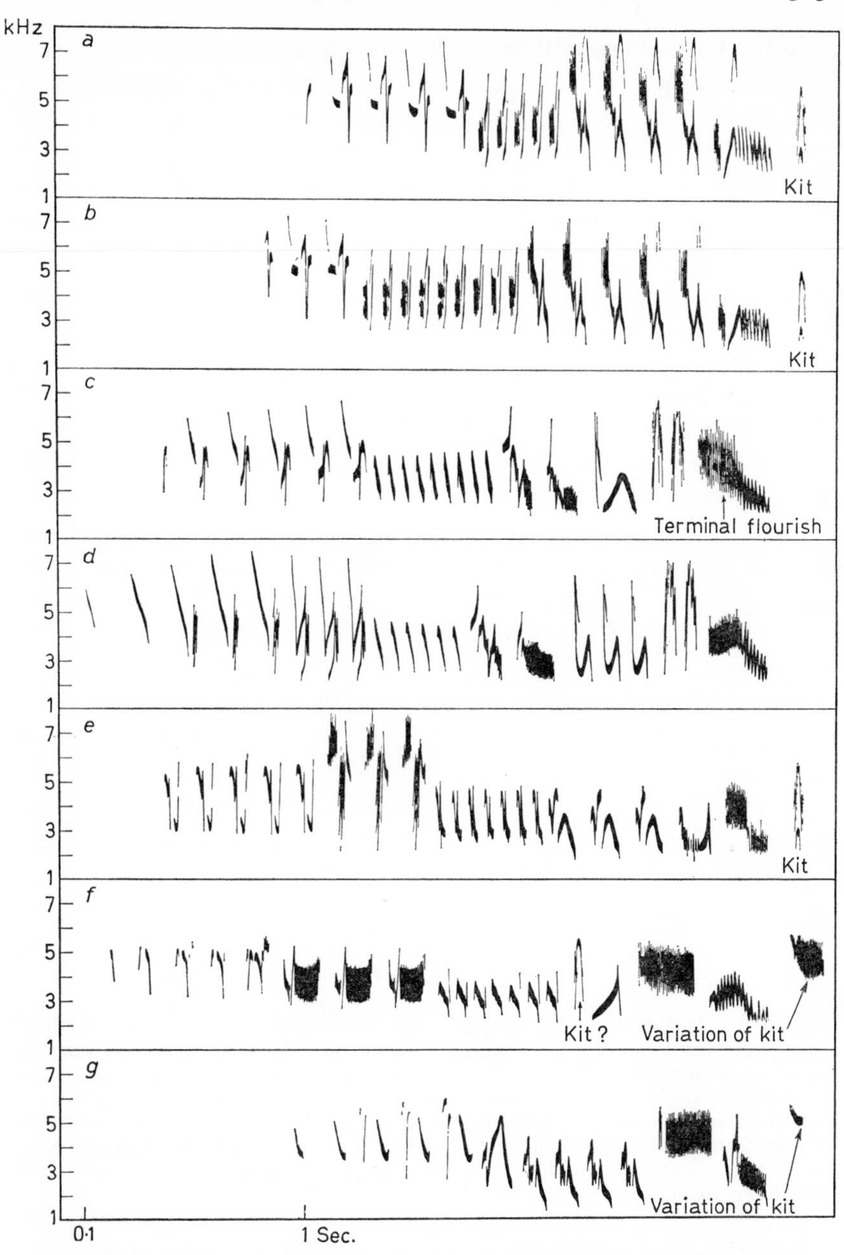

FIGURE 1. Seven songs of chaffinches. (*a* and *d*) and (*b* and *c*) are by the same male. (*a* and *b*) and (*c* and *d*) are songs of the same type. The songs *a*, *b* and *e* have a kit added to them; the songs *f* and *g* have a different kit variant. Recording locations (compare Figure 2): *a*, *b*, *c* and *d* in southwest Germany, *e* near Graz, Austria, *f* and *g* in the Dolomites in northern Italy. The same terminal flourish (*c*, *d*, *e*) and the kit occur in widely spaced regions.

add the kit to some of their song-types but not to others. In the two cases for which I have tape recordings of two song-types with kits by the same male, all the kits of one type differ markedly from those of the other with respect to pitch (Figure 2*d*).

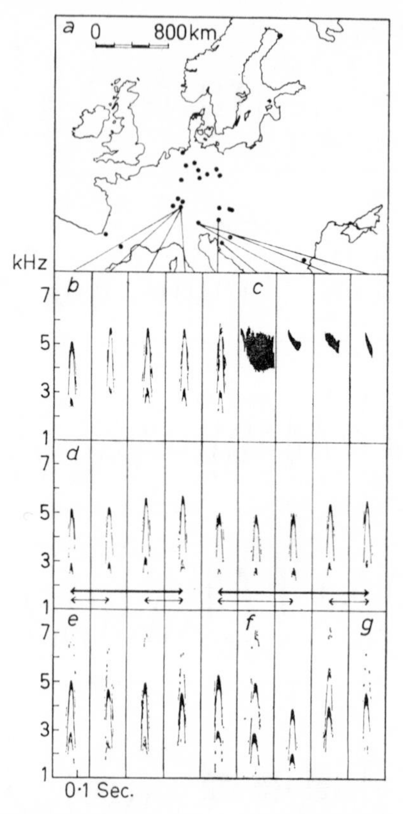

FIGURE 2. *a*, The dots indicate locations where European chaffinches add a kit to their song (according to data by Mauersberger, Conrads and Löhrl, and own observations). *b*, Kit of five different males. *c*, Notes by four males sung instead of the kit; the recording locations are shown by strokes on the map. *d*, Four and five kits by two males (top arrows), kits of songs of the same type (bottom arrows). *e*, Kit of the great spotted woodpecker. *f*, Kit of the middle spotted woodpecker; *g*, kit of the lesser spotted woodpecker. The more pronounced harmonics of woodpeckers, compared to chaffinches, are due to greater loudness.

I had thought (1962) that the kit was the only variant added by the chaffinch to a complete song. V. Dorka and H. Löhrl, however, drew to my notice chaffinches in the Dolomites which are aberrant in this respect. Tape recordings revealed the presence of two new forms

(Figures 1*f* and *g*, and 2*c*). One of these songs (*f*) contains a note at the beginning of the last phrase, i.e. in the middle of the song, which is very similar to a kit except that it lacks the characteristic lower harmonic at about 3 kHz which is typical for all kits. Tembrock (1965) reports a variant ('hweet') by an autumn migrant on Rügen island (Baltic) reminiscent of a warbler (*Phylloscopus*).

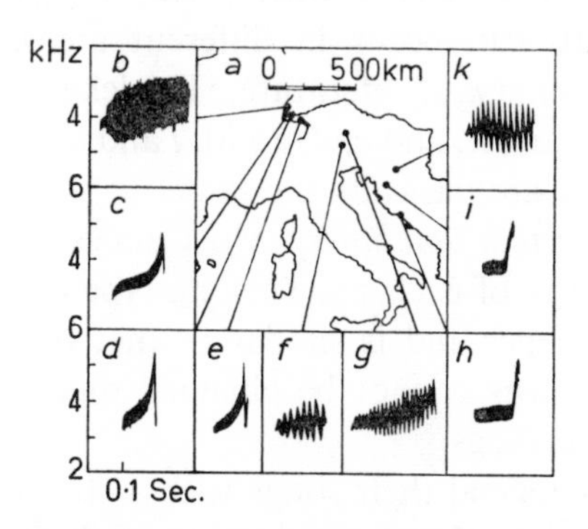

FIGURE 3. So-called rain calls of the chaffinch and recording locations in southwest Germany, Austria, northern Italy and Yugoslavia.

The kit does not necessarily occur with particular song-types nor is it necessarily linked to particular terminal flourishes. This does not, of course, exclude the possibility that a correlation might be demonstrated statistically. The answer to this question is of interest because the entire song, as well as parts thereof, and especially the terminal flourish, are geographically variable.

Chaffinch, rain call. The so-called rain call of the chaffinch is one of the classical examples of mosaic variation on the European continent. As the extreme forms of this call can easily be distinguished by the human ear (see Figure 3), some of its basic features are well known. An accurate sound analysis would nevertheless be of considerable value. In contrast to the examples discussed above, this is not a song or part of a song but a call. It is made only by males. It is linked to the breeding season and is made in various situations (Sick, 1939, 1950; Marler, 1956*a*; Poulsen, 1958). Poulsen's view that this call is a kind of song substitute is in agreement with Bergman (1953) and with my own observations.

The dialect form is very uniform throughout a population (Sick, 1939). This is confirmed by some of my own recordings and tone spectrograms from a 'hweet' area, but further recordings are still required. In one case the boundary went right through a region densely populated with chaffinches, and there was a mixed zone with pure calls of both forms as well as transitions between the two forms. In another case, by contrast, two dialects were sharply divided by a railway line and a few blocks of houses. The uninhabitable zone in this instance was over 400 m wide (Sick, 1939). Whether or not identical rain calls can occur in different areas, as is frequently reported (see summary of literature, Thielcke 1961), is not clear (compare *f* with *g* and *k*, and *c*, *d*, *e* with *i* and *h* in Figure 3). At least the calls *f*, *g* and *k* are basically the same.

Tree creeper (*Certhia familiaris*). Most males have only one song-type. The variability of each male's song-type is greater than in the short-toed tree creeper and is similar to that of the chaffinch. The songs of mature birds cannot be changed qualitatively by making them listen to recordings.

Most tree creepers end their songs with a striking note. In several widely separated areas they have, instead of the terminal flourish, a call sounding like 'teet', or they add the 'teet' to their complete song (with terminal flourish) (Thielcke, 1962). The 'teet' is very probably learnt from the short-toed tree creeper.

Short-toed tree creeper (*Certhia brachydactyla*). All songs are very uniform. Most males have only one song-type. Males with a second song-type sing both songs alternately or in a 'medley sequence'. Making the birds listen to song recordings at mating time has no effect on the quality of the songs.

The highly stereotyped song of a short-toed tree creeper usually consists of six to seven notes (Thielcke, 1964, 1965 *a* and Figure 4). Notes e and E occur in different variants. One variant, occasionally two, is typical for each population. 14·6 per cent of males deviate from the predominant variant. The same variants may occur in different localities separated by regions with different variants. In each of the three cases studied, the notes e and E remained constant for seven to eight years within the population (unpublished). They are not necessarily linked either with other forms of geographic variation or with other mosaically varying features of the songs of the short-toed tree creeper (Thielcke, 1965 *a* and unpublished). I know of one case of a sharp dialect boundary. In this instance the forest is interrupted for 7·5 km by meadows, fields and three small woods

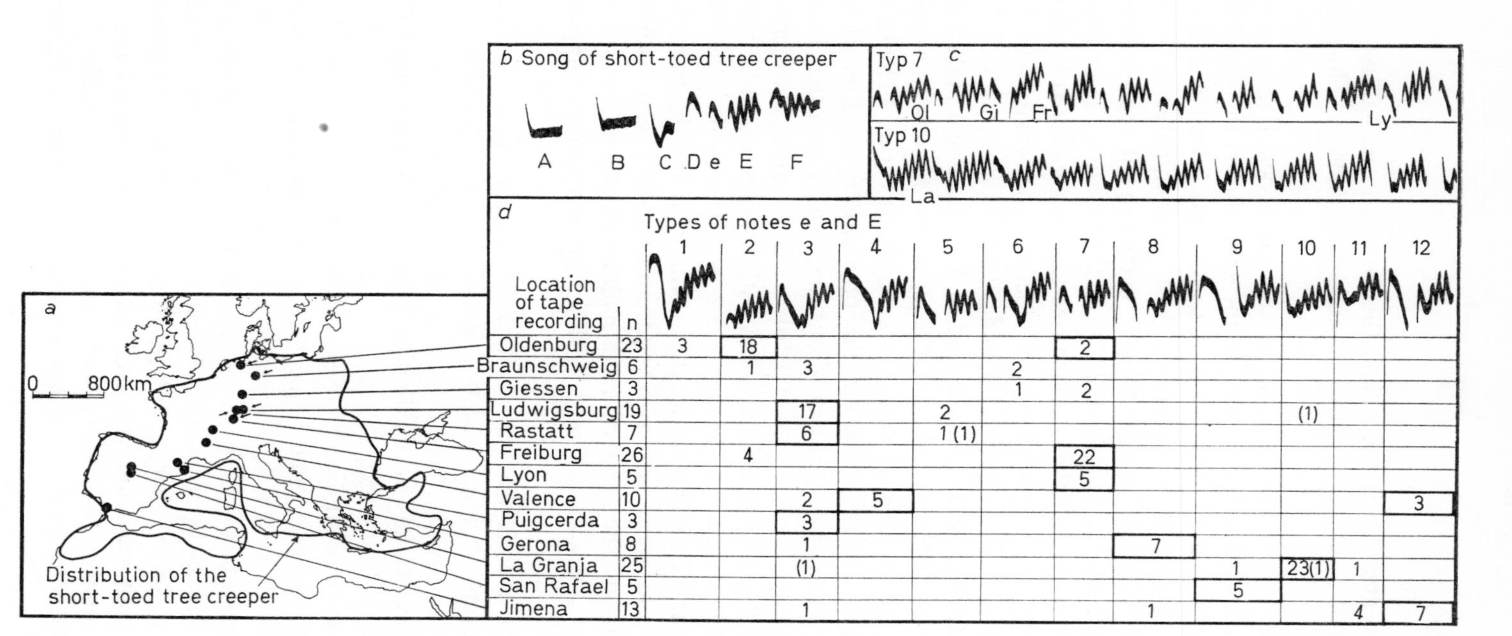

Location of tape recording	n	1	2	3	4	5	6	7	8	9	10	11	12
Oldenburg	23	3	18					2					
Braunschweig	6		1	3			2						
Giessen	3						1	2					
Ludwigsburg	19			17		2					(1)		
Rastatt	7			6		1 (1)							
Freiburg	26		4					22					
Lyon	5							5					
Valence	10			2	5								3
Puigcerda	3			3									
Gerona	8			1					7				
La Granja	25			(1)						1	23(1)	1	
San Rafael	5									5			
Jimena	13			1					1			4	7

FIGURE 4. *a*, Distribution of the short-toed tree creeper and recording locations. Arrows designate locations where both short-toed tree creepers and tree creepers were recorded. *b*, Song of short-toed tree creeper with seven notes. *c*, Notes *e* and E (Type 7 and Type 10) by different males (Ol = Oldenburg; etc.). *d*, Recording locations, number of males (*n*) and frequency of the 12 notes types e and E (after Thielcke, 1965*a*, modified, based on the map by Mauersberger, 1960).

However, even greater distances than this between one habitat and another do not necessarily form a dialect boundary (Thielcke, unpublished).

Notes A, B and C of the song of the short-toed tree creeper descend to different levels of pitch. In the song shown in Figure 4, note C descends to the lowest pitch, A to the next lowest and B to the least low. This pitch sequence C, A, B predominates in many populations in Central Europe; thus, it occurs in all the nineteen recordings made near Ludwigsburg and in all the twenty-six recordings made near Freiburg. In other populations this uniformity is broken (Figure 5). Thus, in a population near Oldenburg, ten males had the pitch sequence C,A,B, as above; six males had the sequence C,B,A; four males had B,C,A; two males had B,A,C; and one male had A,B,C. Higher variability of the pitch sequence is linked with higher variability of the number of notes. I had thought earlier (Thielcke, 1965*a*) that this variability within the population was limited to northern Germany. However recordings made in many other locations in Germany, Austria and Yugoslavia show quite a different picture. Populations with a higher and with a lower variability are also distributed mosaically. This mosaic does not coincide with that of the notes e and E. There are locations with high variability of the pitch sequence, locations with two forms and, finally, locations with no variation at all. The proportions of males with one and with two different songs are also different in different populations (unpublished).

Carolina chickadee (*Parus carolinensis*). Each male has one to four song types. Intra-individual variation is, however, insufficiently well known. Hearing a tape recording or a particular song type of a neighbour probably has no qualitative influence on the songs of a sexually mature bird (Ward, 1966).

At all the locations studied there exists a typical song consisting of four notes in a definite pitch sequence. Conformity with this basic type is particularly strong in four out of nine regions, with 30 to 100 per cent of songs being of this type. The males of each population tend to sing identically or similarly: their song may conform to the basic type or to some other stereotyped dialect form, or it may take the form of high variability of versions of a given song-type. In addition, the number of song-types of a male varies regionally.

Cardinal (*Richmondena cardinalis*). Each male has nine to thirteen

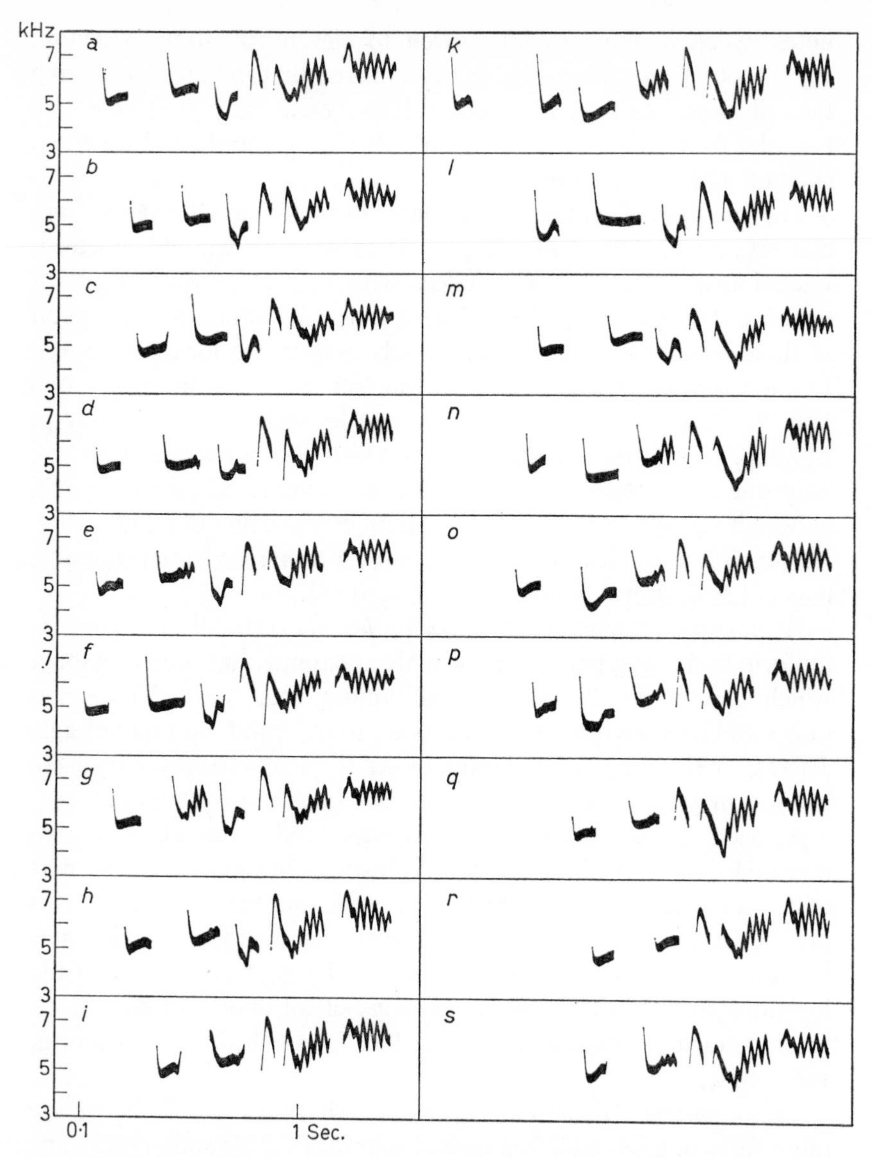

FIGURE 5. (*a* to *i*) Songs by nine short-toed tree creepers from Ludwigsburg. (*k* to *s*) nine songs by eight males from Frankfurt; *k* and *l* are by the same male. The pitch sequence (see text) and the number of notes per song are quite uniform in Ludwigsburg, with only *h* and *i* deviating with respect to the number of notes (five instead of six) and only *i* deviating with respect to the pitch sequence (B, C instead of C, A, B). In Frankfurt, by contrast, the pitch sequence and the number of notes vary considerably. Pitch sequence: 4×B, A, C (*k, n, o, p*), 2×C, A, B (*l, m*), 3×B, C (*q, r, s*); number of notes: 1×7 (*k*), 5×6 (*l, m, n, o, p*) and 3×5 (*q, r, s*). The penultimate note is the same in Ludwigsburg and Frankfurt. Elsewhere it varies mosaically.

notes which it arranges into seven to eleven song-types (Lemon, 1965, 1966). Cardinals probably learn in their first summer and during the following February and March. However, songs played on a tape recorder from mid-February to early April apparently had no effect (Lemon and Scott, 1966).

The number of note-types at each location varied between ten and seventeen, that of song-types varied between ten and nineteen. Each location has its own dialect with respect to the quality of notes' and songs' types. Several song-types are similar and common in three locations, others occur in only one or two locations. Seven London (Ontario) notes occur in fourteen of fifteen locations, one note in six locations, one note in three locations, and one only in London. Of the eleven Ontario notes which do not occur in London, only one is widespread (seven locations). Simple notes and motifs, although varying less with location than complicated ones, nevertheless exhibit local features. Even small barriers, e.g. river systems, may act as dialect boundaries (Lemon, 1966).

White-crowned sparrow (*Zonotrichia leucophrys*). All males have a uniform basic song pattern. Each male apparently has only one song, which it varies slightly (Marler and Tamura, 1962, 1964). The young males and females learn the dialects up to the hundredth day of age. Thereafter the song of the young males is fixed, although they have not yet sung; the females normally do not sing (Konishi, 1965).

At each location (four locations were studied) there is a pronounced dialect. The introductory whistle is the most variable part of the song, the first trills are less variable, and the concluding trills are the least variable. It has so far been possible to assign all males to a given location according to their concluding trills. At one of the locations studied all the parts of songs of all males were in much better agreement than at the other locations (Marler and Tamura, 1962, 1964).

Yellowhammer (*Emberiza citrinella*). All songs are built on the same pattern. Each male has several slightly variable song-types. The males, after a series of songs of one type, change over to another type. In essence, there is a strong similarity to the repertoire of the chaffinch.

The last notes of the song occur in different variants. One or two variants may predominate over large areas (Kaiser, 1965) but there may also be localities with more variants (own observations). Kaiser considers the relatively large extent of dialect regions as being different from mosaic-type variation. This can only be clarified by spectrographic analysis of the songs of many individual yellowhammers. In the boundary zone betweeen two variants there is a concentration of mixed songsters with both song types; the boundary is fairly sharp (Kaiser, 1965).

Discussion. A dialect mosaic consists in the simplest case of one variant which occurs or is absent in several populations. More detailed studies usually lead to the discovery of several variants (the chaffinch kit). A variant may be a call, a note, a group of notes or an entire song qualitatively different from others. Example of a call: the rain call of the chaffinch. Examples of a note: chaffinch, tree creeper, short-toed tree creeper, cardinal, white-crowned sparrow, yellowhammer. Examples of an entire song: chaffinch, Carolina chickadee, cardinal, white-crowned sparrow.

Within a population, the intra-individual or the inter-individual variation may be small or large. Examples: short-toed tree creeper, Carolina chickadee.

We know little about dialect boundaries. They may be quite sharp or they may consist of a mixed zone (rain call of the chaffinch, short-toed tree creeper). A very sharp boundary apparently presupposes an uninhabitable dividing zone.

Constancy of dialect over two to eight years has been definitely proven up to now for only three species (chaffinch, white-crowned sparrow, short-toed tree creeper); in one case (chaffinch) it was probably maintained for at least twenty years.

The predominant variant may be characteristic of nearly all the males or it may only be demonstrable statistically.

It may be assumed with reasonable certainty that the mosaic distribution of bird dialects is not due to adaptation to local factors as is the case, for example, with desert colouration and industrial melanism (Thielcke, 1965*a*). Meanwhile, there is clearcut evidence of a dialect having been learned (white-crowned sparrow). Other findings also point in the same direction (chaffinch: Poulsen, 1951; Thorpe, 1958). In one case it may be assumed with reasonable certainty that a dialect was learned from an alien species because a convergent development of the kit by the spotted woodpeckers (*Dendrocopus major, medius and minor*) and by the chaffinch is most improbable (Thielcke, 1965*a*). There are several examples of specifically preferential imitation of alien vocalizations (Stadler, 1930;

Borror and Reese, 1956; Thielcke, 1963, 1965*a*; Tretzel, 1965, 1966, 1967) so that the acceptance of something alien into a dialect is not as improbable as it may appear at first sight.

The variable preference for certain imitations, reminiscent of changing fashions (Schmidt and Hantge, 1954), may possibly point the way to the historical genesis of dialects. Adaptation of the songster's own song to that of the previous singer (Metfessel, 1940; Thorpe, 1958; Konishi, 1965; Marler, 1967) or to that of a neighbour (Hinde, 1958) explains the genesis of dialects and their transmission to the next generation.

We have as yet no data explaining how mutually independent dialects arise within the same species or even within the same song. We also know nothing about the function of dialects. Dialects occupy far too large areas to be explained in terms of differentiating between roving (dangerous) and harmless (because already in possession of territory) neighbours (see Weeden and Falls, 1959). The function of dialects is perhaps to reduce variability in order to increase the effectiveness of the signal.

Judging by what we know of the settling of young birds, dialect boundaries present no obstacles to the spreading out of young birds. Old birds with the same dialect are therefore not a 'closed group', i.e., do not originate from the same family, and we can thus dismiss altruistic explanations in this context (see Wickler, 1967).

The decisive ontogenetic imprinting of the young birds must therefore occur when they are already settled in their future breeding area. We know in this context that chaffinches are capable of learning songs up to the thirteenth month of age, and white-crowned sparrows up to the hundredth day of age. It follows that white-crowned sparrows must be settled in their home area by that age. This is possible even with migrating species, as shown by Löhrl's (1959) experiments with the collared flycatcher (*Ficedula albicollis*).

Species in which the young males learn the fine points of their repertoire exclusively from the father (bullfinch, *Pyrrhula pyrrhula*, Nicolai, 1959; zebra finch, *Taeniopygia castanotis*, Immelmann, 1966) are not suited for developing mosaic dialects, unless they happen to be extremely sedentary. In the case of the *Viduinae*, close imitation of the vocalizations of the step-father results in a parallel development with the foster species which may lead to species genesis (Nicolai, 1964, 1967).

Male territorial neighbours stimulate each other by their songs. It is possible that this also results in better synchronization with their females (see Brockway, 1962). It is not impossible that dialects may

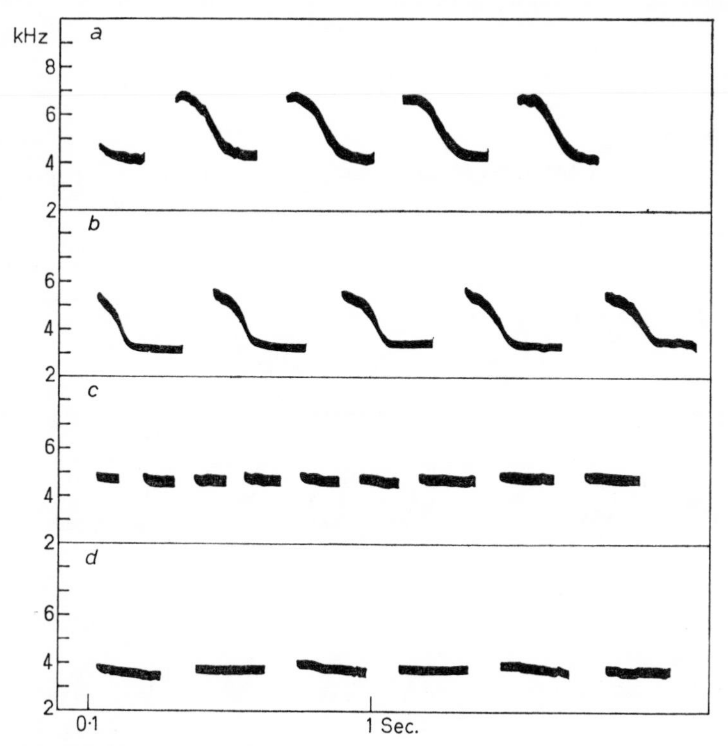

FIGURE 6. Song forms of two willow tit males in the plains (*a*, *b*) and two in the Alps (*c*, *d*); (after Thönen, 1962, modified).

play a role in this context. They may, however, be of purely physiological origin without any social advantage (see Kneutgen, 1964).

Mosaically occurring variants without supplementary isolation can never result in the formation of new species. However, the ability of a species to modify its song repertoire extensively by learning, and to make it uniform within the population, may well be a causative factor in evolution under appropriate conditions. It is possible that the strong species fragmentation of Oscines is linked to their acoustic learning ability.

Different versions in large areas with narrow mixed zones

Willow tit (*Parus montanus*). In Western and Central Europe the repertoire is very simple. Notes of the same form are strung together into a song. Each male apparently has only one song-type (Thönen, 1962).

There are two versions of this basic type (Figure 6). One version

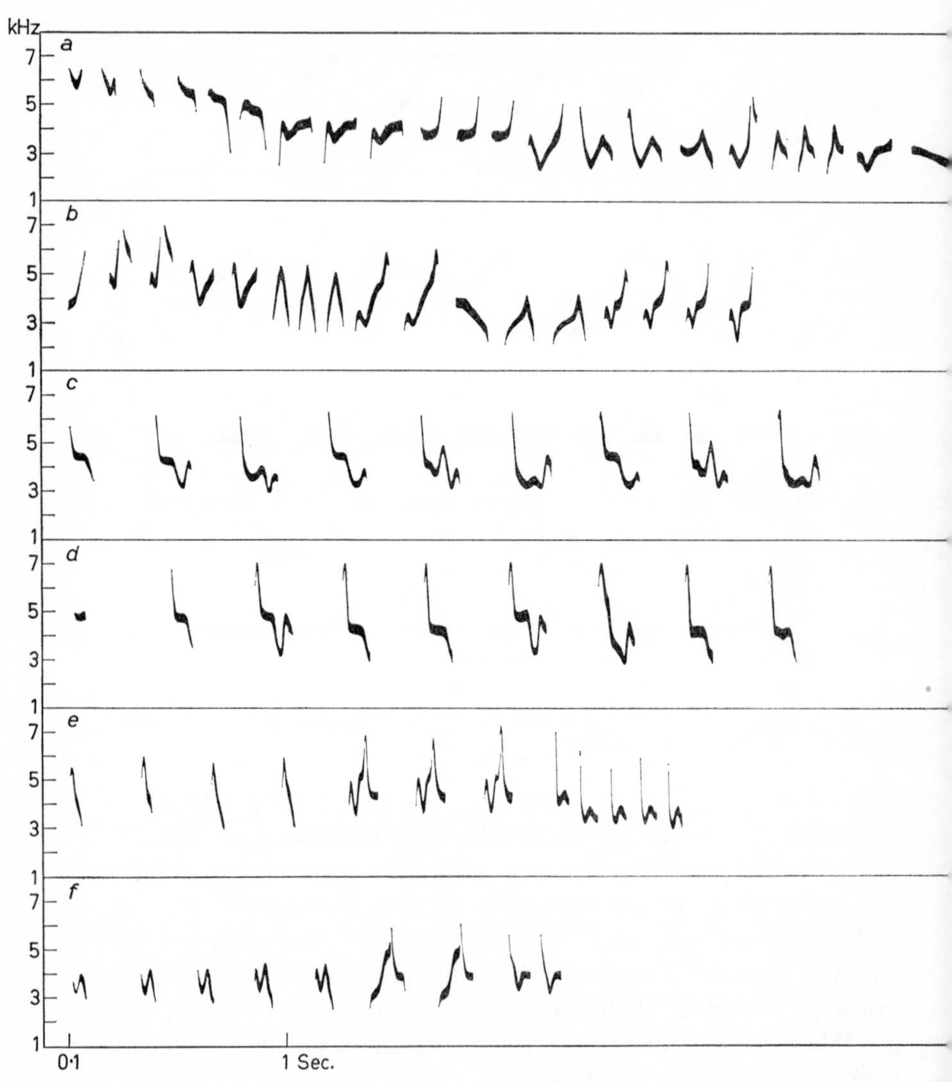

FIGURE 7. Song of a willow warbler from *a* northern Sweden and *b* southwest Germany. Song a chiffchaff from *c* northern Sweden and *d* southwest Germany. Owing to lack of space, songs the same length were chosen, although each male varies considerably the number of notes per so *e* and *f*, songs of two chiffchaffs from Spain.

is found on the central European plain, the other in the Alps. There is no dialect formation within either of these two song regions. Both forms react only to their own song-type. At the boundaries between the two regions there is a narrow mixed zone in which males with both song-types can be found (Thönen, 1962).

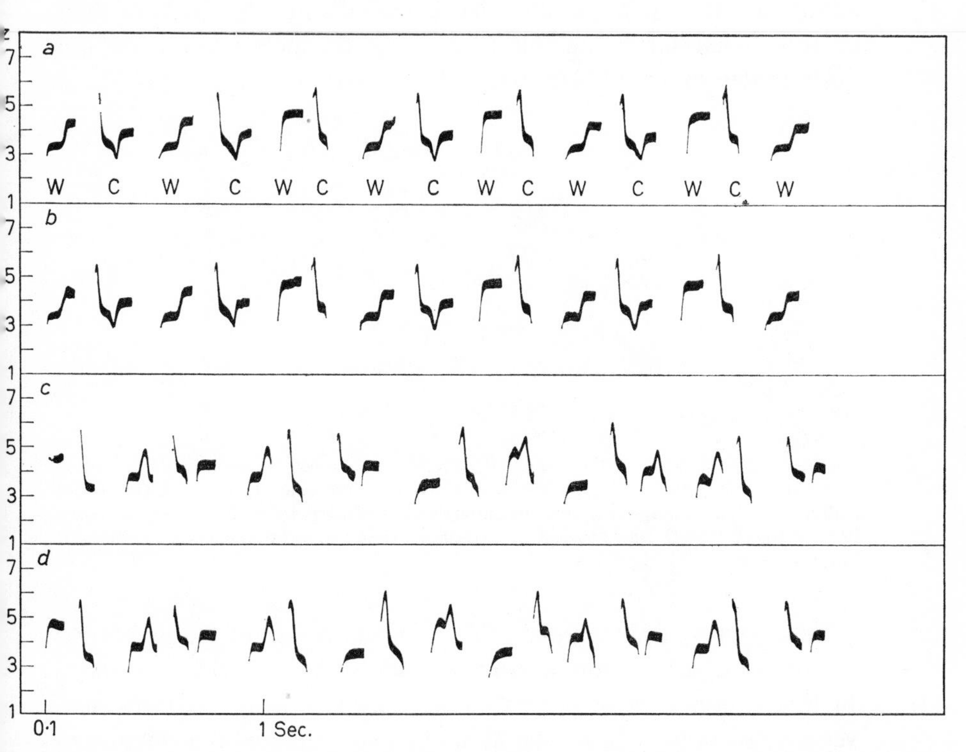

FIGURE 8. *a* to *d*, Four songs of a chiffchaff from Siberia. *a* and *b* and *c* and *d* belong to the same song-types. The songs contain willow warbler type notes W as well as 'normal' chiffchaff type notes C although both species occur in the same region. Spectrograms from gramophone recordings by Veprintsev and Naoomova (1964).

Chiffchaff (*Phylloscopus collybita*). In central, northern and southeastern Europe all the notes are derived from the same basic form. Each male has at least three to nine different notes which it arranges in 'medley sequence' into songs of very variable length (Figure 7). There is apparently no mosaic-type formation of dialects (Thielcke and Linsenmair, 1963; Thielcke, 1965*b*).

In southwest Europe, North Africa and the Canary Islands the notes cannot all be reduced to one basic form. The songs consist of one to three phrases which are repeated stereotypedly. The length of song is relatively constant (Figure 7). Each male has one or two song-types (Marler, 1960; Niethammer, 1963; Thielcke and Linsenmair, 1963).

Judging by the songs recorded by Marler (1960), the Canary Islands chiffchaffs sing as mixed songsters with a strong Spanish admixture. It is quite possible that chiffchaffs in other parts of their southwest habitat sing similarly to those on the Canaries, i.e. not quite as stereotypedly as in Figure 7.

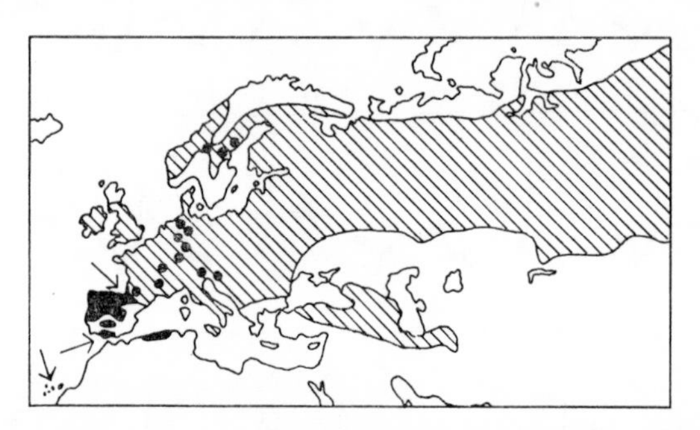

FIGURE 9. Distribution of the chiffchaff (after Voous, 1962). Dots indicate recording locations of 'normally' singing chiffchaffs. Arrows indicate recording locations of chiffchaffs singing 'in Spanish' (on the Canaries after Marler, 1960). Black: Spanish song form of the chiffchaff. In southwest France there exists a mixed region (topmost arrow); (after Thielcke, 1966, modified).

In southwest France there exists a mixed region of the two forms (Figure 9) which is quite narrow in some parts and wider in others. In this region it is possible to find songsters with relatively pure versions of both song forms, as well as songsters with a mixture of the two forms in one song.

Chiffchaffs in southwest Germany react to both song forms, but their reaction to the Spanish form is very markedly the poorer of the two (Thielcke and Linsenmair, 1963).

In Siberia, the song of the chiffchaff is quite different again (Figure 8), as shown by the gramophone recordings by Veprintsev and Naoomova (1964) and in data by Johansen (1954). Song deviations are also reported from Persia (Schüz, 1959), from the Caucasus and for the *sindianus* subspecies (Nicolai and Wolters, 1963).

Discussion. Both Thönen (1962) and I (1963, 1965*a*) consider zones of mixed song forms to be of secondary occurrence—that is to say the populations were separated, developed divergently, and came together again. There are similar patterns of population distribution in terms of morphological features. The best European example of the latter is provided by the carrion crow and the hooded crow (*Corvus corone corone* and *C.c. cornix*) with a very stable hybrid zone (Meise, 1928).

Despite the tendency to geographic variation, songs are quite uniform over large areas. This has been studied best in the 'normal' song form in Europe (Thielcke and Linsenmair, 1963).

The distribution patterns of the willow tit and chiffchaff songs are similar to mosaic variation in two respects. The song forms are largely mutually exclusive, and mixtures of the two forms occur in the contact zones. They differ from mosaics in three respects. The region of each song version is very much larger, and the number of variants within the region is much smaller. The song forms are so different that the males might be thought to belong to different species. Willow tits, and to a lesser extent chiffchaffs, react accordingly when hearing the other song form. The song forms of the great tit may possibly be of the same type. The causal factors for the development of these song types may possibly be the same for all three species.

Radiation and cline

There are indications that several peculiarities of the song of the cardinal (Lemon, 1966) and of the short-toed tree creeper radiate from centres of particularly high frequency of occurrence and become rarer at increasing distances from such centres. More accurate data are required, however, for more precise statements in this context.

Lanyon (1960*a*) found in the calls of *Myiarchus tyrannulus* pitch clines correlated to body size. Borror (1967) also reports on cline-type deviations. Our knowledge of this form of geographic variation is also very scanty.

VARIATION WITH INSULAR DISTRIBUTION

Short-toed tree creeper (*Certhia brachydactyla*)

Apart from mosaically varying peculiarities, the song of the short-toed tree creeper is relatively uniform throughout central Europe. This is particularly noticeable for notes D and F (Figure 10) but is also unmistakable for all the other features of the song. In central and southern Spain, by contrast, there occur large differences from one area to another without any reduction, or with only a very slight reduction, of uniformity within each population. This pattern of increased variability begins to some extent in southern France (Thielcke, 1965*a*).

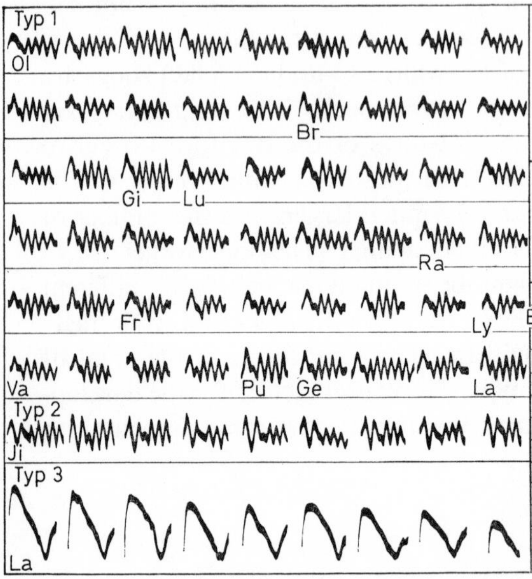

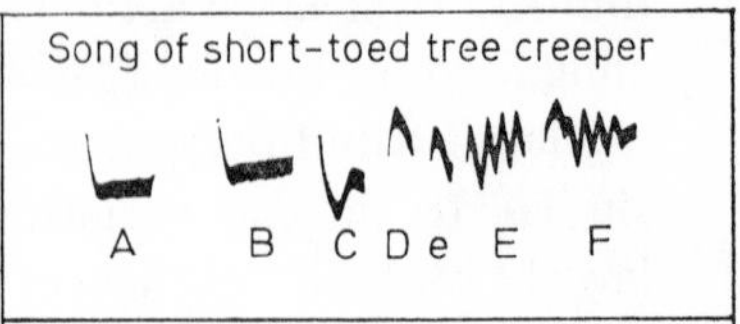

Song of short-toed tree creeper

A B C De E F

Location of tape recording	n	Types of notes F: 1	2	3	4 (fehlt)
Oldenburg	23	23			
Braunschweig	6	6			
Giessen	2	2			
Ludwigsburg	19	19			
Rastatt	7	7			
Freiburg	26	26			
Lyon	5	5			
Valence	10	9			1
Puigcerda	3	3			
Gerona	8	6			2
La Granja	25	1		23	1(1)
San Rafael	5				5
Jimena	14		14		

FIGURE 10. *Left:* The three types of note F by 72 individuals from all locations, selected so as to illustrate the entire individual variation. This representation neglects the slight fluctuations of pitch from one note to another. The arrangement of the notes from top to bottom corresponds to the location sequence from north to south.
Right: Distribution of types of note F. The maximum numbers at each location are framed. In notes of type 1 the first peak is formed as a double peak; in notes of type 2 the second peak is a double peak. The numbers in brackets indicate the number of males singing the corresponding dialect only occasionally or in a second song type (after Thielcke, 1965*a*).

The short-toed tree creeper requires for its habitat woods with at least a small proportion of rough-barked trees, parks or orchards, lanes or gardens. The percentage area of land under forest amounts to 27·1 per cent in Germany, 19·9 per cent in France, and 14·4 per cent in Spain. Commercial timber production in 1950 amounted to 29,723,000 m³ in Germany, 11,665,000 m³ in France, and 2,500,000 m³ in Spain. The timber species *Picea*, *Fagus* and *Abies*, unsuitable for the short-toed tree creeper, account in Germany for less than half of the total area under forest. In Spain, almost all the woods are suitable for the short-toed tree creeper as far as the timber species are concerned but, as can be observed visually and as is confirmed by the commercial timber production figures, the woods consist largely of uninhabitable young stands (see World Forest Atlas, 1951). In addition, the suitable habitats in Spain occur in patches separated by savannahs, fields and citrus orchards. This exerts an isolating effect similar to that of water around islands. There are nevertheless indications in the song that a small degree of interchange does take place between the populations (Thielcke, 1965*a*).

Great tit (*Parus major*)

The songs of almost all great tits are formed on the same principle (Gompertz, 1961; Thielcke, 1968). Each male has four to seven song types. A series of one type is followed by a series of another. Males respond to songs played to them with identical or similar songs (Gompertz, 1961).

Great tits are distributed in a ring extending over the Palearctic and South Asia. The northern part of the ring reaches from western Europe via Siberia to the Pacific; the southern part of the ring extends from Japan via China through southern Asia to Asia Minor and the Mediterranean area (Figure 11). The habitat in the northern part of the ring has very few distribution barriers; by contrast, the impression of a compact habitat in the southern part of the ring, created by the map, is misleading. In reality the southern part of the ring is broken up not only by seas but also by deserts, savannahs and mountains into a large number of large and small regions separated by barriers difficult to overcome. This difference is reflected in the number of varieties occurring in the two parts of the ring; in the northern part of the ring there are only three varieties from Britain to the Pacific whereas in the southern part there are thirty-six (Wolters, 1967). In addition there are in India two species closely related to the great tit (*Parus monticolus* and *P. xanthogenys*) and possibly a third (*bokharensis*) the status of which as a subspecies or a species is in dispute. The splintering of the great tit populations in the southern part of the ring is particularly striking. Similar conditions are found with many other species and genera. On the other hand the less pronounced splitting in the northern part of the ring may equally well be explained by the later settling of these areas.

The songs of European great tits are largely uniform in the basic

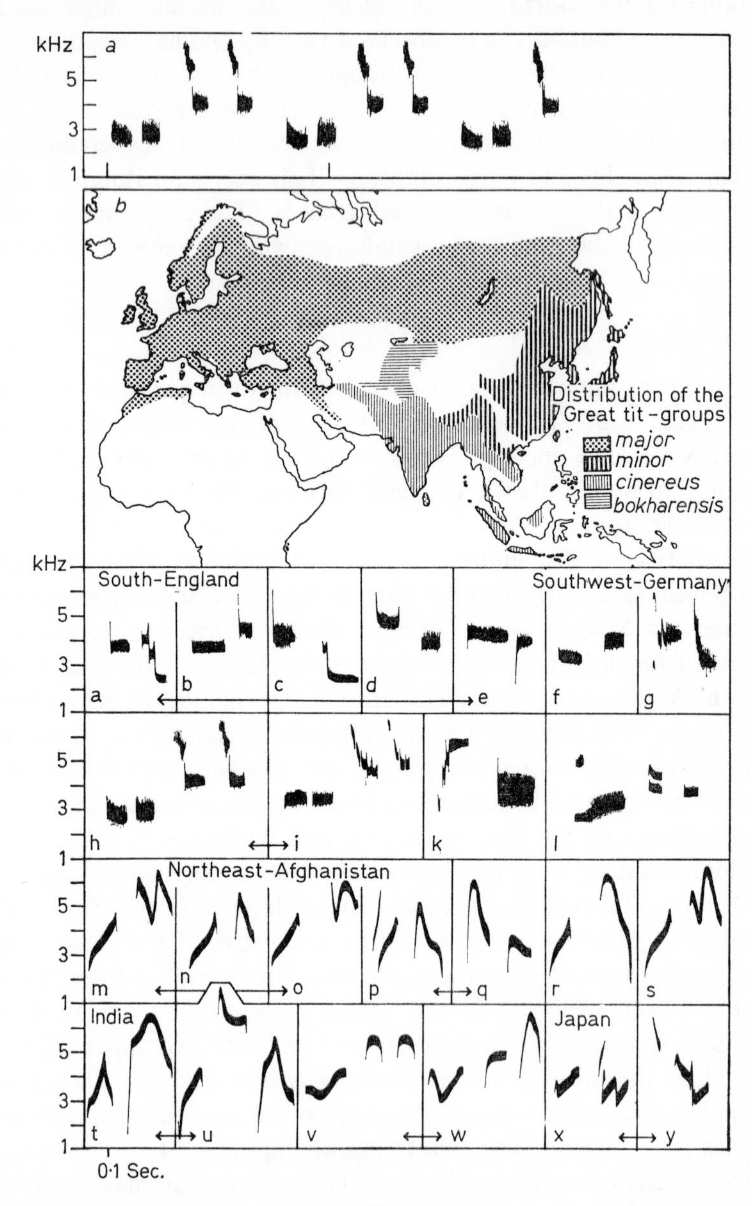

FIGURE 11. *a*. Song of a great tit. *b*. Distribution of the great tit and parts of the song-pattern which are repeated in the song, from different regions. The notes of the European great tit song differ fundamentally from those in southern and eastern Asia (with data by Wolters, 1967; supplemented after Gompertz, 1968).

composition of the notes. By contrast, the Afghan, Indian and Japanese populations differ radically from the European great tits (Gompertz, 1968 and Figure 11)—the Japanese possibly somewhat less than the Indian and the Afghan. There are very probably differences between the Japanese and the other two Asian populations, possibly even between the Indian and the Afghan. However, we have as yet too few tape recordings to be certain on these points.

Discussion

Among the causative factors of greater variation in a fragmented terrain are probably not only isolation but also the smaller number of individuals in each population. Small partial populations do not possess the whole gene complement of a large population. This leads to greater variability from one population to another (Dobzhansky and Pavlovsky, 1957; Mayr, 1963). The same may be expected for the learnt components of song.

Although we have as yet no great tit recordings from Russia, it may be assumed with reasonable certainty that a high degree of splitting in morphological features is paralleled by greater differences in the song.

The differentiation of song has in some instances already reached a degree at least approaching interspecific differences (Gompertz, 1968). Similar observations on the vocalizations of the nuthatch (*Sitta europaea*) were made by Löhrl (unpublished).

Ward (1966) found in his studies on the Carolina chickadee no effect of geographic isolation on the song. Such an effect need not necessarily be expected. It should be borne in mind, however, that in comparing compact and fragmented ranges valid results can only be obtained if the regions studied are sufficiently large. I cannot judge whether or not this condition was fulfilled by Ward.

CONTRAST REINFORCEMENT OF ISOLATION MECHANISMS

According to the hypothesis of contrast reinforcement, the differences acquired in isolation by two geographically separated populations are reinforced when the two populations come into secondary contact, provided the two populations are sufficiently different. This presumes that hybrids are selectively handicapped. According to this hypothesis, the differences between the two forms should be smaller in the allopatric than in the sympatric range. This hypothesis has been verified by model experiments on *Drosophila* (see Mayr, 1963).

However, the examples adduced from ornithology do not stand up to close scrutiny.

Mayr (1963) writes: 'Perhaps the best-substantiated evidence for the reinforcement of isolating mechanisms is the recent overlap in western Russia between the two titmice *Parus caeruleus* and *Parus cyanus* which, when it first occurred (1870–1900), led to frequent production of hybrids ("*Parus pleskei*"). After 50 years of overlap, hybridization has greatly decreased (Vaurie, 1957).' Vaurie, on the other hand, writes: 'As far as I can judge by the statements of Voinstvenski, hybridization takes place still, but it is less frequent than it was in the 1870's and 1880's and apparently occurs only along the northern border of the range of *cyanus*. This suggests that the pressure towards range expansion may still be active to some extent, but the hybrids seem selected against, and eventually complete reproductive isolation and ecological compatibility will probably be achieved.'

Vaurie's authority on the question of frequency of hybrids at the turn of the century is Pleske (1912), whose statements are couched in very general terms (many, frequent, fairly significant, by no means rare). Only twice does Pleske state concrete figures (from another observer): 'In autumn 1902 saw one specimen, in autumn 1903 captured six specimens.' Besides, Pleske was not at all concerned with presenting as complete as possible a picture of the occurrence of hybrids; his main interest was the recognition of the hybrids as a species (*Cyanistes pleskei* Cab.). It is for this reason, as he himself admits, that he picked up his pen again after 15 years of ornithological abstinence.

Vaurie's other source is Voinstvenski (1954) who writes in the Russian Handbook (p. 738) only the following (translated from Russian): 'The name *Parus pleskei* . . . is used to describe specimens with a coloration intermediate between that of *cyanus* and *caeruleus*. These are found on the northern distribution boundary of the two species and are obviously hybrids.'

I consider it rash to draw from the statements of Pleske and Voinstvenski any conclusions regarding differences of frequency, and even rasher to link such conclusions to isolation mechanisms. Assuming that hybrids do occur in fact only on the border of the range of *cyanus*, a simpler explanation would be that, on the border of their range, *cyanus* have fewer chances to find mates of the same species.

Bauer (1957) has demonstrated the probability of this mechanism for the Syrian woodpecker (*Dendrocopus syriacus*).

Lynes (1914) aroused a greater stir by his remarks on the chiffchaffs near Gibraltar: 'The only other breeding *Phylloscopus* . . . was by its song, I think anyone would have agreed, a willow warbler. For a willow warbler, true, the song was unmelodious and disjointed ('tinpotty', if one may use such an expression), the first two notes jerked out, so that for a moment they might have been put down to an eccentric chiffchaff, had they not invariably been followed by the four or five notes in descending scale characteristic of the willow warbler—in short, if it was a poor willow wren's song, it was an impossible chiffchaff'. The chiffchaff was immediately proclaimed a standard example of contrast reinforcement because in Spain there are no willow warblers (*Phylloscopus trochilus*), the sibling species of the chiffchaffs. Wherever both species occur in Europe, their songs are quite different. Marler (1960) published four songs of Tenerife chiffchaffs. He found that 'the individual notes are often similar to those of the European bird, at times they are arranged into a short two-second song . . . reminiscent of the European willow warbler'. He attributes the greater variation on the Canary Islands to loss of contrast owing to the absence of the sibling species. Marler correctly stresses that this conclusion is speculative because other species on the Canary Islands apparently sing exactly as they do on the Continent despite the absence of closely related species on the Canaries. Admittedly, Marler did not have an exact analysis for any of his cases. Our comparison of the Spanish with the western, central, northern and southeast-European chiffchaffs resulted in findings which could be reconciled with the contrast hypothesis, but which nevertheless raised the first doubts (Thielcke, 1963, 1965*a*, and unpublished). This work apparently escaped Marler's notice because, in Marler and Hamilton (1966), the old explanation is repeated without reservations.

The following considerations militate against the contrast hypothesis:

(i) The southern border of the range of the willow warbler does not coincide with the mixed zone of the two chiffchaff song varieties.

(ii) The contact zone between the chiffchaff varieties is very narrow. It is probably secondary.

(iii) The sound spectrograms of Siberian chiffchaffs (Figure 8), in a similar way to those of Spanish chiffchaffs, show coincidences with notes of the willow warbler although the ranges of the two species overlap extensively (Voous, 1962).

Too many supplementary assumptions would be required to reconcile all this with the contrast hypothesis.

Loss of contrast is often considered an indication in favour of, or even as proof of, the contrast hypothesis, as by Marler (1960) for warblers. Marler puts forward in this context the Tenerife blue tit (*Parus caeruleus*) without presenting nearly enough material either from Tenerife or from Britain. The alleged concordance of the song of the Tenerife blue tit with that of *Parus* species absent on the Canary Islands but occurring on the continent with the blue tit, is also not convincing. The song of the blue tit can be variable on the continent also, as I found in my own studies in southwest Germany.

The data of Lack and Southern (1949) and those of Marler and Boatman (1951) on aberrant vocalizations of insular varieties are as subjective as most other data on geographic variation unsupported by sound spectrographic analysis.

Even in cases where losses of genetic or learning inventory do actually occur on islands, there are basic objections against the explanation that all these losses occurred because of cessation of competition by closely related species. Variations occur mainly on small islands located relatively far from the mainland. The founders of the terrestrial bird populations on such islands were in most cases probably only a few individuals who brought with them only a very small part of the genetic and learning inventory of the total population. Perhaps this is one of the reasons for the frequent aberrations found on islands. In order to prove loss or intensification of contrast one should consider ranges of approximately equal size of one species in some of which closely related species do occur and in some of which they do not. Comparisons are valid only in situations differing by only one decisive factor.

No instances are known as yet of contrast reinforcement of isolation mechanisms in birds in the sympatric range. The cases which have been investigated in detail (chiffchaff and willow warbler; short-toed tree creeper and tree creeper) have up to now failed to provide proof. According to Mayr (1963), 'courtship signals' become not only 'general' and 'non-specific' but also variable in the absence of closely related species. I know of no convincing proof of this as far as vocalizations are concerned. Despite the lack of the sibling species, the song of Spanish chiffchaffs is even more stereotyped in its structure than that of chiffchaffs in western, central, south-eastern and northern Europe. The song of short-toed tree creepers is very stereotyped in several populations, and variable in others (Figure 5). This is totally unrelated to the presence or absence of tree creepers, the sibling species of the short-toed creeper. Ward (1966) reports similar findings for the Carolina chickadee. The causes of low or high variability are unknown.

Extreme differences in the songs of closely related sympatric species are very often considered proof of contrast reinforcement. There are, however, examples to the contrary. Thus, the songs of *Parus melanolophus* and *P. major* in Afghanistan are very similar although the two species occur in the same habitat (Thielcke, unpublished). The same is true of the wrens *Thryothorus sinaloa* and *T. felix* in Mexico (Grant, 1966).

LACK OF GEOGRAPHIC VARIATION

Thönen (1962) and I (1962) found a lack of geographic variation in at least one song form of willow tits and chiffchaffs. Lanyon (1960*b*) found no differences in a call and in the song between two allopatric wrens (*Troglodytes bruneicollis* and *T. aedon*), Lanyon (1962) in a call of the meadowlarks (*Sturnella*) and Stein (1963) in the song of Traill's flycatcher (*Empidonax*). It is probable, however, that more detailed studies will reveal a slight degree of geographic variation even in these cases.

CONCLUSIONS

Although the geographic variation of the calls of some species has been analysed, far more precise data are indispensable for further studies. Our knowledge of the constancy, the boundaries and the large scale pattern of dialects is as yet very scanty. Sound spectrographic analysis is indispensable. Comparisons between two or between only a few locations are of little interest. Twenty recording locations are too few rather than too many for arriving at an accurate idea of geographic variation. Publications should contain a number of recordings from each of a number of small regions. Both the authors and the users of published material should differentiate clearly between the actual findings and their interpretation. Up to now only Oscinae have been studied. Just because almost all other bird groups probably lack acoustic learning ability, the non-Oscinae should be studied.

REFERENCES

BAUER, K. (1957) Zur systematischen Stellung des Blutspechtes. *Falke, Sonderheft* **3**, 22–5.

BERGMAN, G. (1953) Über das Revierbesetzen und die Balz des Buchfinken, *Fringilla coelebs* L. *Acta Soc. Fauna et Flora Fenn.* **69**, 1–15.

BORROR, D. J. (1967) Songs of the Yellowthroat. *Living Bird* **6**, 141–61.

BORROR, D. J. and REESE, C. R. (1956) Mockingbird imitations of Carolina Wren. *Bull. Mass. Audubon Soc.* **40**, 245–50 and 309–18.

BROCKWAY, BARBARA F. (1962) The effects of nest-entrance position and male vocalizations on reproduction in Budgerigars. *Living Bird* **1**, 93–101.

CONRADS, K. (1966) Der Egge-Dialekt des Buchfinken (*Fringilla coelebs*)—Ein Beitrag zur geographischen Gesangsvariation. *Vogelwelt* **87**, 176–82.

DOBZHANSKY, TH. and PAVLOVSKY, O. (1957) An experimental study of interaction between genetic drift and natural selection. *Evolution* **11**, 311–19.

GOMPERTZ, T. (1961) The vocabulary of the Great Tit. *Br. Birds* **54**, 369–418.

(1968) Results of bringing individuals of two geographically isolated forms of *Parus major* into contact. Vogelwelt, Beiheft **1**, 63–92.

GRANT, P. R. (1966) The coexistence of two wren species of the genus *Thryothorus. Wilson Bull.* **78**, 266–78.

HINDE, R. A. (1958) Alternative motor patterns in Chaffinch song. *Anim. Beh.* **6**, 211–18.

IMMELMANN, K. (1966) Zur ontogenetischen Gesangsentwicklung bei Prachtfinken. *Verh. dt. zool. Ges., Zool. Anz., Supplement* **30**, 320–32.

JOHANSEN, H. (1954) Die Vogelfauna Westsibiriens. *J. Ornith.* **95**, 64–110.

KAISER, W. (1965) Der Gesang der Goldammer und die Verbreitung ihrer Dialekte. *Falke*, **12**, 40–2, 92–3, 131–5, 169–70, 188–91.

KONISHI, M. (1965) The role of auditory feedback in the control of vocalization in the white-crowned sparrow. *Z. Tierpsychol.* **22**, 770–83.

KNEUTGEN, J. (1964) Beobachtungen über die Anpassung von Verhaltensweisen an gleichförmige akustische Reize. *Z. Tierpsychol.* **21**, 763–79.

LACK, D. and SOUTHERN, H. N. (1949) Birds of Tenerife. *Ibis* **91**, 607–26.

LANYON, W. E. (1960*a*) The middle American populations of the Crested Flycatcher *Myiarchus tyrannulus. Condor* **62**, 341–50.

(1960*b*) Relationship of the House Wren (*Troglodytes aedon*) of North America and the Brown-Throated Wren (*Troglodytes brunneicollis*) of Mexico. *Proc. 12th Int. Orn. Congr.* 450–8.

(1962) Specific limits and distribution of meadowlarks of the desert grassland. *Auk*, **79**, 183–207.

LEMON, R. E. (1965) The song repertoires of Cardinals (*Richmondena cardinalis*) at London, Ontario. *Canad. J. Zool.* **43**, 559–69.

(1966) Geographic variation in the song of Cardinals. *Canad. J. Zool.* **44**, 413–28.

LEMON, R. E. and SCOTT, D. M. (1966) On the development of song in young Cardinals. *Canad. J. Zool.* **44**, 191–7.

LÖHRL, H. (1959) Zur Frage des Zeitpunktes einer Prägung auf die Heimatregion beim Halsbandschnäpper (*Ficedula albicollis*). *J. Ornith.* **100**, 132–40.

LYNES, H. (1914) Remarks on the geographical distribution of the Chiffchaff and Willow-Warblers. *Ibis* **10**, 304–14. Series 2.

MARLER, P. (1952) Variation in the song of the Chaffinch *Fringilla coelebs*. *Ibis* **94**, 458–72.

(1956*a*) The voice of the Chaffinch and its function as a language. *Ibis* **98**, 231–61.

(1956*b*) Behaviour of the Chaffinch, *Fringilla coelebs*. *Behaviour, Supplement* **5**.

(1960) Bird songs and mate selection. Animal sounds and communication. *Am. Inst. Biol. Sci.* no. 7, 348–67.

(1967) Comparative study of song development in sparrows. *Proc. 14th Int. Orn. Congr.* 231–44.

MARLER, P. and BOATMAN, D. J. (1951) Observations in the birds of Pico, Azores. *Ibis* **93**, 90–9.

MARLER, P. and HAMILTON, W. J. III (1966) *Mechanisms of Animal Behaviour*. New York, London, Sydney.

MARLER, P. and TAMURA, M. (1962) Song 'dialects' in three populations of White-crowned Sparrows. *Condor* **64**, 368–77.

(1964) Culturally transmitted patterns of vocal behaviour in Sparrows. *Science* **146**, 1483–6.

MAUERSBERGER, G. (1960) In Stresemann und Portenko. *Atlas der Verbreitung palaärktischer Vögel*. Berlin.

MAYR, E. (1963) *Animal Species and Evolution*. Cambridge, Massachusetts.

MEISE, W. (1928) Die Verbreitung der Aaskrähe. *J. Ornith.* **76**, 1–203.

METFESSEL, M. (1940) Relationship of the heredity and environment in behaviour. *J. Psychol.* **10**, 177–98.

NICOLAI, J. (1959) Familientradition in der Gesangsentwicklung des Gimpels (*Pyrrhula pyrrhula* L.) *J. Ornith.* **100**, 39–46.

(1964) Der Brutparasitismus der Viduinae als ethologisches Problem. *Z. Tierpychol.* **21**, 129–204.

(1967) Rassen-und Artbildung in der Viduinengattung Hypochera. *J. Ornith.* **108**, 309–19.

NICOLAI, J. and WOLTERS, H. E. (1963) *Vögel in Käfig und Voliere*. Aachen.

NIETHAMMER, G. (1963) Zur Kennzeichnung des Zilpzalps der Iberischen Halbinsel. *J. Ornith.* **104**, 403–11.

PLESKE, T. (1912) Zur Lösung der Frage, ob *Cyanistes pleskei* Cab. eine selbständige Art darstellt, oder für einen Bastard von *Cyanistes caeruleus* (Linn.) und *Cyanistes cyanus* (Pallas) angesprochen werden muß. *J. Ornith.* **60**, 96–109.

POULSEN, H. (1951) Inheritance and learning in the song of the Chaffinch (*Fringilla coelebs*). *Behaviour* **3**, 216–28.

(1958) The calls of the Chaffinch (*Fringilla coelebs* L.) in Denmark. *Dansk orn. Foren. Tidsskr.* **52**, 89–105.

PROMPTOFF, A. (1930) Die geographische Variabiltät des Buchfinkenschlages (*Fringilla coelebs* L.). *Biol. Zentralbl.* **50**, 478–503.

SCHMIDT, K. and HANTGE, E. (1954) Studien an einer farbig beringten Population des Braunkehlchens (*Saxicola rubetra*). *J. Ornith* **95**, 130–73.

SCHÜZ, E. (1959) *Die Vogelwelt des Südkaspischen Tieflandes*. Stuttgart.

SICK, H. (1939) Über die Dialektbildung beim Regenruf des Buchfinken. *J. Ornith.* **87**, 568–92.

(1950) Der Regenruf des Buchfinkens (*Fringilla coelebs*). *Vogelwarte* **15**, 236–7.

STADLER, H. (1930) Vogeldialekte. Alauda **2**, Suppl. 1–66.

STEIN, R. C. (1963) Isolating mechanisms between populations of Traill's flycatchers. *Proc. Am. Phil. Soc.*, **107**, 27–50.

TEMBROCK, G. (1965) Beobachtungen zum Gesang des Buchfinken (*Fringilla coelebs* L.) *J. Ornith.* **106**, 313–17.

THIELCKE, G. (1961) Stammesgeschichte und geographische Variation des Gesanges unserer Baumläufer (*Certhia familiaris* L. und *Certhia brachydactyla Brehm*). *Z. Tierpsychol.* **18**, 188–204.

(1962) Die geographische Variation eines erlernten Elementes im Gesang des Buchfinken (*Fringilla coelebs*) und des Waldbaumläufers (*Certhia familiaris*). *Vogelwarte*, **21**, 199–202.

(1964) Zur Phylogenese einiger Lautäuβerungen der europäischen Baumläufer (*Certhia brachydactyla* Brehm und *Certhia familiaris* L.). *Z. zool. Syst. Evolut.-forsch.* **2**, 383–413.

(1965*a*) Gesangsgeographische Variation des Gartenbaumlaufers (*Certhia brachydactyla*) im Hinblick auf das Artbildungsproblem. *Z. Tierpsychol.* **22**, 542–566.

(1965*b*) Der Gesang schwedischer Zilpzalpe (*Phylloscopus collybita*). *J. Ornith.* **106**, 352–4.

(1966) Ritualized distinctiveness of song in closely related sympatric species. *Philos. Trans. R. Soc. Lond. B*, **251**, 493–7.

(1968) Gemeinsames der Gattung Parus. Ein bioakustischer Beitrag zur Systematik. *Vogelwelt, Beiheft* **1**, 147–64.

THIELCKE, G. and LINSENMAIR, K. E. (1963) Zur geographischen Variation des Gesanges des Zilpzalps, *Phylloscopus collybita*, in Mittel-und Südwesteuropa mit einem Vergleich des Gesanges der Fitis, *Phylloscopus trochilus*. *J. Ornith.* **104**, 372–402.

THODAY, J. M. and BOAM, T. B. (1959) Effects of disruptive selection. II. Polymorphism and divergence without isolation. *Heredity* **13**, 205–18.

THÖNEN, W. (1962) Stimmgeographische, ökologische und verbreitungsgeschichtliche Studien über die Mönchsmeise (*Parus montanus* Conrad). *Orn. Beob.* **59**, 101–72.

THORPE, W. H. (1958) The learning of song patterns by birds, with especial reference to the song of the Chaffinch *Fringilla coelebs*. *Ibis* **100**, 535–70.

TRETZEL, E. (1965) Imitation und Variation von Schäferpfiffen durch Haubenlerchen (*Galerida c. cristata* (L)). Ein Beispiel für spezielle Spottmotiv-Prädisposition. *Z. Tierpsychol.* **22**, 784–809.

(1966) Spottmotivprädisposition und akustische Abstraktion bei Gartengrasmücken (*Sylvia borin borin* (Bodd)). *Verh. dt. Zool. Ges., Zool. Anz., Supplement* **30**, 333–43.

(1967) Imitation und Transposition menschlicher Pfiffe durch Amseln (*Turdus m. merula* L.). Ein weiterer Nachweis relativen Lernens und akustischer Abstraktion bei Vögeln. *Z. Tierpsychol.* **24**, 137–61.

VAURIE, C. (1957) Systematic notes on palearctic birds. No. 26. Paridae: The *Parus caeruleus* complex. *Am. Mus. Novit.* 1833.

(1959) *The Birds of the Palearctic Fauna*. London.

VEPRINSTEV, B. N. and NAOOMOVA, Z. R. (1964) *The Voices of Wild Nature: Siberian Birds*. Gramophone record.

VOINSTVENSKI, M. A. (1954) In Dementiev *et al.*, *Die Vögel der Sowjetunion*, Bd.5. Moskau (Russian).

VOOUS, K. H. (1962) *Die Vogelwelt Europas und ihre Verbreitung*. Hamburg and Berlin.

WARD, R. (1966) Regional variation in the song of the Carolina Chickadee. *Living Bird* **5**, 127–50.

WEEDEN, J. S. and FALLS, J. B. (1959) Differential responses of male Ovenbirds to recorded songs of neighboring and more distant individuals. *Auk* **78**, 343–51.

WELTFORSTATLAS (1951) *Herausgegeben von der Bundesanstalt fur Forst-und Holzwirtschaft, bearbeitet von R. Torunsky*. Haller Verlag o.O.

WICKLER, W. (1967) Vergleichende Verhaltensforschung und Phylogenetik. In *Die Evolution der Organismen*, Bd. 1. 3. Auflage. Stuttgart.

WOLTERS, H. E. (1967) In J. Nicolai und H. E. Wolters. *Vögel in Käfig und Voliere*. Aachen.

PART F
LITERARY AND AESTHETIC ASPECTS

INTRODUCTION

In the preceding sections some recent advances in our understanding of the development, causation, function and evolutionary significance of avian vocalizations have been discussed. Man's appreciation of bird-song far preceded his scientific interest in it, and we may therefore ask whether the latter helps us to understand the former. In the next essay Armstrong (chapter 15) describes various aspects of man's pre-scientific associations with the sounds of birds—their role in magic and ritual, the means man has used to imitate them. the use of bird song in literature, and man's changing attitudes to them. It is worth noting that the interest is not solely one way, as the occasional records of birds imitating human music demonstrate (Thorpe, 1967). Indeed, Tretzel's (1967) description of a small population of blackbirds (*Turdus merula*) which imitated a man's whistle (or imitated each other's imitations of that whistle) lends a little perspective to any discussion of whether bird song should or should not really count as music—'The simple whistle of the model consisted of four notes in the average frequency of 1·19, 2·07, 2·19 and 1·74 Kc. . . . The whistle of the model varied considerably in pitch and rhythm, however. The blackbirds transposed the whole motif upwards by approximately a fifth and ornamented it: along with a short grace-note before the first note, they provided the second note with a long, steeply rising slur which stresses the transposition. The ornamentation . . . is equated in principle with that customary in classical music. While the pitch of the motif (for instance of its two high notes) fluctuated inside a whole octave in 45 of the model's whistles on two consecutive days, . . . the imitations of one of the blackbirds did not fluctuate by more than a semi-tone.'

This problem of the relation between human and avian music is taken up again in the last essay, where Hall-Craggs (chapter 16) comes to grips with what is perhaps the most difficult problem of all—the nature of the aesthetic content of bird song. Her suggestion, that the resemblances between bird song and human music can be understood in terms of similar functional requirements, gives us new understanding of both their basic properties and the ways in which these have been elaborated. But the link between the first essay in this book, concerned with the tonal quality of bird song, and this discussion of aesthetics in the last, is one which remains to be forged.

REFERENCES

THORPE, W. H. (1967) Vocal imitation and antiphonal song and its implications. In *Proc. 14th Int. Orn. Congr.* (D. W. Snow, ed.). Blackwell, Oxford. pp. 245–63.

TRETZEL, E. (1967) Imitation und Transposition menschlicher Pfiffe durch Amseln (*Turdus m. merula* L.). *Z. Tierpsychol.* **24**, 137–61.

15. ASPECTS OF THE EVOLUTION OF MAN'S APPRECIATION OF BIRD SONG*

by **EDWARD A. ARMSTRONG**
23, Leys Road, Cambridge

> The bird's song is certainly the same
> The change is in the emotions of the man.
> Po Chü-i (fl. AD 772–846)[1]

THE IMITATION AND INTERPRETATION OF BIRD UTTERANCES IN LANGUAGE, MAGIC AND RITUAL

Representations by Palaeolithic artists testify to early man's interest in birds, and remains found in Mesolithic and Neolithic deposits indicate that prehistoric fowlers sought them and their eggs as food. Their feathers were also used as adornments. We may be sure that these hunters, in common with those of more recent times, attended to bird utterances as a means of locating and identifying their prey.

In almost the earliest literature surviving from the past the significance of bird calls receives comment. The Sumerian Tilmun myth, describing a time when the world was young and neither men nor animals had yet acquired their distinctive features, lays stress on bird calls as characteristic of particular species once a more stable state of affairs had been attained:

> The raven in Tilmun did not croak (as the raven does nowadays),
> The kite did not utter the cries of a kite (as the kite does nowadays).
> Kramer (1944); Frankfort *et al.* (1949).

It was thus recognized that it is of the essential nature of a species to utter its distinctive call.

The onomatopoeic naming of animals, especially birds, is so

* We should like to thank all those who gave permission to reproduce material in this chapter.

[1] 'Hearing the Early Oriole', trans. F. Ayscough and A. Lowell (1922). *Fir-Flower Tablets* (London: Constable and Co.; New York: Houghton Mifflin & Co.).

widespread that it must be an ancient practice. In nursery language we say 'dicky-bird', 'bow-wow' and 'moo-cow':

> "For Mimicry is inborn in man from childhood up:
> "and in this differeth he from other animals,
> "being the most imitativ: and his first approach
> "to learning maketh he in mimicry, and hath delight
> "in imitations of all kinds."[1]

Among peoples of lower culture, as well as in more sophisticated societies, this nomenclatorial device is employed. In determining the species of birds mentioned in the Old Testament, scholars have based identifications on a comparison of the sounds of the Hebrew words with the utterances of the most likely species (Driver, 1955), and in Chinese (Cantonese) the words for bird, dog and buffalo—*tiu*, *gao* and *gnau* respectively—represent a chirrup, growl and moo. The English, Greek and Latin names of the hoopoe, ἔποψ and *upupa*, are onomatopoeic.

The great importance attached to birds by many hunting peoples is indicated by the extent to which their movements and cries are imitated in ceremonial, especially in the dance (Armstrong, 1965). During the initiation ceremony conducted by some South American tribes, the neophytes are taught how to mimic the calls of animals (Loeb, 1929). The imitation of birds is given great prominence in shamanistic cultures, and has frequently been described in detail by those acquainted with the Ural-Altaic peoples. The Kirghiz-Tatar shaman mimics various creatures as he bounds and prances around his tent: 'He whinnies, coos, imitating with remarkable accuracy the cries of animals, the songs of birds, the sound of their flight, and so on' (Castagné, 1930). Among the Yakut the shaman, during a séance in which he visits the sky realm, flaps his arms as if they were wings and honks like a goose (Radloff, 1884; Éliade, 1964). The adornments of the shaman's costume confirm that he identifies himself with a bird—sometimes of a particular species, such as the eagle, owl or diver (Shirokogorov, 1923). Moreover, models of birds erected on poles close to the scene of the shaman's activities are regarded as essential to the correct performance of the ritual (Harva, 1938). Early bronzes show that the shamans of ancient China also wore bird costumes (Hentze, 1941; Waterbury, 1952).

It has been suggested that the most intriguing of all Palaeolithic

[1] Robert Bridges (1929), *The Testament of Beauty* Bk iv (Oxford: The Clarendon Press).

man's works of art, a painting in the cave of Lascaux, may be interpreted as a shamanistic scene (Kirchner, 1952; Narr, 1959; Armstrong, 1969[1]). The painting can be seen only by those descending a fissure in the cavern floor—an indication of its special esoteric significance. A bird-masked man is shown leaning backwards, possibly indicating a trance state. He is confronted by a wounded bison, and beside him is a bird on a post. The assimilation of his appearance to a bird and the presence of the bird on the post—probably a model bird—agree with shamanistic practice. Furthermore, the bird-man and accompanying bird are depicted with open beaks as if calling. Visual imitation is certainly represented and, possibly, vocal imitation also.[2]

Non-industrialized folk in many areas mimic bird utterances and activities in their ceremonies, and even in modern Europe such performances survive. The *Schuhplattler* dance may originally have been inspired by the lek display of the bird whose feathers the dancers wear in their hats—the blackcock. In Slovenia masqueraders still appear in villages at Shrovetide dressed to represent birds. They enact a fertility ritual, dancing and pirouetting in front of cottages, crowing like cocks (Armstrong, 1969). Whether or not such practices are survivals from shamanistic rites, shamanism is very ancient and was much more widespread than the area in which it is still, or was until recently, practised. Evidence of its earlier prevalence may be detected by probing beneath the surface of Germanic, Greek, Hungarian, Irish and Mongolian culture (Davidson, 1968; Dodds, 1951; Roheim, 1954; Ross, 1966; Boyle, J. A., *in litt.*).

Very ancient, too, is the belief that song is magically potent and

[1] The first edition of the latter (1958) includes a reproduction in colour.

[2] The imitation of animal utterances may have influenced the development of language. A major difficulty is to account for the transition from the condition found in most mammals other than man, where each species has a restricted repertoire of signal sounds with an almost complete inability to imitate new ones. The situation in which we could most readily envisage this transition being achieved would be one in which (*a*) its adaptive value would directly favour survival, and (*b*) the extraneous sounds imitated would be (*i*) attention-provoking, (*ii*) distinctive and (*iii*) easily mimicked. Imitation of a snake's hiss would fulfil these requirements, making it possible for a companion to take avoiding action appropriate to a venomous reptile rather than a large mammal. The ability to imitate a hiss could thus have become an adaptation for naming a particular class of potentially highly dangerous predators, and the way would be open for adaptations establishing a nexus—sound, extraneous object, appropriate reaction. The occurrence of an aspirate in the name for a snake in many languages as various as German, Somali and Chinese should be taken into account. Further development of this hypothesis must be reserved for elsewhere.

creative. The Germanic term for a magic charm, *galdr*, is derived from *galan*, to sing, the term applied to the vocalizations of birds—an indication that no firm distinction was made between the song magic of birds and men (de Vries, 1956–7). Singing birds were said to charm men asleep for centuries (Armstrong, 1946; Ross, 1966). Songs could build cities and music subdue fierce animals. Apollo thus raised the towers of Troy, and Orpheus with his lyre charmed beasts and even trees, as Lorenzo reminded Jessica.[1] It was very widely believed that birds were able to convey esoteric knowledge through being able to speak human language, and that favoured individuals could acquire the ability to understand or speak bird language, sometimes by eating snake's flesh or other devices (Frazer, 1933). Folktales concerning a long-distant age begin: 'When animals could still speak'. From early times birds were assumed to be oracular, and even today in industrialized Britain such powers are sometimes attributed to the raven, crow, magpie and cuckoo. Some persons are disquieted when a magpie flies by, or may turn a coin for luck when the cuckoo is first heard in spring. In Ireland common sayings are that the raven and the grey crow 'tell the truth' and that to have 'raven's knowledge' is 'to see all and know all' (Armstrong, 1969).

THE INSTRUMENTAL, VOCAL AND POETIC IMITATION OF BIRD SOUNDS

A discussion of the possibility that attention to bird song may have stimulated man's attainment of the art of music is beyond the scope of this essay (Szöke, 1962). Some anthropologists consider bones with regular perforations, about 20,000 years old, to have been musical instruments (Leroi-Gourhan, 1964–5), and the sounds emitted by some of these prehistoric bone tubes are notes close to the pitch of some birds. The present writer has presented detailed evidence elsewhere in support of the view that Bronze Age men drummed to call down rain in imitation of the magical thunderbird, the woodpecker (Armstrong, 1969). Early man may have decoyed birds by whistling or vocal calls, as both primitive and sophisticated fowlers do today (Hartley, 1950).[2]

[1] Alfred, Lord Tennyson, 'Tithonus', *Poetical Works*, Oxford Standard Authors (London, 1953); William Shakespeare, *The Merchant of Venice*, v. i. 58–82.

[2] A small collection of traditional European bird decoy calls is exhibited in the Musée de l'Homme, Paris. Mr J. Boswall informs me (*in litt.*) that such decoy calls are used in Laos. Tribespeople in Malaya use a wind instrument to imitate the drongo.

The elaborate artistic decoration of some chattels fashioned by Old Stone Age man indicates that at times he enjoyed considerable leisure and thus had opportunities to develop arts other than the visual. While it is hardly to be expected that much evidence concerning the use of instrumental music to call birds should survive in early literature, we find this comment in the Chou Li, the Chinese Book of Ritual[1]:

> To call birds a single change of melody is necessary.
> Thus *rapport* is established with the spirits of the
> mountains and the forests.

In 1650 Athanasius Kircher set the nightingale's song in music and wrote the calls of the cuckoo and the quail in musical notation (Cf. Armstrong, 1969). Bird calls have been included in their compositions by a succession of composers, major and minor, culminating in the *Catalogue d'oiseaux* of Messiaen (Fisher, 1966; Scholes, 1938).[2]

A considerable number of poets have attempted to render bird songs in syllables or words. Aristophanes transcribed the nightingale's song in nonsense syllables as:

> τιὸ τιὸ τιὸ τιὸ τιὸ τιὸ τιὸ τιὸ τριοτὸ τριοτὸ τοτοβρίξ κτλ.
>
> (*Aves*, 237 ff.)

According to a traditional Italian saying, quoted by Dante,[3] the foolish blackbird sings when sunny days encourage it in February: *Omai più non ti temo*—'Lord, I fear Thee no more', supposing that spring has arrived. Mrs Joan Hall-Craggs confirms that this is an onomatopoeic representation of a phrase of the bird's song. The translator of an ancient poem in Japanese suggests that the author chose his words to represent the birds singing—*zuri, mari, wari, mari*:

> *Saezuri no takamari owari shizămarinu.*
> The singing of the birds
> Louder and louder, then softer and softer
> To silence.[4]

Charles Kingsley (1884) pointed out that some of the old German *Minnelieder* seemed to be copied from the songs of birds. Elizabethan

[1] Trans, E. Biot (1951), *Le Tcheou-li*. Paris. The text does not make clear to what extent the music was imitative. Through the kindness of Dr L. Picken I have been enabled to hear recordings of Chinese instrumental music including realistic imitations of bird vocalizations.

[2] Miss Gladys E. Page-Wood has transcribed in musical notation phrases of the songs of many British birds in a manuscript deposited in the Newton Library, Cambridge.

[3] *Purgatorio*, xiii. 122.

[4] Kyoshi (fl. 1874–1959) (Blyth, 1963–4).

poets, including Shakespeare,[1] occasionally sprinkled their verse with 'wood-notes wild'—'jug-jug', 'tirra lirra', 'cuckoo', 'tu-witta-woo', and so forth, to represent the songs of nightingale, skylark, cuckoo, tawny owl, and other birds. Browning's[2] reiterative song thrush is well known, and Tennyson[3] attributed to the bird:

> Summer is coming, summer is coming,
> I know it, I know it, I know it,
> Light again, leaf again, life again, love again,
> Yes, my wild poet.

Charles d'Orléans[4] represented the skylark's song-flight thus:

> La gentille alouette
> Avec son tire-lire-à-lire,
> Et tire-lire-à-lire
> Tirelirant tire
> Vers la voûte du ciel,
> Puis s'envol vers ce lieu
> Vire,
> Et désire dire:
> Adieu Dieu!
> Adieu Dieu!

Among other minor poets writing in this vein, seeking to transcribe song as well as to capture the mood evoked, William Allingham[5] attained considerable success, including five birds of the Irish countryside—chaffinch, blackbird, song thrush, skylark and robin—in one poem. This technique culminated in Walter Garstang's *Songs of the Birds*.[6] Now that the electronic recording of bird song has been brought to such perfection, there is little to encourage further attempts to render bird vocalizations in this picturesque way.

NUMINOUS EXPERIENCE AND EMOTIONAL AMBIGUITY IN THE APPRECIATION OF BIRD SONG

As far back in history as information is available, men enjoyed numinous experiences in contact with nature. Such experiences have

[1] William Shakespeare, *The Winter's Tale*, IV. ii. 9; *Love's Labour's Lost*, V. vii. 928; John Lyly (1584), 'Spring's Welcome', *Alexander and Campaspe*; Thomas Nashe (1592), 'Spring', *Summer's Last Will and Testament*.

[2] Robert Browning, 'Home thoughts from abroad', *Poetical Works*, Oxford Standard Edn., (London, 1940).

[3] Alfred, Lord Tennyson, 'The gardener's daughter', Oxford Standard Authors, (London, 1953.)

[4] Charles d'Orléans, cit. in Massingham H. J. (1922), *Poems about Birds* (London.)

[5] William Allingham (1854) 'The lover and birds'. *Flower Pieces and Day and Night Songs* (London); see also Armstrong (1923).

[6] Walter Garstang (1922) *Songs of the Birds* (London).

been described as involving a sense of a *mysterium tremendum*, and in them feelings of fascination, awe, exaltation, fear or wonder may mingle. Normally the thrill is not readily characterized since emotions blend or are touched with ambivalence (Otto, 1923). When intense, these experiences may be regarded as religious: indeed, numinous experience is often an element in mystical states and, to those who sustain it, may be permanently life-enriching. It can also be an important factor in the appreciation of nature. Here we must confine ourselves to a few illustrations of its significance in relation to the appreciation of bird songs and calls.

A hymn in the Rigveda entitled *Aranyani, Goddess of the Forest*[1] palpitates with the eeriness of night in the primeval jungle:

> Aranyani, Aranyani, who wanderest as if lost, why dost thou ask for the village? Does not fear overtake thee?
>
> When to the cry of the *vrishavara* the *chichika* responds, Aranyani goes proudly...
>
> It is as though oxen were feeding, and as though one saw a house. And at evening Aranyani creaks like a wagon.
>
> The dweller in the forest thinks at night, 'Someone shouted'.
>
> Aranyani smites not, unless another comes...

Passing from the poetry of three thousand years ago to the prose of our own time, Edmund Selous (1901), a man not normally emotionally expansive, thus describes cooing call-notes heard in Shetland. For a time they mystified him:

> One of the strangest sounds that came to me on that lonely island was the courting-note of the male eider-duck. This varies a good deal, not in the sound, which is always the same, but in the duration and division of it... The sound seems always to be on the point of catching, yet just to miss the human intonation, sometimes suggesting a soft (though often loud) mocking laugh, at others a slightly ironical or surprised ejaculation. But this element only trembles upon it and is gone. Rousing for a moment the sense of man's proximity with its attendant associations, these vanish almost in the forming, and are replaced by a feeling of unutterable loneliness and wildness. For what recalls, yet is far other, enforces the sense of the absence of that which it recalls... If not quite music, it was most softly harmonious, and always, from first to last, brought into my mind with strange insistency, those lines in the *Tempest*:
>
> Sitting on a bank,
> Weeping again the king my father's wrack,
> This music crept by me upon the waters,
> Allaying both their fury and my passion
> With its sweet air.

Fully to appreciate Selous's description, it must be read in full by

[1] E. J. Thomas (1923), *Vedic Hymns* (London: John Murray, Ltd.).

one familiar with eiders in their breeding haunts. A like impression is conveyed by the lovely musical clamour of a large herd of Bewick swans—a plangent chorus, which together with the wing-music of mute swans may have given rise to the traditions of singing swans in Celtic mythology.

The clanging note of the bell-bird in the South American forest stirs the emotions in a similar way, as a number of travellers have remarked (Armstrong, 1946). For a moment the surprised wayfarer is transported by the bell-note out of the wilderness to the Sunday calm of an English hamlet. Almost simultaneously, conscious of the enveloping rain forest, disillusionment and a sense of desolation assail him.

Robert Burns[1] refers in one of his letters to related experiences of a less disturbing nature:

> I never hear the loud solitary whistle of the curlew in a summer noon, or the wild cadence of a troop of grey plovers in an autumn morning, without feeling an elevation of soul like the enthusiasm of devotion or poetry.

There is a human quality in the notes of eider duck, Bewick swan and curlew, and the bell-bird's tolling recalls civilized society; yet these birds belong to the wilds. On hearing them imagination falters and mixed emotions are aroused. Creatures with ambiguous characteristics which recall human qualities, such as owls whose faces and voices seem vaguely human, are particularly prone to arouse disquiet, as folklore testifies (Armstrong, 1969). Among mammals, gibbons, which look like a grotesque parody of humanity and whose weird halloos have the timbre of a woman's voice, figure in Chinese and Siamese literature as emotive, ambivalent symbols associated with human and scenic desolation. Po Chü-i[2] links the gibbon with a bird which in the Far East has ambivalent, mingled tragic and pleasant associations:

> I left the imperial city. Last year banished far
> To this plague-stricken spot, where desolation
> Broods on from year to heavy year, nor lute
> Nor love's guitar is heard. By marshy bank
> Girt with tall yellow reeds and dwarf bamboos
> I dwell. Night long and day no stir, no sound,
> Only the lurking cuckoo's blood-stained note,
> The gibbon's mournful wail.

[1] Robert Burns, Letter to Mrs Dunlop, New Year's Day, 1789, *Letters*, Masonic edn. (London, 1928).

[2] Po Chü-i (fl. AD 772–846) 'The Lute Girl', trans. L. Cranmer-Byng (1911). *A Lute of Jade*, London. p. 79; cf. Li T'ai-po (fl. AD 702–762) 'The Terraced Road of the Two-edged Sword Mountains', trans. Ayscough and Lowell (1922) *op. cit.*

By such subtleties poets reinforce the impact of their verse, calling in aid responses stemming from our animal ancestry and awakening in us inchoate, ambivalent emotions.

ANIMISM AND EMPATHY IN THE INTERPRETATION OF BIRD SONG

Animism is among the oldest and most widespread thought-forms in which empathy with nature is manifested. If we exclude Palaeolithic, and possibly a few other types of prehistoric and Egyptian, art, such empathy appears to have found artistic expression in the Far East, especially in China, earlier and more effectively than in the West, as is evident in the long history of Chinese landscape painting and illustration of bird life. Throughout centuries animistic empathy has inspired Japanese art and literature. As Shinto, animism survives as a highly organized cult.

The Manyōshū, an anthology of seventh and eighth century poetry—much of it concerned with nature, contains remarkably few allusions to the visual characteristics of the animals mentioned but lays stress on their utterances (Keene, 1965). This may be due in part to an interest in the alliance of poetry and music, but it is also related to the implicit assumption which underlies some western as well as other eastern nature poetry, that the more animate makes the less animate articulate. In the *Haiku*, a brief animistic poem, neither mystical nor symbolic, the Japanese throughout the centuries during which it has been used as a form of expression have sought to capture fleeting moments of experience. Authorities on Japanese poetry emphasize that sensation rather than emotion is expressed:

A mountain path;
Wild geese in the clouds,
The voice of mandarin ducks in the ravine.[1]

In the leafy tree-tops
Of the summer mountain
The cuckoo calls—
Oh, how far off his echoing voice.[2]

There are affinities between such poetry and the Middle English

[1] Sogi (fl. A.D. 1421–1502) (Blyth, 1963–4).
[2] Ōtoma Yakamochi (fl. A.D. 718–85) (Keene, 1965).

song, 'Sumer is icumen in, Lhude sing cuccu', and also with Early Irish poetry, such as the lines in 'First winter song':[1]

> Dull red the fern;
> Shapes are shadows;
> Wild geese mourn
> O'er misty meadows

THE PERSONALIZING OF NATURE IN THE INTERPRETATION OF BIRD SONG

Many folk still in the food-gathering and hunting stages of culture do not make our clear distinctions between men and animals. They assimilate themselves to animals by wearing antlers, skins, bird masks or plumes. Paintings and engravings from the Palaeolithic to the Iron Age show that earlier peoples followed similar customs. Folk belonging to lower cultures also tend to attribute human characteristics to animals. The Ural-Altaic tribespeople, already mentioned, not only behaved in their ceremonies as if they were animals, but sometimes treated animals as if they were human beings. Before sacrificing a bear the Gilyaks led him through the village and feasted him on fish and brandy (von Schrenk, 1891). Ainu women suckled bear cubs which were to be sacrificed as if they were human children (Batchelor, 1901). In Greek mythology we have refinements of ambivalent beliefs in which there is no absolute demarcation between the divine, the human and the animal. Less familiar, perhaps, are the related conceptions underlying our own early poetry. Alluding to the riddles with which our ancestors amused themselves, a recent writer remarks: 'People in Anglo-Saxon times, living uncomfortably close to the natural world, were well aware that . . . creation is animate, and that every created thing, every *wiht*, had its own personality. . . . The riddle is a sophisticated and harmless form of invocation by imitation; the essence of it is that the poet, by an act of imaginative identification to which Vernon Lee gave the name "empathy", assumes the personality of some created thing—an animal, a plant, a natural force . . . by performing this particular ventriloquism the poet extends and diversifies our understanding of—or at least our acquaintance with—the noumenous (*sic*) natural world, of whose life, or even existence, modern men are becoming

[1] A. P. Graves, trans, (n.d.) *The Book of Irish Poetry* (Dublin: The Educational Company of Ireland, Ltd.). Reproduced with the permission of John Graves.

progressively more unaware.' On these lines must a conundrum such as this be interpreted:

> When it is earth I tread, make tracks upon water
> or keep the houses, hushed is my clothing,
> clothing that can hoist me above house-ridges
> at times toss me into the high heaven
> where the strong cloud-wind carries me on
> over cities and countries;
> accoutrements that throb out sound, thrilling strokes
> deep-soughing song, as I sail alone
> over field and flood, faring on,
> resting nowhere. My name is—.

The solution, as all who have heard the bell-beat of their wings will recognize, is—the mute swan (Alexander, 1966).

Nature personified is sometimes portrayed enjoying bird song, as in Blake's poem[1] in which he refers to the skylark:

> All Nature listens silent to him, and the awful Sun
> Stands still upon the Mountain looking on this little Bird
> With eyes of soft humility, and wonder, love and awe.
> Then loud from their green covert all the Birds begin their Song:
> The Thrush, the Linnet and the Goldfinch, Robin and the Wren
> Awake the Sun from his sweet reverie upon the Mountain . . .

THE JUBILANT BIRDS

Through his capacity for empathy man not only finds in nature spirits which often bear a close resemblance to himself, but also divinizes and humanizes, sometimes simultaneously, aspects of, or powers manifested in, nature. According to Shinto doctrine the sea is 'a Mighty Being'. The metaphors of poetry, or even common speech, often preserve the memory of earlier convictions or embody beliefs or half-beliefs which have their own emotional validity. Vaughan[2] could refer to stones as 'deep in contemplation', and write:

> the quick world
> Awakes and sings;
> The very winds,
> And falling springs,
> Birds, beasts, all things
> Adore Him in their kinds.

[1] William Blake (1804), *Milton* (London).
[2] Henry Vaughan (1957), *Works* (Oxford: The Clarendon Press).

Wordsworth[1] declared:

> Ye blessed Creatures, I have heard the call
> Ye to each other make; I see
> The heavens laugh with you in your jubilee;
> My heart is at your festival.

Throughout the centuries people of very different cultures have expressed in verse and song the belief that the birds, and all nature, join with them in rejoicing and singing praise. This belief has heightened man's spiritual experience and enriched literature. Creation, including man, may be conceived as adoring and praising the Creator. In a hymn ascribed to Akhenaten the worshippers chant:

> All the birds rise up from their nests
> And flap their wings with joy
> And circle round in praise of the living Aten.

(Baikie, 1926)

Tertullian[2] wrote:

> Nay, the birds too, rising out of the nest, upraise themselves heavenward, and instead of hands, extend the cross of their wings, and say somewhat to seem like prayer.

Entering the Earthly Paradise, Dante[3] hears verses of Psalm xcii 'Lord, thou hast made me glad through Thy works' and listens to the dawn chorus:

> A delicate air, that no inconstancies
> Knows in its motion, on my forehead played,
> With force no greater than a gentle breeze,
>
> And quivering at its touch the branches swayed,
> All toward that quarter where the holy hill
> With the first daylight stretches out its shade;
>
> Yet ne'er swayed from the upright so, but still
> The little birds the topmost twigs among
> Spared not to practise all their tiny skills;
>
> Rather they welcomed with rejoicing song
> The dawn-wind to the leaves, which constantly
> To their sweet chant the burden bore along.[4]

[1] William Wordsworth, *Intimations of Immortality* iv. 1–4.

[2] Tertullian. *De spectaculis*.

[3] *Purgatorio*, xxviii. 7–18. Trans D. Sayers (1955) (Harmondsworth: Penguin Books, Ltd.).

[4] The Psalmist (cxlviii.10) called on the birds to praise the Lord and the theme recurs in Christian literature from the second century to the present day. Throughout Lent it is expressed in Anglican churches when the Benedicite is sung. There is a distinction between the Jewish-Christian orientation to nature and that of the animist or pantheist. The approach of the latter is from the variety and virtues of nature to God or Reality, however conceived; whereas the Christian, apprised by prophets and mystics and through his own experience of the transcendent glory of God, finds in nature innumerable expressions of His beauty, bounty and power. Compromise may be reached between the two points of view but the difference between them remains important (Armstrong, in prep.).

THE EMOTIVE USE OF BIRD SONGS AND CALLS IN LITERATURE

Birds have always intrigued mankind, in part because like human beings they go on two legs, sing, wear bright adornments, dance and build; hence empathy with them has not been so difficult to achieve as with, say, insects or fish. In spite of this, comparatively few species have acquired sufficient emotive valency to figure prominently in folklore or literature. These reappear again and again in legend, poetry and visual art throughout centuries. Some owe this distinction to association with times and seasons: dawn and dusk, spring and autumn. In many countries, from ancient to modern times, the cock's crowing has been connected not only with dawn but with the parting of lovers (Hatto, 1965). In Europe, the skylark, mounting at daybreak in song to the heavens, became a symbol of joy. Some conspicuous migratory birds, such as the goose and crane, acquired contrary associations in literature because it was observed that their appearance heralded increasing sunshine and warmth but their departure foretold dark days and chilly winds. The dove, sacred to Astarte and Aphrodite, early became associated with love in the Middle East. It is a harbinger of spring, its courtship is conspicuous and its voice is interpreted as having a tender human quality. The Hebrew youth urged his beloved[1]:

> Rise up, my love, my fair one, and come away.
> For lo, the winter is past, the rain is over and gone.
> The flowers appear on the earth; the time of
> the singing of birds is come, and the voice of the turtle
> is heard in our land.

The birds do more than illustrate man's *joie de vivre*: they augment it and may even be thought to share it. There is a path wide open from the shaman's self-identification with birds to their symbolic treatment by poets, prophets and priests. Imagination, dwelling on their powers of flight, migration, songfulness, and certain aspects of their breeding behaviour, readily appropriated them as symbols.

Some poets, notably Shakespeare, refining the use of symbolism, have used emotive contrast effectively in connection with their references to birds. In *Macbeth* 'the temple-haunting martlet' flitting where 'the air smells wooingly' contrasts with the sinister crow making wing to 'the rooky wood'; and in *Romeo and Juliet* the lark, bird of dawn and joy, is opposed to Philomel, bird of darkness and

[1] *Song of Songs*, ii. 10–12.

passion.[1] Dante[2] employed this device of avian contrast even more subtly by making different species and their vocalizations symbolic of Hell and Paradise. He compares the ever-homeless souls in Hell with the cranes he had seen and heard flying over the yellowing plains of Lombardy:

> And as the cranes in long lines streak the sky
> And in procession chant their mournful call,
> So I saw come with sound of wailing by
> The shadows fluttering in the tempest's brawl...

When the repose of the souls in Paradise is described, the comparison is—significantly—not with the lark's song but with the silence which ensues. Tranquil in making God's purpose their own the waiting souls are

> Like the small lark who wantons free in air,
> First singing and then silent, as possessed
> By the last sweetness that contenteth her.

The emotive use of natural sounds, particularly bird vocalizations in association with scenery to induce the appropriate mood in the reader, is a familiar poetic device. The author of the Anglo-Saxon *The Seafarer* describes a lonely mariner sailing along a rugged coast while the cries of sea-birds are carried on the wind:

> To the storm striking the stone cliffs
> gull would answer, eagle scream
> from throats frost-feathered. No friend or brother
> by to speak the despairing mind.
>
> (Alexander, 1966)

The screaming birds accentuate the impression of nature launching attacks against the lonely, distraught seaman. About the same period a Japanese poet, Kamo Taruhito, viewing the ruins of the palace on Mount Kagu, linked bird calls with another desolate scene:

Mount Kagu

> As spring has come when the mist trails
> On the heaven-descended hill of Kagu,
> The lake is rippled by the breeze through the pines
> And cherry flowers fill the leafy boughs.
> The mallards call to their mates
> Far off on the water,
> And the teal cry and whirr
> About the beach.

[1] *Macbeth*, I.vi.4; III.ii.52; *Romeo and Juliet*, III.v.1–11; Armstrong (1963*b*).
[2] *Inferno*, V. 46–49; *Paradiso*, XX. 73–75. Trans L. Binyon (1943) (London: Macmillan and Co., Ltd.).

But the pleasure-barges, which of old
The courtiers, retiring from the palace rowed,
Now lie desolate,
Oarless, poleless and unmanned.

Envoys

It is plain that no one comes boating:
The diving mandarin-ducks and teal
Dwell upon the barges.
How timeworn all has grown, unnoticed!
On the hill of Kagu the moss lies green
At the ancient roots
Of the speary cryptomerias.

(Keene, 1965)

The bright flowers and calls of the mated ducks, symbols of marital fidelity, disporting themselves where once courtiers dallied, contrast with the derelict pleasure-boats and sombre conifers.

A similar nostalgic sadness is evoked by the fluting of Tennyson's curlews[1]:

'Tis the place, and all around it, as of old, the curlews call,
Dreary gleams about the moorland, flying over Locksley Hall.

In verses by Robert Frost,[2] the poet sadly accepts that, though there are times when we would like to think the birds' songs reflected our feelings, we are all the more desolate because we can no longer believe it to be so. As he stands by the ruined farmhouse he notices that:

The birds that came to it through the air
At broken windows flew out and in,
Their murmur more like the sigh we sigh
From too much dwelling on what has been.
. . . .
For them there was really nothing sad,
But though they rejoiced in the nest they kept,
One had to be versed in country things
Not to believe the phoebes wept.

THE CONTRAST BETWEEN THE PAGAN CLASSICAL AND CELTIC CHRISTIAN RESPONSE TO BIRD SONG

Although in Hesiod and Aristophanes we find traces of the empathic attitude to birds, Greek factual curiosity led the poets, as the Anthology shows, to interest in them, not for their own sake but to the extent that they could be thought of anthropomorphically. Thus they wrote about species, such as the jay and parrot, which are able to

[1] Alfred, Lord Tennyson, 'Locksley Hall', *op. cit.*
[2] Robert Frost (1955), 'The need of being versed in country things', *Selected Poems* (Harmondsworth: Penguin Books Ltd.).

acquire human speech. Crinagoras facetiously versified an imaginary situation in which an escaped parrot taught all the birds around to cry 'Hail' to Caesar:

> Orpheus on the mountains wild
> The animals made mild,
> But now, unbidden all
> The birds on thee, Caesar, melodious call.
>
> (Douglas, 1928)

The Romans, materialistic and power-minded, had little sympathy with nature. Lucretius[1] movingly expressed his sympathy for a cow bereft of her calf, Pliny[2] has a surprisingly eloquent appreciation of the nightingale's song, and Horace decorated his verse with allusions to nature; but Virgil[3] stands out, almost alone, as the Roman poet who expressed sympathetic interest in, and feeling for, birds. In his verse he distinguishes the calls of three species of dove, and as we recite his description of the nesting rock dove taking flight we hear the clatter of the bird's wings:

> *Qualis spelunca subito commota columba,*
> *Cui domus, et dulces latebroso in pumice nidi,*
> *Fertur in arva volans, plausumque exterrita pennis*
> *Dat tecto ingentem—mox aëre lapsa quieto,*
> *Radit iter liquidum, celeris neque commovet alas.*

> As when a dove her rocky hold forsakes
> Rous'd, in a fright her sounding wings she shakes;
> The cavern rings with clattering; out she flies,
> And leaves her callow care, and cleaves the skies;
> At first she flutters: but at length she springs
> To smoother flight, and shoots upon her wings.

In Europe, intense feeling for nature and enjoyment of bird song first blossomed in Ireland, where the clamour of the marching legions was never heard. There Christendom burst forth into carols extolling nature. Typical of this joy in nature are the verses which a ninth-century Irish scribe[4] paused to write in the margin of the manuscript he was engaged in copying:

> A leafy grove surrounds me quite;
>   For my delight the blackbirds flute;
> While o'er my little book's lined words
>   Sweet warbling birds their Scribe salute.

[1] Lucretius, *De ver. nat.* ii. 352–66. [2] Pliny, *Nat. Hist.* x (29), 43.

[3] Virgil, *Georgics*, i. 375–89; *Eclogues*, i. 57 f.; *Aeneid*, v. 213–17, tr. Dryden. Cf. G. White, *The Natural History of Selborne* (1789). Letter xliv to W. Pennant.

[4] A. P. Graves (n.d.) *op. cit.* 'The Scribe'. Reproduced with the permission of John Graves.

The cuckoo in his mantle grey
  Cries on all day through lush tree tops,
And verily—God shield me still!
  Well speeds my quill beneath the copse.

Chaucer, in *The Manciple's Tale*, commenting with pity on the plight of caged birds, maintained and introduced into English poetry the tradition of Christian compassion for wild creatures prominent in Christian hagiography throughout the centuries (Armstrong, in prep.). It is ironical, but understandable and commendable, that enlightened humanitarian concern for birds is becoming more ardent and widespread in our time, when man's activities throughout the world are reducing the numbers of many species and menacing others with extinction.

FROM MEDIEVAL SYMBOLISM TO THE POETIC-SCIENTIFIC DESCRIPTION OF BIRD SONG

In England the observational study of birds initiated by William Turner made slow headway. Shakespeare, countryman as he was, thought of nature largely in medieval terms, and many of his birds are more symbolic than real (Armstrong, 1963*b*). But gradually, as observation supplanted tradition, the poets paid more attention to what the birds did and sang, so that lists of names and clichés were no longer regarded as elegant. Drayton paraded his ornithological knowledge in verse, and birds began to take their place in poetry, not as incidental embellishments, but in their own right. Increasing affluence and the spread of education created a situation in which a poet such as Clare, turning to nature for solace as many have done before and since, could find a public. The growth of a more objective outlook appears in the scepticism of a poet who was no enthusiast for wild nature—Pope[1]:

Is it for thee the lark ascends and sings
Joy tunes his voice, joy elevates his wings?
Is it for thee the linnet pours his throat?
Loves of his own and raptures swell the note.

More than half a century later Coleridge[2] directly challenged traditions of two millennia and proclaimed that the nightingale sang merrily:

A melancholy bird! Oh, idle thought!
In nature there is nothing melancholy.

[1] Alexander Pope (1734). *An Essay on Man* (London).

[2] Samuel Taylor Coleridge (1912). 'The Nightingale', *Poems* (Oxford: The Clarendon Press) but cf. *Phaedo* 85a, where Socrates refers to the birds singing for joy.

> But some night-wandering man whose heart was pierced
> With the remembrance of a grievous wrong
> Or slow distemper, or neglected love,
> (And so, poor wretch, filled all things with himself
> And made all gentle sounds tell the tale
> Of his own sorrow) he, and such as he,
> First named these notes a melancholy strain.

Inspired by the English countryside, perhaps at its loveliest during the eighteenth and nineteenth centuries, and sometimes also by mystical experience of nature or Platonic influence, poets such as Wordsworth and Coleridge themselves sustained and nurtured in their readers an attention to and love for nature which, in Burns' phrase, was akin to 'the enthusiasm of devotion'. Poems were written about individual song birds and eventually about their songs. Interest awakened in species little honoured by earlier writers. Ralph Hodgson[1] depicted the missel thrush singing as is his wont on a tempestuous day:

> I heard the gray thrush piping loud
> From the wheezing chestnut-tree;
> The gray thrush gripped the spray that bowed
>
> Beneath the storm, and brave sang he—
> O he sang brave as he were one
> Who hailed a people newly free.

Winifred Letts,[2] continuing the long tradition of Irish countryside poetry in which nature is viewed with religious delight, described the willow warbler's song:

> Among the alders of the glen
> Flitting and singing from each tree,
> The willow-wren
> Murmurs the decades of his rosary.
> He tells his beads, each one a minor note;
> They hang, they fall, descending to Amen.
> Unceasing, wistfully he makes his plea
> That restless souls may find tranquillity.
>
> Now little bedesman, lead my thoughts apart,
> Sing in the quiet places of my heart.

Thus, particularly during the last hundred years, naturalists, poets and people in general have become increasingly fascinated by bird music. Electronic recordings now enable us to listen to the songs of

[1] Ralph Hodgson (1907), 'The Missel Thrush', *The Last Blackbird* (London: Allen and Unwin, Ltd.).

[2] Mrs Verschoyle, cit. in Armstrong (1946).

birds in far-distant places, and the sound spectrograph has made possible the visible transcription and scientific analysis of bird vocalizations: but perhaps the delights of hearing a bird's song in the field may be somewhat tarnished by first making its acquaintance at second-hand.

If poets have received most prominence in this survey it is because quotations in verse illustrate more effectively the changing attitudes to birds than the work of prose writers. But among these W. H. Hudson, of whom Conrad said 'He writes as the grass grows', described bird songs with simplicity, imagination, charm and accuracy not surpassed by any other naturalist. Of the redstart's song he wrote:

> The opening rapidly warbled notes are so charming that the attention is instantly attracted by them. They are composed of two sounds, both beautiful—the bright pure gushing robin-like note, and the more tender expressive swallow-like note. And that is all; the song scarcely begins before it ends, or collapses, for in most cases the pure sweet opening strain is followed by a curious little farrago of gurgling and squeaking sounds, and little fragments of varied notes, often so low as to be audible only at a few yards' distance. It is curious that these slight fragments of notes at the end vary in different individuals, in strength and character and in number, from a single faintest squeal to half a dozen or a dozen distinct sounds. In all cases they are emitted with apparent effort, as if the bird strained its pipe in the vain attempt to continue the song.

Poets have become wary of venturing into a province once almost exclusively theirs but now invaded and like to be pre-empted by the scientist. We await a new Lucretius. But the modern poet can unite compassion with accurate observation and a scientific outlook as in Robert Frost's[1] lines describing the subsong of a sleepy bird:

> A bird half wakened in the lunar noon
> Sang halfway through its little inborn tune.
> Partly because it sang but once all night
> And that from no especial bush's height;
> Partly because it sang ventriloquist
> And had the inspiration to desist
> Almost before the prick of hostile ears,
> It ventured less in peril than appears.
> It could not have come down to us so far
> Through the interstices of things ajar
> On the long chain of repeated birth
> To be a bird while we are men on earth
> If singing out of sleep that way
> Had made it much more easily a prey.

[1] Robert Frost (1955) 'On a bird singing in its sleep', *op. cit.*

THE DESECRATION OF THE WORLD OF NATURAL SOUND

Some of us can still recall the time when only in the streets of the great wens, such as London, were folk out of earshot of singing birds, the hedged roads were ribbons of song, and the sky belonged to the lark. Now, in this country and many others it is increasingly difficult to listen to birds without the intrusion of man's cacophonies. The internal combustion engine breaks the stillness of the Antarctic; in African forests and Arabian deserts the voice of the transistor is heard. The preservation of areas where those yet unborn may be able to hear, study and enjoy natural sounds secure from interruption by machine noises is of vital importance but presents such difficulties, because amenity is subordinated to profit, that conservationists, in a world with its priorities astray, appear too daunted even to contemplate its possibility. Man's desecration of his environment by noise is the most pervasive and gratuitous of his many outrages against nature.

THE ENDURING SPIRITUAL SIGNIFICANCE OF BIRD SONG

It is not only for the intrinsic interest, inspiration and beauty of bird song that we should esteem it, but also because a sense of continuity with the past is important for our spiritual health as life becomes more complex. Despite the changes in man's attitude, his response to the utterances of birds has retained so much from the past that in appreciating bird song and what has been written about it we become alive to insights and sentiments widely shared. It is an achievement of great music, visual art and literature that they alleviate our loneliness and enable us to realize that, although the centuries have brought many changes, others have stood where we stand and been inspired by universal, enduring things—not least by the songs of birds. Delight in nature wedded with art enables us to commune with the past and live more intensely in the present. The twittering swallows outside our windows in the early morning will be more to us than a pleasant background to our waking thoughts if we recall that, while man has passed from food gathering to civilization and empires have flourished and declined, the birds have remained substantially faithful to their songs. This generalization is not contradicted by what is known of song learning in birds (Thorpe, 1961, 1963; Armstrong, 1963*a*). Long ago the Greeks welcomed the swallow with a song, depicted on a vase the greeting accorded the first to be seen, and complained in verse of chattering swallows which kept lovers awake in the early morning.[1] According to a papyrus note written many centuries earlier a love-sick Egyptian girl, aroused by the swallows, heard them inviting her into the countryside but refused to go without her lover, who was tardily awakening beside her (Wilson, 1965). Keats[2] was not mistaken when, listening to the nightingale, he reflected:

> The voice I hear this passing night was heard
> In ancient days by emperor and clown,
> Perhaps the self-same song that found a path
> Through the sad heart of Ruth, when, sick for home,
> She stood in tears amid the alien corn.

Li T'ai-po[3] hails us across the centuries:

> Now the long, wailing flight of geesebrings autumn in its train,
> So to the view-tower cup in hand to fill and drink again,
>
> And dream of the great singers of the past,
> Their fadeless lines of fire and beauty cast.
> I too have felt the wild-bird thrill of song behind the bars,
> But these have brushed the world aside and walked amid the stars.
>
> In vain we cleave the torrent's thread with steel,
> In vain we drink to drown the grief we feel;
> When man's desire with fate doth war, this, this avails alone—
> To hoist the sail and let the gale and waters bear us on.

ACKNOWLEDGEMENT

After nearly fifty years friendship it is a great pleasure to add my essay to this volume as a tribute to Professor W. H. Thorpe.

REFERENCES

ALEXANDER, M. (1966) *The earliest English Poems*. Harmondsworth: Penguin Books, Ltd.

ARMSTRONG, E. A. (1923) What the birds sing. *The Pageant of Nature* (P. C. Mitchell, ed.). London: Cassells. **2**, 614–20.

(1946) *Birds of the Grey Wind*. London: Lindsay Drummond. (1st edn. Oxford: The University Press 1940.)

(1963*a*) *A Study of Bird Song*. Oxford: The University Press.

[1] Athenaeus, 360c; S. Reinach (1899), *Répertoire des Vases peints*, **1**, 96; N. Douglas, *op. cit.* (Chapman and Hall, Ltd.).

[2] John Keats (1820) 'To a nightingale', *Poems*. London.

[3] Li T'ai-po (1911), 'Drifting', Trans. L. Cranmer-Byng. *Op. cit.*

(1963*b*) *Shakespeare's Imagination*. Nebraska University Press. (1st edn. London: Lindsay Drummond, 1946.)

(1965) *Bird Display and Behaviour*. New York: Dover Publ. Inc. (1st edn. Cambridge: The University Press, 1942, under the title, *Bird Display*.)

(1969) *The Folklore of Birds*. New York: Dover Publ. Inc. (1st edn. London: Collins, 1958.)

(in prep.) *From Galilee to Assisi: The Christian tradition of delight in nature.*

BAIKIE, J. (1926) *The Amarna Age*. London: A. and C. Black, Ltd.

BATCHELOR, J. (1901) *The Ainu and their Folklore*. London: Religious Tract Society.

BLYTH, R. H. (1963–4) *A History of Haiku*. Tokyo: Hokuseido Press. **1**, 17, 47; **2**, 130.

CASTAGNÉ, J. (1930) Magie et exorcisme chez les Kazak-Kirghizes et autres peuples turcs orientaux. *Rev. des études islamiques, Paris*, 53–131.

DAVIDSON, H. R. E. (1968) *Gods and Myths of Northern Europe*. Harmondsworth: Penguin Books, Ltd.

DODDS, E. R. (1951) *The Greeks and the Irrational*. Berkeley and Los Angeles: University of California Press.

DOUGLAS, N. (1928) *Birds and Beasts of the Greek Anthology*, London: Chapman Hall, Ltd.

DRIVER, G. R. (1955) Birds in the Old Testament. *Palestine Exploration Quarterly* **86–7**: 5–20, 129–40.

ÉLIADE, M. (1964) *Shamanism*. London: Routledge and Kegan Paul, Ltd.

FISHER, J. (1966) *The Shell Bird Book*. London: Ebury Press and Michael Joseph.

FRANKFORT, H. and H. A., WILSON, J. A. and JACOBSEN, T. (1949) *Before Philosophy*. Harmondsworth: Penguin Books, Ltd.

FRAZER, SIR J. G. (1933) *Spirits of the Corn and of the Wild*. London: Macmillan and Co., **2**, 146.

HARTLEY, P. H. T. (1950) An experimental analysis of interspecific recognition. *Symp. Soc. exp. Biol.* **4**, 313–36.

HARVA, U. (1938) *Die religiösen Vorstellungen der Altaischen Völker*. F F Communications No. 125 Suomalainen Tiedeakatemia Helsinki. Werner Söderstrom Osakeyhtio, Porvoo, Helsinki.

HATTO, A. T. (1965) *Eos*. London, The Hague and Paris: Mouton and Co.

HENTZE, K. (1941) *Die Sakralbronzen und ihre Bedeutung in den fruechinesischen Kulturen*. Antwerp: De Sikkel.

HUDSON, W. H. (n.d.) *Afoot in England*. London: J. M. Dent and Sons, Ltd.

KEENE, D. (1965) *The Manyōshū*. New York and London: Columbia University Press.

KINGSLEY, C. (1884) *Prose Idylls*. London: Macmillan and Co.

KIRCHER, A. (1650) *Musurgia universalis, sive ars magna consoni et dissoni*. Rome: *Ex typographia Haenedium Francisci Corbelletti*.

KIRCHNER, H. (1952) Ein Archäologischen Beitrag zur Vorgeschichte des Schamanismus. *Anthropos*, **47**, 244–86.

KRAMER, S. N. (1944) Sumerian mythology: A study of spiritual and literary achievement in the third millenium B.C. *Mem. Amer. phil. Soc.* **21**, 55.

LEROI-GOURHAN, A. (1964–65) *Le Geste et la Parole*. Paris: Albin Michel. **2**, 217.

LOEB, E. M. (1929) *Tribal Initiations and Secret Societies*. Berkeley.

NARR, K. (1959) Bärenzeremoniall und Schamanismus in der alteren Steinzeit Europas. *Saeculum* (Freiburg and Munich) **10**, 233–72.

OTTO, R. (1923) *The Idea of the Holy*. Oxford: Humphrey Milford.

RADLOFF, W. (1884) *Aus Sibirien*. Leipzig: T. O. Weigel, **2**, 20–50.

ROHEIM, G. (1954) *Hungarian and Vogul Mythology*. New York: J. Augustin, Locust Valley.

ROSS, A. (1966) *Pagan Celtic Britain*. London: Routledge and Kegan Paul, Ltd.

SCHOLES, P. (1938) *The Oxford Companion to Music*. Oxford: Humphrey Milford.

SCHRENCK, L. VON (1891) *Reisen und Forschungen im Amur-lande*. St. Petersburg: Kaiserl, Akademie der Wissenschaften Za.

SELOUS, E. (1901) *Bird Watching*. London: J. M. Dent and Sons, Ltd.

SHIROKOGROV, S. M. (1923) General theory of shamanism among the Tungus. *J. Roy. Asiatic Soc.* (N. China Branch) Shanghai, **54**, 296.

SZÖKE, P. (1962) Zur Entstehung und Entwicklungsgeschichte der Musik. *Studia Musicologica* **2**, 33–85.

THORPE, W. H. (1961) *Bird-Song*. Cambridge: The University Press.

THORPE, W. H. (1963) *Learning and Instinct in Animals*. London: Methuen. (1st edn. 1956).

VRIES, J. DE (1956–7) *Altgermanische Religionsgeschichte*. Berlin: Walter de Gruyter and Co. **1**, 304.

WATERBURY, F. (1952) *Bird-deities in China*. Ascona. *Artibus Asiae Curat. editionem Alfred Salmony. Suppl. X.*

WILSON, J. A. (1965) In Hatto, A. T. *Eos*. London, The Hague and Paris: Mouton and Co. 105–6.

16. THE AESTHETIC CONTENT OF BIRD SONG

by **JOAN HALL-CRAGGS**
Little Holt, Woodcote, Reading

'There are two musical races in the world—the birds and the humans.'[1]

INTRODUCTION

To speak of an aesthetic content implies recognition of some quality in bird song which gives pleasure to man. That many of the sound sequences made by birds do please us in some way is self-evident; if they did not, the term 'song' would never have been used to describe them. We distinguish quite sharply between the song and the calls of one and the same species; this dichotomy is acceptable in terms of function. But we also distinguish between the territorial proclamations of species which 'sing' and those which do not. Superficially, the vocal utterances of some species in certain situations are called 'song' because they exhibit the temporal and tonal patterning characteristic of human music. If we accept that music appeals to the aesthetic sensibilities then its criteria must be applicable to bird song when, indeed, the latter has aesthetic content.

Both music and aesthetics have been defined in many different ways but there is a marked lack of accord among authors. This is partly due to preoccupation with the cultural approach to music which varies in detail according to the conventions of the particular society and becomes modified as that society develops. The human response to aesthetic stimuli is subject inevitably to conditioning and this adds to the proliferation of definitions of what is meant by 'aesthetic'.

Aristotle (Politics, I, 2) advocated taking into account the origin and first growth of phenomena in order to be clear about their meaning. This view would enable us to modify the definitions of

[1] Percy Scholes. *The Listener's History of Music* (1947).

music according to the stages of its subsequent development. Authorities such as Bullough (1957) and Langer (1957) regard the consideration of origin, as distinct from import, as likely to lead to errors, it being inadmissible to argue from the original function to functions at a later stage of development. Sachs (1965) held the view that the origin of music cannot be solved. Nevertheless, from the numerous definitions of music which have been advanced it is clear that bird song is only excluded from those which include the qualifying term 'human'; it therefore seems reasonable to assume that the attributes common to both are due to corresponding origins of function. Biologists will not disagree with Newton (1950) when he states: 'Function creates form'. If there is adequate evidence for formal correspondence in the music of man and the songs of birds there is some justification for suspecting convergent evolution.

POSSIBLE ORIGINS OF MUSIC

It may be assumed that man's immediate ancestors, in common with most other animals, had a repertoire of innate cries and that, in the emergence of articulate speech, these were modified to suit the requirements of particular circumstances. Since man is a social animal living generally in close contact with his fellows, his most economical and therefore most usual method of vocal communication was probably a kind of speech and not, as has been suggested, a kind of song. After forty years as laryngologist to Covent Garden, Sir Milsom Ress (in Newman, 1956) concluded that 'the vocal cords were not put into the human throat for the purpose of singing'. If this is so, song must have evolved through the selection and accentuation of those characteristics of speech which proved to be advantageous in certain circumstances.

Articulate speech, formed by a combination of vowels and consonants and usually pitched at a level prescribed by the anatomy of the individual, suffices for short range communication. If we wish to communicate vocally over distances greater than those we cover in the field of conversation we can increase the intensity and raise the fundamental frequencies to approach the range of greatest aural sensitivity, and in addition we may increase the duration of each word unit in order to overcome interference. Increase in the duration of a word unit can only be achieved by prolonging the vowel sounds within it and, as the vowel content is extended, the consonantal content will become relatively ineffective until a point may be reached where the original word loses its meaning. If primitive man wished to signal vocally over distances which exceeded the carrying power of ordinary speech and, at the same time, to overcome interference from extraneous sounds, he would probably have found most effective some technique which incorporated all or most of such modifications of speech; the result would have been a type of yodelling or hooting. Communicatory yodelling is still practised in many parts of the world today as instanced by A. E. Cherbuliez in the discussion following 'Zu den Jodlertheorien' (Graf, 1961). The most rudimentary kind of pitch patterning or unmeasured music could have evolved from such a signalling system. Extensive modification of speech might reduce the precise meaning of word units and, at the same time, would demand more physical effort from the signaller than would ordinary speech but, as meaningful signals became established, compensation would be afforded by the saving of time—an important factor in emergencies—and by reduction of gross physical effort in the form of travel. At a later stage when musical instruments were constructed the signalling range could be extended and further economy of time and effort achieved. Long range signalling might well have had decided survival value for man.

The evolution of language gave primitive man advantages over his less articulate predecessors, as did his greater capacity to alter or manipulate his environment to suit his requirements. When handling large objects he must have found at an early stage that coordinated, cooperative effort reduced individual or uncoordinated mass effort. The only way to coordinate mass physical effort is to apply it in time. Time measurement may have been effected in a number of ways but eventually some kind of chanting might emerge, a rudimentary antecedent of labour songs so well exemplified in the sea shanty. This could account for the regularly recurring strong beat which has revealed itself in much of the music of different races and ages. But metronomic regularity is not, and has never been, essential to music in the broad sense; it served merely to coordinate physical effort, not to convey information.

If these suppositions concerning the origins of music are acceptable, it is possible to trace an analogy with bird song. The functional aspect of vocal utterances in relation to the territorial requirements of birds has been established (Thorpe, 1961; Armstrong, 1963;

Marler and Hamilton, 1966). Those species of bird which, by traditional musical standards, are considered to have the best songs are usually those with the most marked territorial behaviour; the greater the territorial requirements the more musical we judge the song to be. Conversely, highly social species with a minimal territory range have no true song although they may be extremely vocal; hence, in many instances the degree of musicality may correlate with the distance over which the voice must be projected. The many types of quiet singing which are included in the category 'subsong' may be communicatory or noncommunicatory; in either case it is clear that there is no functional requirement for the voice to carry far and the acoustic characteristics of subsong resemble those of speech in so far as both suffice, when necessary, for short range communication.

DISTINGUISHING CHARACTERISTICS OF LONG RANGE SIGNALLING

If a long range vocal communication system is to be serviceable it must contain sounds which will carry the required distance in a particular habitat and, within certain limits and prescribed situations, it must advertise the location of the signaller; in addition it must serve purposes of identification at increasingly finer levels and, above all, it must overcome interference. Those characteristics which distinguish human song from speech are similar to those which appear when long range territorial bird song develops out of subsong in the same species. These are as follows:

(i) Fundamental frequencies raised to approach the range of greatest aural sensitivity.
(ii) Greater concentration of energy around the main frequency of the individual sound units.
(iii) As a result of (i) and (ii) there is overall decrease in bandwidth.
(iv) Increase in intensity of most of the sound units.
(v) The sound units tend to become discrete and prolonged or rapidly reiterated; their duration and frequency become fixed.
(vi) The rest periods between the sound units are of fixed duration.
(vii) As a result of (v) and (vi), temporal patterning is evident.
(viii) As a result of (ii) and (v), tonal patterning is evident.
(ix) As a result of (vii) and (viii), the song form is evident; only when a song form is established can exact repetition occur.
(x) The auditory impression is of rhythm, and analysis of highly rhythmic songs often gives a visual impression of symmetry.
(xi) There is a much greater degree of predictability.

Of these modifications of speech and subsong numbers (i) to (v) indicate decrease in bandwidth and lengthening of the duration of the sounds; vocal carrying power is increased and the sound units may be of more readily discernible pitch. The pitch of a tone is discernible when 70 per cent of the energy lies between ±5 per cent of the principal frequency (Bürck, Kotowski and Lichte, 1936, in Stevens and Davis, 1938). Numbers (v) to (ix) establish the song form. As the signalling range of the voice is extended interference is likely to increase so that 'redundancy' in the form of repetition must be introduced; thus distinct and exactly reproducible forms are essential to long range signalling. This may account for the repetitive nature of many territorial bird songs and for the fact that in human art music, which appears to have developed out of a primitive long range communication system, we do not merely tolerate but need repetition of an order that would be quite intolerable in the spoken word. We still, however, concede to verbal repetition when we wish to emphasize distance in time or space: 'far, far away', 'long, long ago', 'wide, wide world', and the poet often creates an impression of distance by the same means. Van der Post (1961), with true insight into the nature of music, writes of the Bushman's ability to express remoteness:

> The plain was, as they put it in their tongue, 'far, far, far away' to the east. It was lovely how the 'far' came out of their mouths. At each 'far' a musician's instinct made the voices themselves more elongated with distance, the pitch higher with remoteness, until the last 'far' of the series vanished on a needle-point of sound into the silence beyond the reach of the human scale.

It seems justifiable, therefore, to argue that increasing precision of pitch, the establishment of temporal forms and the repetitive nature of the phenomenon, as used by birds and men, may be due to similar functional requirements.

RHYTHM AS THE BASIS OF FORM IN BIRD SONG AND MUSIC

Sachs (1965) points out that the concepts of rhythm and form overlap and cannot be strictly kept apart. In musical parlance 'rhythm' is generally used when discussing small sections of a work (motifs and phrases), 'form' is used when referring to the larger sections. This is an arbitrary distinction, useful for analytical purposes but tending to obscure the crucial fact that all musical forms, whether brief or lengthy, depend upon rhythm. Rhythm is in time what symmetry is in space and equilibrium is in matter (Schopenhauer, 1859).

Musical form may be said to begin at that point of extreme simplicity where no more than two sounds occur in a determined temporal and tonal relationship. In the total absence of competition, indeterminate series of sounds might suffice for communicatory purposes but in the process of speciation distinct song forms must emerge. It is necessary, therefore, to seek some basic principle of organization which may become manifest without conscious intervention. The most notable attribute of long range bird song is temporal organization of a kind that is not to be found in subsong of the same species. The organization, like the quality of the individual units, appears to depend upon distance. As the range of the voice is extended the effort involved in producing the sounds must increase (the energy of a sound is negligible compared with the energy expended in producing the sound); and when an organism the size of a song bird must project the voice over several hundred feet in an open environment, the effort involved is likely to be considerable. Such organization as tends to minimize effort must be advantageous, and a song form will be most efficient when the units are so ordered as to balance expenditure of effort and rest. The simplest manifestation of orderly distribution in time is a pulse of given duration, frequency and intensity, energy being expended on the beat and recuperation occurring between beats; auditory stimulation of this kind, however, would lead rapidly to habituation and allow little scope for specific diversification.

When temporal organization becomes complex the need to balance effort and rest can be met by transforming the pulse into impulse and repulse—a more usual term for the latter is recoil. Units of sound become organized on a scale of gradually increasing expenditure of energy followed by a recuperative period when effort is gradually relaxed; this is the rhythmic ebb and flow of music. Long continued multiphrased bird songs with minimal pauses between phrases exhibit such rhythmic organization; discontinuous songs often exhibit either impulse, as in the song of the chaffinch *Fringilla coelebs*, or recoil, illustrated by the song of the blue tit *Parus caeruleus*. The greater the rhythmic organization the more persistent the song, hence the latent order of music and bird song, rhythm, appears to derive from the organism's unconscious tendency to minimize effort. Subsong, like speech, may continue for long periods without evincing any definite rhythmic structure but the voice does not have to carry and the effort involved is slight compared with that required in long range signalling.

If impulse and recoil constitute rhythm there must be a climactic point towards which the remaining sound units progress or away from which they recede. The climax of a musical phrase can occur only once and, like the most rarely-used word in a sentence, it is the point which rivets the attention. But whereas the pivotal word in a sentence cannot usually be anticipated, the musical climax may be—namely by the gradual build-up or rhythmic impulse towards it. The intensified organization of musical units (as compared with language units) may be associated with the greater need to minimize effort when signalling over increasing distance, because it requires less effort to approach the climax of a phrase by degrees than by a sudden leap. Such organization—the rhythm—appears to be a distinction between music and language at the syntactic level; there is an obvious transition between the two by way of poetry and drama but the rhythm of these must coincide with semantic indications which would be unduly restricted by the exactly prescribed minutiae of musical temporal patterning.

It should be possible to apply Zipf's (1949) Principle of Least Effort to music and bird song; this is an empirical rule that the number of occurrences of a word in a long stretch of text is the reciprocal of the rank order of frequency of occurrence (Pierce, 1962). Mandelbrot (1953), who put forward a theoretical explication of Zipf's law, does not assume it but aims to show that it follows from simple premisses (Cherry, 1957).

Thorpe (1966) refers to the work of the psychologists Wheeler (1929) and Tolman (1932) who formulated 'laws' similar to that of Zipf which appeared to account for the refinement of behaviour patterns; with practice, an animal tends to minimize the effort expended in achieving a goal. These laws failed to come up to expectations as instruments for exact prediction. However, in music, whether it is considered as an art or as a long range communication system, it is the rhythm which may be held to account for the curiously predictable nature of the phenomenon and which led Puffer (1905) to define music as 'the art of auditory implications'.

SELECTIVE PRESSURE AND RHYTHMIC ORGANIZATION

The songs of the three sympatric species of the genus *Phylloscopus*: chiffchaff, *P. collybita*, willow warbler, *P. trochilus* and wood warbler, *P. sibilatrix* illustrate the correlation of increasing complexity with intensification of rhythmic organization resulting in specific distinctiveness of song. But for slight variation in pitch the chiffchaff's song is virtually an expression of the simplest, pulse-like form. Variety of form is expressed in the song of the wood warbler by gradual increase in duration and intensity of the units combined with decrease in the duration of the rest periods between units, followed by the reverse process; organization is achieved through powerful rhythmic impulse to a climax followed by recoil. Variety in the willow warbler's song takes the form of a step-wise descent in pitch combined with an overall increase in intensity; taken together, these give an organized rhythmic drive towards the final sound.

If it were possible to assess the actual effort involved in singing the three songs, it would probably be found to be about the same, bearing in mind the fact that the duration of a chiffchaff's song bout usually exceeds that of the song form of either wood warbler or willow warbler.

By musical criteria the song of the chiffchaff might be considered to be of greatest antiquity. It would be less easy to place the other songs in time for the wood warbler, being more exclusively a bird of woodland habitat, appears to have developed a song which is acoustically better adapted to that environment.

It is interesting to note that where the chiffchaff is not in competition with the willow warbler, the pulse-like aspect of its song deteriorates and the specific formality is lost (Marler, 1960). It would seem that the process of speciation may have initiated a trend from formlessness to simple formality of song in an existing species thereby requiring equally formal songs as isolating mechanisms in the divergent species.

EVOLUTIONARY TRENDS IN FORM

It has been argued that the emergence of musical form in the vocal communication systems of birds and men stems from the corresponding origin of function. It is possible to support this claim by pointing to parallel evolutionary trends in form.

For the purpose of analysis musical form has been divided into many categories but these may be reduced to two main lines of development: one is the limiting process of variant formation—the 'maqam' of the folk musician, and the other embraces all additive forms, binary, ternary and their derivatives. Scholes (1955) sums up the difference between these two lines of development by saying that change and repetition appear in alternation in the extended forms, whereas they appear in combination in the 'air with variations'.

Szabolcsi (1965) describes 'maqam': 'General outline and basic features are all that is laid down; the rest is left to performance . . . .' The outline may indicate no more than the general direction in which the melody shall move and its approximate duration; the performer is left to express his individuality *within* the prescribed pattern. Thus the folk musician starts with rather more than the young chaffinch which has inborn recognition of the duration and tonal quality of the specific song together with a tendency to sing the song at intervals of about ten to twenty seconds (Thorpe, 1958). Although chaffinches kept in close proximity tend to sing similar variants of the song type and wild birds reply to the song of a rival with the most nearly resembling variant in their own repertoire (Hinde, 1958) it is unlikely that any two individuals in the wild have identical repertoires or sing identical variants. Marler and Hamilton (1966) write: 'individuality is as common in birds with small repertoires as in those with large ones' and Borror (1961, in Marler, 1966) concludes that two individuals rarely share identical songs. Clearly variant formation is a limiting process; its basis is repetition within which there may be just so much change as shall reveal the identity of the singer (or the individuality of the musician) without concealing the identity of the form. It requires no further extension in time which is advantageous to both song-bird and prenotational man who have to rely upon memory. In avian communication adherence to the form ensures specific recognition, while diversification within the form, which appears to be limitless, ensures individual recognition.

Formal development by extension—the addition of similar yet contrasting material to an existing song type—is less usual in bird song but where it occurs, as in the songs of many of the Turdines, it may reach immense proportions. In the writer's experience the repertoire of the European blackbird, *Turdus merula*, may comprise from 15 to 40 motifs at the beginning of the breeding season from which material, mostly by a process of redistribution and combination, from 12 to 95 phrases may be created. These figures are taken from the analyses of the songs of seven individuals, each recorded daily throughout their song seasons; the average for the seven is 28·8 motifs or basic song types and 24·4 'developed' forms.

The total duration of the musical material used by blackbirds with extensive repertoires may exceed three minutes yet contiguous repetition is rare in the dawn song except during the first few days of the season. At the outset the repertoire comprises an undisciplined jumble of motifs but as the season progresses the material is organized into phrases and compound phrases, sometimes by what appears to be a painstaking transitional process, and some degree of predictability in the order of occurrence of the phrases emerges (Hall-Craggs, 1962). The significant factor in the organization of the material is that, in the songs of some of these birds, phrase construction by the addition of motif to motif exemplifies the principle of rhythmic impulse and recoil; brief tonal and temporal patterns are matched to form 'symmetrical' wholes. The constructional basis appears to be identical to that which we find in our own music. At the unit level there may be some lack of tonal 'purity' and some tones may tend to merge rather than to be discrete, but this is also true of much folk music and a good deal of contemporary art music.

Figure 1 shows a blackbird phrase (sound spectrogram in Hall-Craggs, 1962) formed by combining two basic motifs. This is compared with as widely divergent types of European music as possible: the opening phrase of the Gigue from Bach's Suite No. 3 in D and an entire sea shanty: 'Blow the man down.' To facilitate comparison the three examples have been transposed to the same octave c′ to c″ (261 to 522 Hz) and into the same key: C. Each example is divided into two sections, irrespective of total content, because the material compressed into two subphrases in the Bach and the bird examples is expanded to encompass four phrases in the shanty.

The following points of correspondence occur:

(i) Equality of duration in sections 1 and 2.
(ii) Change of tonal and temporal patterning in section 2 (much more pronounced in Bach).
(iii) Section 1 ends at a point of anticipation.
(iv) Section 2 ends at a point of finality.
(v) Melodic outlines similar and incorporating a suggestion of pentatonicism:
Sections 1: C′ E G A (Bach omits the E).
Sections 2: C″ G E C′ (bird omits final C).
(vi) Climactic points the same:
Sections 1: the single high A.
Sections 2: the single high C.

It is noteworthy, too, that in the Bach and the bird phrases the climax occurs at the end of an ascending figure in section 1 (impulse) and, following an anacrusis,[1] at the beginning of a descending figure in section 2 (recoil). The underlying rhythm is the same and the bird's 'instinct' for balancing motif against motif in extension is as sure as man's. The bird's phrase was recorded 83 times but on no occasion was the order of the motifs reversed; when the constituent motifs were sung in other contexts section 1 was retained as an introductory figure and section 2 as a concluding figure.

FIGURE 1. A blackbird's phrase compared with two widely divergent types of European music. Each has been transposed to the same octave and key.

The duration of the bird's phrase is three seconds, and this is only one of a chain of phrases; its approximate rate of occurrence over the whole season is about once in 6½ minutes of song. The duration of the shanty is twelve seconds at Terry's (1921) metronome speed of ♩. = 88 and there are seven verses which may be repeated as required. Repetition is as common in human music as in bird song and the discrepancy in the musical memories of man and bird is more apparent than real. Until the advent of musical notation it is unlikely that exactly reproducible melodies (i.e. not improvisatory) exceeded thirty seconds in duration; familiarity gained by the aid of a written record is man's advantage.

[1] An unstressed note or group of notes at the beginning of a phrase (Scholes, 1955).

MUSIC, COMMUNICATION AND AESTHETICS

Form in art music and efficiency in long range communication are dependent upon the same principles. The main principle of musical form is to achieve a mean between the extremes of all change and all repetition (Scholes, 1955). In order to arrest the attention of the listener a composer must introduce novelty; but all novelty induces the psychological fatigue of sustained effort which must be averted by reference to the familiar. Scholes calls this the dual principle of musical form: unity plus variety.

Cherry (1957) writes: 'To set up communication the signals must have at least some surprise value, some degree of unexpectedness, or it is a waste of time to transmit them.' And, in the mathematical sense, 'information is measured in terms of the *statistical rarity* of signs'. When an organism is signalling over a distance in a noisy and competitive environment, a considerable amount of repetition (redundancy) must occur if the message is to be received and correctly interpreted, but some form of variety must be introduced or the system ceases to be informative; a compromise between change and repetition is necessary.

Both music and long range communication have, therefore, not only similar qualitative units and the same temporal organization—rhythm—but the psychological requirements of the former are a reflection of the functional requirements of the latter. The objective distinction between the two systems appears to be one of degree only: the extent to which the organization of the parts within the whole reflects the organization of the units within the parts. A musical work should epitomize the highest level of such organization; it must become a unified whole. It may be argued that when this stage is reached there is an essential difference between music and communication in that the musical work becomes a fully predictable series of events and ceases, therefore, to communicate in a practical manner; yet our appreciation tends not to decrease but to increase with familiarity. The unpredictable factor remains in the nature of the performance.

If a specific bird song is of a monotonously repetitive kind, that is, if the species has only one song type and the individual but its own single variant of that type, then the unpredictable factor—necessary to set up communication—can only be introduced by varying the time intervals between utterances or by changing the song post; the precise 'when' and 'where' cannot be predicted. The multiphrased singer can sing persistently and for long periods without changing its perch or varying the duration of the interphrase pauses but, so far as we know, the precise 'what' cannot be predicted. Sotavalta (1956, in Armstrong, 1963) claims that there is sufficient organization in the song of the Sprosser nightingale *L. luscinia* to estimate a probable order of occurrence of the phrases, and Craig (1943) indicates that the song phrases of the eastern wood pewee *Myiochanes virens* may be musically adapted to their mode of occurrence. The same may be said of the song of the blackbird but in this the phrases tend to become associated predominantly and not exclusively with others: thus completely random distribution may be eliminated but some degree of unpredictability always remains. Predictability increases as the season progresses; this might be expected for the song must be most efficient as a communication system at the outset when the bird is establishing or enlarging the breeding territory. When boundaries are settled and the song serves to maintain, rather than to gain, a territory, the functional aspect may be relaxed and organization of the song material is intensified. The attention of the bird appears to be directed towards the song rather than to the function of the song. Attention to form rather than to function would imply that, for the bird, the activity is becoming aesthetic.

A form may be considered objectively beautiful if it perfectly fulfils its function. This is the view of Sir Herbert Read (1963) who even maintains that such an object will be automatically a work of art. If this is so, then bird song, having the nature of organized form, is beautiful. Indeed, if Read's dictum is applied to the evolutionary process the inescapable conclusion is that natural selection operates in favour of the beautiful. But subjectively we may assess the value of an object according to its beauty or according to its usefulness and Newton (1950) asserts that only when form is perceived purged of all ideas of function can it be thought of as beautiful.

If bird song must remain operative in its primary function it cannot suffer complete differentiation; but this indissolubility of form and function may enhance rather than impair its aesthetic content. As Santayana (1896) points out '. . . we accept the forms imposed upon us by utility, and train ourselves to apperceive their potential beauty' and 'Disaster follows rebellion against tradition or against utility, which are the basis and root of our taste and progress.'

A song bird's identity is expressed in musical form by limited deviation from a norm; the human musician expresses his individuality by the same means, whether creatively or re-creatively. If the form of art music should become totally divorced from function it will cease to communicate and will, therefore, cease to exist. The need for a form to fulfil its original function may retard its rate of evolutionary progress, keeping it in step with organic evolution and precluding wide, experimental deviation: if human music is a hare beginning now to jink wildly in order to escape the mechanization which will destroy it, bird music is the tortoise advancing imperceptibly along a straight course. Pre-taped musical works and the frequent mechanical repetition of a single interpretation of a work must inevitably induce habituation in man. Only those works which permit of interpretative variety within a unified whole, as those bird songs which may be freely adapted by many members of a single species, will survive. Survival is the only infallible test of quality in music.

Adaptability is as essential to music as it is to life which is, perhaps, only to be expected of a form which is the prerogative of life. All other sense data can be found in inorganic matter while, so far as is known, the form of music remains the privilege of birds and men.

REFERENCES

ARMSTRONG, E. A. (1963) *A Study of Bird Song*. London.

BORROR, D. J. (1961) Intraspecific variation in passerine bird songs. *Wilson Bull.* **73**, 57–78.

BULLOUGH, E. (1957) *Aesthetics*. London.

BÜRCK, W., KOTOWSKI, P. and LICHTE, H. (1936) Frequenzspektrum und Tonerkennen. *Ann. d. Physik*. **25**, 433–49.

CHERRY, C. (1957) *On Human Communication: A Review, a Survey and a Criticism*. New York and London.

CRAIG, W. (1943) The song of the wood pewee (*Myiochanes virens*). *N.Y. State Mus. Bull.* **334**, 6–186.

GRAF, W. (1961) Zu den Jodlertheorein. *J. Int. Folk Music Council*, **13**, 39–42.

HALL-CRAGGS, J. (1962) The development of song in the blackbird (*Turdus merula*). *Ibis* **104**, 277–300.

HINDE, R. A. (1958) Alternative motor patterns in chaffinch song. *Anim. Behav.* **6**, 211–18.

LANGER, S. K. (1957) *Philosophy in a New Key*. Cambridge, Mass.

MANDLELBROT, B. (1953) Contribution à la théorie mathématique des jeux de communication. *Publ. inst. statist.* 2. University of Paris.

MARLER, P. (1960) Bird songs and mate selection. In Lanyon and Tavolga *Animal sounds and communication*, Washington, D.C., pp. 348–67.

MARLER, P. and HAMILTON, W. J. (1966) *Mechanisms of Animal Behaviour.* New York.
NEWMAN, E. (1956) *From the World of Music.* London.
NEWTON, E. (1950) *The Meaning of Beauty.* London.
PIERCE, J. R. (1962) *Symbols, Signals and Noise: The Nature and Process of Communication.* London.
PUFFER, E. D. (1905) *The Psychology of Beauty.* Cambridge, Mass.
READ, H. (1963) *To Hell with Culture.* London.
SACHS, C. (1965) *The Wellsprings of Music.* The Hague and New York.
SANTAYANA, G. (1896 and 1955) *The Sense of Beauty: Being the Outline of Aesthetic Theory.* New York.
SCHOLES, P. (1947) *The Listener's History of Music.* London.
(1955) *The Oxford Companion to Music.* London.
SCHOPENHAUER, A. (1859 and 1966) *The World as Will and Representation.* New York.
SOTAVALTA, O. (1956) Analysis of the song patterns of two Sprosser nightingales (*Luscinia luscinia* L.) *Ann. Soc. Zool. Botan. Fenn. Vanamo,* **17,** 1–31.
STEVENS, S. S. and DAVIS, H. (1938) *Hearing: Its Psychology and Physiology.* New York.
SZABOLCSI, B. (1965) *A History of Melody.* Budapest and London.
TERRY, R. R. (1921) *The Shanty Book.* London.
THORPE, W. H. (1958) The learning of song patterns by birds, with especial reference to the song of the chaffinch, *Fringilla coelebs. Ibis* **100,** 535–70.
(1961) *Bird Song: The Biology of Vocal Communication and Expression in Birds.* Cambridge.
(1966) Ritualization in ontogeny: Animal play. *Phil. Trans. R. Soc.* **251,** 311–19.
TOLMAN, E. C. (1932) *Purposive Behaviour in Animals and Men.* New York.
VAN DER POST, L. (1961) *The Heart of the Hunter.* London.
WHEELER, R. H. (1929) *The Science of Psychology.* New York.
ZIPF, G. K. (1949) *Human Behaviour and the Principle of Least Effort.* Cambridge, Mass.

NAME INDEX

Author's contributions listed in bold type.

SUBJECT AND SPECIES INDEX

Species are listed under both common and scientific name unless the former is not generally used or the latter is uncertain.